U0112362

Food Immunochemistry and Immunology

Libing Wang
Chuanlai Xu

 **SCIENCE PRESS**
Beijing

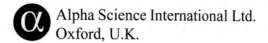 Alpha Science International Ltd.
Oxford, U.K.

Authors
Libing Wang
Chuanlai Xu

Responsible Editors
Haiguang Wang
Hao Wang

Co-Published by:

Science Press
16 Donghuangchenggen North Street
Beijing 100717, China

and

Alpha Science International Ltd.
7200 The Quorum, Oxford Business Park North
Garsington Road, Oxford OX4 2JZ, U. K.

www. sciencep. com

ISBN 978-7-03-033817-4 (Science Press, Beijing)

Printed in Beijing, China

Preface

Rapid urbanization and corresponding lifestyle changes mean that more and more people purchase processed food, eat in restaurants, or buy meals from street vendors. Food production is increasingly industrialized. This introduces multiple opportunities for contamination to occur and makes populations more vulnerable to lapses in food safety at any point along the production chain.

China is a big country for food production and consumption, and also a big trading partner for food. So far, there are 448,000 food enterprises of various varieties in China with a huge annual output. The food industry faces great pressure in developing systems such as total quality management and hazard analysis to identify and manage key steps in food production. Food safety authorities in force a major control. The Chinese government, as it has always done, attaches great importance to food quality and safety. It has been making unremitting efforts to improve food quality and to ensure food safety since the "10th National Food Safety Management 5-year Plan". Quality control analysis is an essential tool in food production. Immunoassay, and related immunochemical analytical procedures [e. g. , ELISA, fluorescence polarization immunoassay (FPIA) and immunochromatographic assay (ICA)] have been widely used to detect various residues in foods. *Food Immunochemistry and Immunology* summarizes the most representative immunochemical technologies applied in food detection. It consists of 13 chapters. Chapter 1 provides the reader with a concise overview of immunology. Chapter 2 describes the binding properties of an antibody to an antigen. Chapters 3 and 4 introduce the common procedure for antibody preparation, purification and modification. Chapter 5 summarizes the labelling technology for antibody with tracers. Chapters 6-9 outline the primary knowledge of time-resolved fluorescence immunoassay, techniques in the molecular immunology and immunogenetics, biomimics (molecular imprinting technique) and immuno-electron microscopy technique. Chapters 10-13 give state-of-the-art information about the application of immunoassay in the following topics: pesticide, veterinary drug, biological toxin and other persistent pollutes detections. This book is of special benefit to researchers on antibody production of chemicals and immunoassay development in food analysis.

I greatly appreciat the help from my teammates (Dr. Chifang Peng, Dr. Hua Kuang, Dr. Wei Ma, Dr. Liguang Xu, Dr. Yingyue Zhu, Dr. Yuan Zhao, Dr. Liqiang Liu and others in preparing this edition of the book). Many more people collaborated in the project, directly and indirectly, and I would like to acknowledge my gratitude to all of them.

Libing Wang

Contents

Colored Figures

Chapter 1   Introduction

1.1   History of immunology and immunological techniques

Immunology is a subject which deals with the answer process and mechanisms of the immune system of the host, when the immune system recognizes and eliminates the deleterious organisms and their compositions. The important physiological function of immune system is to recognize the "auto" and "others" antigens and responsd to them. Provided that the immune system has normal immune functions, it will reject the "others" antigens, which is called immune protection such as anti-infection and anti-tumor immunity. However, the immune response will possibly damage tissues and bring about allergic diseases if the immune system is in immunological disorder. If the immune tolerance against autoantigens is disrupted, the immune system will have immune response to autoantigens, that is autoimmune phenomenon which will bring tissues damage and autoimmune diseases. Thus, the immune system play the roles of rejecting things alien to it and maintaining autotolerance through recognizing and distinguishing the "auto" and "others" antigens. One of the important research fields of modern immunology is to prevent, diagnose and treat diseases through immunology principle and methods[1].

Over the past few decades, the dazzling achievements and technical advantages of immunology have given it a big push to infiltrate into other disciplines and generate many other branches and interdisciplinary aseas of immunology. Nowadays, the researches in immunology at cellular and molecular levels have pushed it to a new level and it has become a new industry with huge market potential.

As did the as other subjects, immunology appeared, developed and mellowed step by step with the development of society and the advancement of science. The development of immunology is divided into four stages, the empirical immunology, the classical immunology, the recent immunology, and the modern immunology[2].

1.1.1   The empirical period of immunology

Anti-infectious immunity was the first immunologic function firstly known by human beings. The medical scientists in ancient China achieved a great deal of experience during timeless clinical diagnosis of smallpox(Figure 1-1). They found that the rehabilitated smallpox

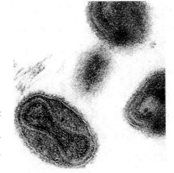

Figure 1-1   Smallpox Virus

patients, their caregiver and the people wearing the clothes that were tainted with smallpox scab wouldn't be infected by smallpox again. Consequently, they created an adventurous method of blowing the smallpox scab powder into no-strisl for safeguarding against infection. This is the earliest vaccine in the world. It is a great contribution before the development of medical science and the discovery of smallpox virus.

The time when the human-pox vaccination(Figure 1-2) was first used is not known for certain. Based on a medical book it was considered that it began during the Kaiyuan era of the Tang Dynasty (during A. D. 713-741). It was also considered that the record of human-pox vaccination began in Emperor Song Times, i. e. A. D. in 11th century, according to the medical book in china. But actually a lot of medical books prove that the human-pox vaccination had been greatly improved by the Longqing era of the Ming Dynasty in the 16th century. The book "Experiential Smallpox Vaccination Method" recorded the difference between *Variola major* and *Variola minor* and the latter was considered safer.

Figure 1-2   Human-pox vaccination

Human-pox vaccination was developed and began to be used in Qing Dynasty. And spread widely. It was also woticed by neighbor countries and soon spread to Russia, North Korea, Japan, Turkey, England etc. After human-pox vaccination had spread to Middle East in the 15th century, the local people developed intradermal inoculation on the basis of the nose-blowing method. In 1721, Mary Montagu, the Wife of the British Ambassador to Turkey, introduced the inoculation method to England and the method spread rapidly in Europe. However, the empirical variolation could bring the danger of artificial infection although it had immune effects, so it hadn't been accepted universally. It is called the period of empirical immunology. It was a start for the mankind to know about the immunity of the organisms. Certainly, human-pox vaccination provided invaluable experience for the English doctor, E. Jenner, who invented vaccinia and French immunologist Louis-Pasteur who invented attenuated vaccine. During this period, people found much immunity phenomena which are really great contributions for medical science.

Figure 1-3   Jenner inoculated the boy

1. 1. 2   The period of classical immunology

The period of classical immunology was from the 18th century to the mid-20th century. During this period, people worked on science experiments instead of phenomena observations of the research of immunology[3, 4]. The important results in this period were as stated below.

(1) The discovery of cowpox vaccine

Cowpox vaccine was an important development of immunology after the appearance of human-pox vaccination. It not only compensated for the lack of human-pox vaccination and but also could be mass-produced in the laboratory. The cowpox vaccination was introduced into China in 1804 and used instead of human-pox vaccination, rapidly. The invention of cowpox vaccine should be attributed to the British doctor, Jenner.

In the 18th century, a rural doctor in England Edward Jenner (1749～1823) observed that the milkmaids who suffered from cowpox would no longer be infected by smallpox. He completed a famous experiment successfully on May 14, 1796. Jenner inoculated a healthy boy with cowbox (Figure 1-3) and when, subsequently, the same boy was inoculated with smallbox the boy didn't contract it. After a series of experiments, Jenner published the famous report of cowpox vaccine.

The creative invention was named as the Jenner cowpox vaccine inoculation, which was the most important weapon during past 200 years against smallpox. On May 8 in 1980, people announced the global eradication of smallpox at the 33rd World Health Assembly (WHA) in Geneva. Jenner's contribution was enormous, so people always attribute the origin of immunology to the creation of cowpox vaccine and the invention of cowpox inoculation. This was the world's first successful vaccine and it contributed greatly to defeat smallpox ultimately. After inoculation, the vaccine would only cause local reaction and did not cause serious damage, but could be effective in preventing smallpox. It not only compensated for the lack of human-pox vaccination, but could also be mass-produced in the laboratory. So the cowpox vaccine replaced human-pox vaccine rapidly and it was accepted by the medical world. But the microbiology hadn't been developed at that time and people didn't know about the pathogens of smallpox and cowpox.

(2) The invention of attenuated live vaccine

Immunology remained at a standstill for nearly a whole century since the cowpox was invented by Jenner. In 19th century, microbiology developed rapidly with the great efforts of French immunologist Louis Pasteur and German bacteriologist Hobert Koch, and other. Pasteur was a French chemist, microbiologist and immunologist. He quantitatively analyzed the relationship between the conversion of various substances and microbial growth and reproduction in the process of alcoholic fermentation. He proved that lactic acid and alcohol were the results of microbial life activities and proposed the method of heat sterilization, which laid the foundation of industrial microbiology and medical microbiology. In his methodology, he fixed the problem of the isolation culture of bacteria creatively and then obtained the pure bacteria, which provided conditions for the preparation of artificial vaccine.

Pasteur researched the ways to obtain attenuated strains consciously through a series of scientific experiments. He finally found that it was available to obtain attenuated strains through physics, chemistry and biology methods. In 1880, Pasteur discovered the inoculation method of chicken bacillus; in 1881, he prepared bacillus anthracis attenuated vaccine by high temperature culture; in 1885, he obtained attenuated strain by continuous passage in rabbits and obtained the rabies virus by continuous passage of rabies vaccine in rabbits and made the active immunization method popular. To commemorate the exploits of Jenner in the past century, he named the method as 'vaccination' and called the preparations 'vaccine'. The preparation of attenuated vaccine laid the foundation for experimental immunology and also opened a new phase for the development of vaccine. The development of microbiology laid the foundation for the formation of immunology.

In 1883, Russian zoologist Metchnikoff found that the cells can swallow foreign bodies outside when he studied the migration effect of starfish larval cells. He also found daphnia blood cells can kill mold spores. Afterwards he found that leukocyte played the role of swallowing of bacteria in experiments with cells of rabbits and humans. He considered that the phagocytosis of leukocyte played an important role in the immune mechanism, which was called cell immunity theory. In 1908, he and Ehrlich shared the Nobel Prize in 1908 for physiology and medicine because of the formation of side chain theory of antibody.

(3) The discovery of antibody

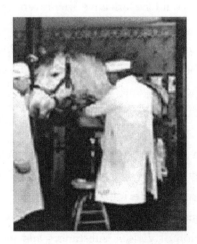

Figure 1-4   Behring and Kitasato

In 1890, German scholar Behring (1854-1917) and Japanese scholar Kitasato immunized animals (horse) with detoxified diphtheria exotoxin in Koch laboratory (Figure 1-4) and found a substance which could neutralize diphtheria exotoxin and was called anti-toxin. The normal animals which were administrated with the immune serum had the ability of neutralizing exotoxin. So the method was called artificial passive immunization. In 1891, Behring cured a patient with diphtheria by animal immune serum successfully. This was the first case that used the passive immunization. In the same year, the passive immunization was rapidly applied in clinical treatment. Behring and Kitasato firstly applied diphtheria antitoxin in the treatment of diphtheria and created the precedent of artificial passive immunotherapy. After that, people found many compounds, which could combine with microorganisms or their products in the serum of immune animal and patients with infectious diseases, such as agglutinin, precipitin, etc. they all were called antibodies and the substance which caused the antibody production was called antigen. The discovery of the specific binding of antigen and antibody laid the foundation for serological diagnosis of infectious diseases. Eventually they both shared the Nobel Prize of Physiology and Medicine in 1901.

(4) The discovery of complement

In the late 19th century, the phenomenon of immune bacteriolysis was soon followed with the anti-toxin. In 1894, Pfeiffer found the phenomenon of immune bacteriolysis. He injected vibrio cholera bacteria into abdominal tissues of a guinea pig which had been immunized by the bacteria, and found that the new injected bacteria dissolved rapidly. In addition, he injected immune serum and the corresponding bacteria simultaneously into a normal guinea pig and obtained the same result. In 1895, Bordet heated fresh serum at 60℃ for 30 min, and then added corresponding bacteria into it. He found that only the agglutination occurred and yet the bacteriolytic capacity was lost. So he considered that there were two kinds of substances in immune serum that related to bacteriolysis, one substance which was thermally stable and called antibody, could combine specifically with the corresponding bacteria or cells and cause agglutination; the other one, which was a thermally unstable substance and called complement, was a normal component of serum without specificity and could assist antibody in dissolving bacteria or cells and showed its effects only in the presence of antibody.

(5) The establishment of serological methods

In the 10 years after the discovery of antitoxin, the specific components such as bacteriolysin, agglutinin and precipitin were discovered in immune serum. They could react with corresponding cells or bacteria. In 1901, the word "immunology" first appeared in the book *Index Medicus* and the publication *Journal of Immunology* started in 1916. As a discipline, henceforward immunology was formally recognized by the people. Meanwhile, the subject serology which researched the reaction of antigen and antibody began to form and developed gradually. In 1896, Durham and others found agglutination reaction. In 1897, Kraus found precipitation reaction and Landsteiner discovered ABO blood group of humans. Since then, these experiments gradually

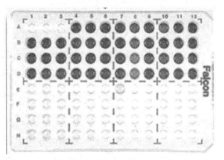

Figure 1-5　Serological reaction

made inroads into the clinical laboratory. In the following decades, serology research represented the mainstream development of the immunology. Many serological detection methods in vitro were found based on the reaction of antibody and antigen to detect antibody or antigen. These methods included agglutination, precipitation and complement fixation, etc. and provided important tools for the diagnosis and epidemiological investigation of infectious diseases (Figure 1-5).

(6) The studies on immunochemistry

After the concept of antigen and antibody were established, their physical and chemical properties and the chemical basis of the specific combining aroused interest. In the early 20th century, Landsteiner and others prepared an artificial azoprotein antigen by coupling aromatic organic molecule to protein and studied the material basis of specific reaction of antibody and antigen. His results indicated that the antigen specificity was determined by the structure and conformation of the small molecules.

In 1929, Heidelberger studied the antigen-antibody reaction quantitatively. In 1934, Marrack put forward the lattice theory of antigen-antibody reaction and theoretically explained the phenomenon of serological response. Tiselius and Kabat established the serum protein electrophoresis in 1938. They proved that antibodies were gamma globulin and studied the structure and function of antibody through some separation and purification methods. In 1941, Coons invented immunofluorescence technique. In 1942, Freund invented adjuvant. In 1955, Grabar and others established the immunity electrophoresis and found the heterogeneity of antibody molecules, which made significant progress in the research of antibody and antibody structure.

(7) The proposals of antibody production theory

In 1897, Ehrlich put forward a theory on antibody production, that was side-chain theory. He considered that antitoxin molecules were present in the cell surface and the combination with external toxins would stimulate cells to produce more anti-toxin molecules and these molecules shed from the cell surface into the blood. He, as an advocate of humeral immunity, confirmed that both toxin and non-toxin could induce antibody production in vivo and antibody could neutralize corresponding induction in vitro to undergo agglutination and precipitation. So he indicated that the formation of antibody was an immunity response phenomenon of the body and the corresponding antibodies were mainly produced in body fluids. Consequently, he established the humeral immune theory, which was not accepted by immunity academics. But his experiment provided important theoretical support for Behring's anti-toxin treatment and made it reasonable. In 1897, he published an important article about diphtheria antitoxin. In this article, the quantitative study of antigen-antibody reaction explained the relationship between antibody specificity and its chemical structure as well as the essence of complement binding to antigen-antibody complex. He made important contributions to immunochemistry and serology. He and Ehrlich shared the Nobel Prize for physiology and medicine in 1908 because of the side chain theory of the antibody production.

In the 1930s, Haurowitz and Pauling successively proposed the direct template theory and the indirect template theory of antibody production. They agreed that the specific structure of the antibody depended on antigen and denied the existence of antigen receptor in the membrane of antibody producing cells. The theory laid one-sided stress on the role of immunity response to antigen and yet ignored the nature of antigen recognition by body's immunity system, which transgressed the basic law of immune response and hindered the research on antibody production.

(8) The discussion on mechanism of body protective immunity

In 1890, German doctor Behring and Japanese scholar Kitasato found diphtheria antitoxin. Behring was awarded the Nobel Prize in Medicine in 1901 because of his research on antitoxin serum, particularly for the contribution of utilizing serum for therapy and prevention of diphtheria and tetanus, etc.

The study of body protective immunity mechanism drew much attention. During this period two schools came into being. One was the cellular immunity school represented by Metchnikoff and they considered that the anti-infection immunity was determined by phagocytosis cells in vivo; the other was humeral immunity school represented by Ehrlich and they believed that serum antibodies were the major factors in anti-infection immunity. Bordet found complement, which provided support for humeral immunity. These schools all adhered to their own views, but each school only addressed different side faces of complex immunity mechanism and all were one-sided. In 1903, Wright and Douglas found the existence of a substance (opsonin) which could raise the phagocytosis greatly in serum and other body fluids when they studied phagocytosis. German scholar Ehrlich put forward the side-chain theory, which unified the two schools initially and led people to realize that the body's immunity mechanism included two aspects: humeral and cellular immunity.

1.1.3 The period of early modern immunology

It is the period of modern Immunology from the mid-20th century to 1960s. In this period, people broke through template theory of anti-infective immunity, and had a relatively comprehensive understanding of the immunity response of organism. Meanwhile, many new findings frequently challenged the traditional concept of immunology and some new theories of immunity emerged. So actually, it is the period of biological immunology during which major achievements included the following aspects.

(1) The discovery of delayed-type hypersensitive reaction

After discovering mycobacterium tuberculosis, Koch attempted to achieve the purpose of immunotherapy by re-infecting subcutaneously patients with mycobacterium

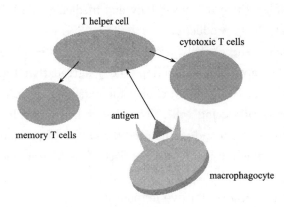

Figure 1-6   Cellular immunity

tuberculosis. Instead, it caused local tissue necrosis, which called Koch phenomenon. This phenomenon was specific and independent of antibody. Until 1942, Chase and others made a deeper study of Koch phenomenon. They transferred the serum of sensitized guinea pig to normal animals, which didn't cause tuberculin reactions; when the lymphocytes were transferred, the reaction was positive. Thus it proved that the tuberculin reaction was not inspired by the antibody, but by the sensitized cells. In addition it proved that the body could produce not only humeral immunity but also cell immunity (Figure 1-6).

(2) The discovery of immunity tolerance

In 1945, Owen found that there were two red blood cells present simultaneously in two fraternal twins and the phenomenon was known as mosaic blood cells. The different blood cells in the body don't cause immune response with each other and the phenomenon is known as natural tolerance. Since then, Medawar and others successfully induced alien artificial antigen tolerance in neonatal mice and revealed that the body's immune cells could lead to tolerance of the corresponding antigen in the developmental stage when exposed to autoantigen or alien antigen. In 1953, Billingham and Medawar successfully induced tolerance in mice, which strongly supported Burnet theory. Since the classical view of immunology was seriously challenged, people began to pay attention to the problem of immune biology, which promoted immunology to enter a new period: the period of immune biology.

(3) The proposal of clone selection theory

In 1958, inspired by side chain theory of Ehrlich and Jerne's "natural selection", Australian immunologist Burnet proposed that there was a variety of cell lines (clone) in vivo that correspond to all kinds of antigens. Antigen entering body would select the appropriate cell lines to combine with it, then activated the cells, and made it to proliferate and produce specific antibody. These views were called clone selection theory.

In the embryonic period, if an antigen was exposed to selected appropriate cell lines, these cell lines would be excluded or lose activity in the inhibitory state. The body lost the reactivity against this antigen and tolerance was formed. Thus the theory explained the body's tolerance for autoantigens. This assumption can not only explain the mechanism of antibody formation but also many phenomena of immune biology, such as antigen recognition, immune memory, immune tolerance and autoimmunity and so on. It boosted greatly the development of modern immunology. In 1960, Burnet and Medawar shared the Nobel Prize for their research on acquired immune tolerance.

The basic points of clone selection theory are as follows: ① There is a variety of cell lines identifying different antigens, and the cell surface of each cell line expresses and recognizes the same receptor of corresponding antigen; ② After antigens enter body, they will react with corresponding cells of receptor cell lines, which make the cells to activate, proliferate and differentiate into effector cells or memory cells; ③ In embryonic period, the cells will be damaged, removed or be in inhibitory state when autoantigen immune cells contact with the autoantigen, and the cell was called a forbidden clone; ④Although the theory isn't perfect, it explains most of the immune phenomena, such as antigen recognition, immune memory formation, the establishment of self-tolerance and the occurrence of autoimmunity. This theory is accepted by most scholars and was evidenced by the later experiments. This epochal immunological theory promoted the development of modern immunology greatly.

In addition, Snell found histocompatibility antigens in 1948. Witebsky established the animal model of autoimmune disease in 1956. Meanwhile, immunology technology developed rapidly, for instance, indirect agglutination reaction and immune label technology were established, which further promoted the research and application of immunological basic theory. Biological immune phenomena forced people to jump out of the anti-infection field or even stand outside medical field to survey immunology.

1.1.4   The period of modern immunology

Period of modern immunology is from 1960s onwasds. During this period, the position of lymphocyte cell lines in the immune response was confrmed and the structure and function of immunoglobulin were clarified; a great deal of research were carried out about immune system, particularly about cytokines and adhesion molecules; valuable discussions were carried out about diversity and type of conversion of immunoglobulin and breakthrough achievements were made in many aspects. In the 1980s, significant progresses were made in molecular immunology. Firstly, a breakthrough was made in the genetic control of antibody diversity, which greatly promoted the development of immunology. Because of the development of biotechnology, antigen-specific T cell clones in vitro could be established and made use of cell, which with cellular and molecular hybridization technology, all together, provided good conditions for studying T cell receptor property at the molecule and gene level.

(1) The studies on immune system

In 1957, Click found that extirpating chicken bursa would cause the drawback of antibody in chicken. It was considered that the bursa was the main place for antibody production. Meanwhile, the cell producing antibody was called B cell. Miller and Good found that removing of the early thymus would cause immune defects and a serious decline in antibody production in mammalian. The result proved that the immune cells in thymus executed cellular immunity, and the cells were called T cells. Claman and Mitchell put forward the concept of T cell subsets in 1969. Since then, the distribution of the T and B lymphocytes after maturing and differentiating in thymus and bursa were further confirmed in peripheral lymphoid tissue. The synergy of T, B cell in antibody production was also confirmed. These achievements established the basis of the immune histology and cytology.

(2) The structure and function of antibody

In 1960s, Porter obtained active fragment (Fab) and crystallized fragment (Fc) by papain hydrolysis. Edelman obtained the antibody polypeptide chains by chemical fracture method. They demonstrated the molecular structure of antibodies by proving antibody consisted of polypeptide chains by chemical reduction method and demonstrating heterogeneity of antibody molecules through antigenic analysis. Since then, people unified name of the antibody globulin and established the classification of immunoglobulin: IgG, IgM, IgA, IgD and IgE (Figure 1-7).

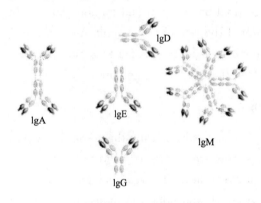

Figure 1-7　Immunoglobulin

In 1950s, Edelman (1929 -) and Porter (1917-1986) studied the basic chemical structure of antibodies with serum and urine of the multiple myeloma through enzyme digestion and other chemical methods. World Health Organization named it immunoglobulin in 1964. In 1969, Edelman completed the determination of primary structure of the human immunoglobulin. In 1972 he won the Nobel Prize for Physiology and Medicine.

(3) The proposal of immune network theory

In 1972, Jerne put forward immune network theory. The doctrine considered that: antibodies and antigen receptor and antibody on lymphocyte surface have unique specificity; they kept a relatively stable state before antigen entering; when the antigen went into the body, the balance was broken, which would cause the production of specific antibody. When the specific antibody reached a certain amount, anti-idiotypic antibody would be produced. Idiotypic determinant of a group antibody could be recognized by anti-idiotypic antibody molecule in body and a lymphocyte surface antigen receptor may be recognized by another group of lymphocyte anti-idiotypic surface receptor. Thus, an idiotypic-anti-idiotypic immune network formed between lymphocytes and antibodies. Network theory discussed the immune regulation mechanism and considered that the immune response proposed by antigen stimulation was not endless but constrained by the idiotypic antibody to maintained the body's physiological stability and balance. Jerne received the Nobel Prize in 1984.

(4) The studies on antibody diversity

As early as 1960s, Dreyer and Benner and others proposed a hypothesis that the peptide chain of encoding immunoglobulin (Ig) gene was composed by two genes. In the embryonic period, they are separated from each other and rearranged and spliced together during the differentiation and development of B cell. They speculated that gene of eukaryotic cells may be separated from each other firstly and before expression they must be rearranged and spliced together during cell differentiation. Japanese scholars Tonegawa and others cloned the encoding genes of IgG V region and C region by molecule hybridization techniques and demonstrated that B cells encoded gene structure of IgG during the period of differentiation and development by utilizing cloned cDNA probes. These findings illustrated the genetic control of the diversity of antigen binding sites and thus provided a basis for genetic control of immunoglobulin diversity. Tonegawa won the 1987 Nobel Prize in Medicine for it.

Thus, immunological gradually developed into an independent subject with its own theoretical system and special research methods.

(5) The studies on cytokines and immune cell membrane molecules

In the recent 20 years, immunology research focused on cytokines and immune cell membrane molecules. At first cytokinse were extracted from the cell culture medium and their function and structure were studied. Interleukins (IL), interferon (IFN), tumor necrosis factor (TNF), colony stimulating factor (CSF) and other cytokines were discovered successively and then one had a better understanding on their biological functions and behaviowr characteristics. On this basis, cytokines could be mass-produced through genetic engineering, which promoted the applications of cytokine in clinical therapy and experimental research.

Immune cells have many types of membrane molecules including T, B cell antigen recognition receptors (TCR / BCR), major histocompatibility antigen, leukocyte differentiation antigen (CD), mitogen receptor, cytokine receptors and immunoglobulin receptors, etc. In the early 20th century, it was found that the normal tissue or tumor transplant rejection between different species or individuals was determined by cell surface major histocompatibility molecules (MHC I / II molecules). Since then, it was noted that a restriction of MHC existed. That was when T cell antigen receptor (TCR) recognized antigen, they identified their own MHC molecules at the same time. Great efforts have been for the study of the leukocyte differentiation antigen (CD) to reveal the function of T cell subsets, cell activation pathway, membrane signal transduction and cell differentiation regulatory mechanisms. In addition, during the study of cytotoxic T cells (CTL) killing effect, it was found that the FasL expressed by CTL cells would combine with Fas which is expressed by target, consequently caused cascade activation of caspase in target cells, then cleaved DNA, and led to the death of target cell. These processes are called programmed cell death (PCD) or apoptosis.

(6) The development of application immunology

In 1956, Glick found the role of the bursa; in 1961, Miller discovered the function of the thymus. In 1966, Claman and others distinguished B cells and T cells and found their immune synergy simultaneously. Afterwards, they established a methods to identify different T cell subsets and found the interaction mechanisms between immune cells and the major histocompatibility complex restriction. Meanwhile, the humeral immunity field continued to develop in-depth.

In 1975, Kohler and Milstein invented hybridoma technique. They fused mouse myeloma cells and B cells sensitized by sheep red blood cells (SRBC) in vitro to form hybridoma. The hybridoma cells not only maintained the unlimited growth features of myeloma cell, but also had the ability to synthesize and secrete antibody. With the application of this technology, uniformitarian antibodies which only recognized a single epitope could be produced. This antibody was called monoclonal antibodies (McAb). There are many advantages of McAb, such as high purity, high specificity and that can be mass produced, etc. It has been widely used in serological diagnosis and the detections of immune cells and other cell surface molecule. Meanwhile, it was also applied to researching on directed therapy of cancer by chemically coupling to nuclide, toxins or drugs. It was developed rapidly on molecular probes for immune identification and immune therapy as well as directed delivery of drug with monoclonal antibody, which promoted the development of molecular biology, cytology, immunology and medicine, etc.

It was also a breakthrough achievement to apply molecular biology techniques in immunological research. Genetically engineered antibodies, such as fully human antibodies, single chain antibodies and bispecific antibodies, were prepared by molecular hybridization and molecular genetics theory. They have more advantages than the McAb. At present, the molecular hybridization technique has also been used to study gene structure, function and expression mechanism of immunoglobulin molecules, T cell receptor molecules, complement, cytokines and MHC molecules. In the 1980s, the polymerase chain reaction (PCR), an in vitro nucleic acid amplification technology, appeared. The technology was applied to preparing recombinant vaccines, DNA vaccines and transgenic plant vaccines, thus that opened up a new prospect for immunoprophylaxis. It has yielded great economic and social benefits because of wide preparation of recombinant cytokines with genetically engineering.

Contributing to the development of RT-PCR technology, functional fragment of antibody molecules was direct or expressed successfully by *E. coli* and the phage display technology (phage display) came into being. In the early 1990s, phage antibody library technology appeared, which made the development possible from the application of DNA recombinant technology in reforming monoclonal antibody to cloning a new monoclonal antibody with genetic engineering technology. Henceforward antibody engineering went into a new era.

Transgenic animals played an important role in scientific research. Integrating a foreign gene into the animal genome is known as transgene. The transgene means integrating a foreign gene into the genome of some cells or the genome of all organizations. The animals with only a portion of the genome of cells integrated with foreign genes are called chimeras animals (chimera mosaic animal). If all the cells in animals were integrated with exogenous genes, the animals would have the ability to transmit the foreign gene to successors. This animals are usually called transgenic animals.

In contemporary life sciences, the progress of stem cell research drew a great attention. Stem cell research and its application was ranged as the top ten scientific results twice by U. S. *Science* magazine, so that the stem cell has become the most influential front-line science following the Human Genome Project and has become a hotspot of contemporary life sciences.

Stem cells (stem cell) are early undifferentiated cells with multi-differentiation potential and self-replication function. Under certain conditions, they can differentiate into a variety of functional cells and formed into a variety of tissues and organs. If the stem cells can be separated and differentiated along a specific direction, then healthy tissues will replace the damaged tissue cells of patients to treat diseases.

Scientists believe that stem cell research is of incalculable medical value. Theoretically, stem cells can be used to treat various human diseases. It has been discovered that stem cells from human embryonic or bone marrow can be used for cultivation of different human cells, tissues or organs, so it and is expected to become a new source of organs for transplant. Tissue and organ transplantation would be the fundamental measures for capturing the human cardiovascular disease, cancer and other major diseases. The wide range of clinical applications of stem cells and their derivative tissues and organs would lead to a new medical technology revolution[5, 6].

1.2   The extensive application of immunological techniques

1. 2. 1   Immunology and medicine

The development of immunology and its penetration to the medical disciplines resulted into a number of branches and cross-immune subjects, such as immunopathology, immunogenetics, immunopharmacology, immunotoxicology, neural immunology, tumor immunology, transplantation immunology science, reproductive immunology and clinical immunology, etc. Studies on these branches have greatly promoted the development of modern biology and medicine. The development of immunology would promote medical advances in cancer prevention, organ transplantation, prevention of infectious diseases, prevention and treatment of autoimmune diseases, reproductive control and anti-aging, etc. [7-9].

1. 2. 2　Immunology and biology

Immune system recognizes self and non-self, and immune tolerance to self components as well as immune response to non-self components, which are all related with the transmission of information between cells, intracellular signal transduction and energy conversion and other basic characteristics of life processes. The function of immune system was controlled by heredity. Till recently, little was known about the genetic control of a variety of physiological functions of the body. Study on immunogeneticsis uncovered the genetic control mechanisms of physiological function system of the body firstly. It is significant for the research on physiological function of body at gene level.

The process of development and maturation of immune cells are accompanied by changes in membrane surface markers. Malignant changes in immune cells at any stage of its development have inherent, specific membrane markers. These malignant cells in different stages are ideal models for the study on malignant transformation mechanisms of cells and of great importance for studying the development of malignant tumos.

The studies on the structure and functions of MHC gene complex, allelic exclusion of gene expression of immunoglobulin, genes of immunoglobulin and other immune molecular and DNA binding proteins regulate the expression of cytokines and that greatly enriched the molecular biology and promoted the recognition of gene structure and expression of eukaryotic. The development of immunological technology provided powerful tools for life science research. Applications of monoclonal antibody have brought about a breakthrough change for the biological sciences and the combination of immunohistochemical and molecular hybridization techniques have enabled the studies on genes and their expression to the quantitative, qualitative and location extent. Clearly, the development of immunology played an important role in biology[10].

1. 2. 3　Immunology and the development of biotechnology

Recalling the history of immunology, it is clear that every step of important progresses in immunology is driving the development of biotechnology. Between the late 20th century and the early 21st century, the tremendous success of application of immunology on infection-fighting promoted the development of biological products industry. Applications of artificial active immunization and passive immunization effectively controlled the spread of many infectious diseases. In the past 30 years, tremendous progress in immunology promoted the development of biological high-tech industry at a dee-per level and broadened its scope. The application of monoclonal antibodies produced by cell engineering and cytokines produced by genetic engineering provided a large class of new drugs with immunological regulation effects for clinical medicine. These new drugs primarily focused on regulating the body's immune function with fewer side effects, and thus had an irreplaceable role in the treatment of various diseases compared with tradi-tional medicines. Biological high-tech industry with cytokines and monoclonal antibodies as the main products has become an emerging industry with a huge market poten-tial[11-13].

1.3　Present situation and prospects of immunology

1.3.1　The in-depth development of immunology at molecule level

One after another breakthrough in molecular immunology was made in recent years. For example, people found that the natural immune system could recognize the pattern shared by pathogens through specific receptor which was called pattern recogni-tion receptors (PRR). Recently, it has been studied thoroughly on the molecular struc-ture and the signal transduction pathways of PRR and it has been found the molecular mechanism has a leading role on innate immune system in guiding acquired immune re-sponse. In addition, a large number of membrane molecules were discovered by using monoclonal antibodies and molecular biology technology. There have been as many as 250 membrane molecules which were uniformly named leuckocyte surface differentiation antigens (CD molecules). Meanwhile, it quite thorough studies on the molecular struc-ture, biological functions, gene structure of major histocompatibility complex (MHC) molecules, T cell and NK cell recognition receptors, complement molecules, cytokines and chemokines[14] have been conducted.

In recent years, some breakthroughs have been made in the study of the molecular mechanism of lymphocyte. For examples, it was found that PU.1/Ikaros could adjust the development of T and B cell, GATA-3 influenced the growth of T lines, EIA / EBF / Pax could regulate the growth of B cell, and so on. Molecular mechanisms studies on the differentiation and development of T cells in the thymus showed that the thymus cell membrane molecules, pTα / TCR molecules, Bortch molecules, CD30/CD153 and CD69 molecules all were related with their differentiation. It was also found that Ras-MAPK signal transduction was associated with positive selection and yet was unrelated to the negative selection.

The studies on the molecular mechanisms of activation, proliferation, differentia-tion and apoptosis of T and B cell showed that the process of T and B cells immune re-sponse was related with a variety of membrane receptor molecules, such as the TCR /

BCR complex molecules, adhesion molecules, chemokine receptor, complement receptor, Fc receptor and Fas, etc. These molecules play different roles in immune response.

Mouse and human helper T cell (Th) could develop and differentiate to different functional Th cells (such as Th1, Th2, Th3 etc.) under different environmental conditions (such as APC type, antigen type, cytokines). Currently these T cell subsets are mainly distinguished by the characteristics of their cytokine secretions. It can be predicted that the membrane surface molecules which can be used to distinguish different Th cell subsets will be found in the near future. Th1/Th2 balance adjustment can direct the types of immune response and has drawn much attention in autoimmune diseases.

1.3.2 The interaction of immune system and neuroendocrine system

In early 1980s, it was found that immune cells could synthesize opioid peptides. Since then a rapid progress has been made in the field of the relationship between immune system and neuroendocrine system, which greatly enriched the understanding on the stabilized mechanisms of internal environment. It has been demonstrated that there are neurotransmitter and endocrine hormone receptor on the surface of immune cells and the surface of nerve cells also contains cytokine receptor. Thus, the two systems can exchange information and regulate functions through their surface receptors and release media to maintain stable internal environment.

1.3.3 The infiltration of immunology into biology, basic medicine, clinical medicine and preventive medicine

The infiltration of immunology into the various disciplines of biology and medicine spawned a lot of subdisciplines, such as the immune biology, molecular immunology, immunogenetics, immunopathology, immune pharmacology, immunotoxicology, neuroimmunology, tumor immunology, transplantation immunology, reproductive immunology, blood immunology, immunology of skin and the elderly immunology, etc. The development of these branches has greatly promoted the progress of modern biology and medicine, which produced a great impact on the prevention and treatment of cancer, organ transplantation, infectious disease prevention, treatment of autoimmune diseases, reproductive control and anti-aging, etc. [15-17].

1.3.4 The applied research of immunology promoted the development of biotechnology

The development of immunology at each stage has pushed the development of biotechnology. In the period of classical immunology, the contribution of immunology in anti-infections promoted the advent of biological products. During the period of modern immunology development, the monoclonal antibody produced by cell engineering, and recombinant cytokines as well as genetically engineered antibodies produced by genetic engineering has provided a large class of new treatment drugs with immune regulation, neutralizing toxins or other treatment fuctions. The studies on genetically engineered vaccines and DNA vaccines have opened new avenues for the prevention of infectious diseases. Currently, research and development of biotechnology products has become an emerging industry with a huge market potential. The 21st century will be the era of the rapid development of biological engineering. The results of cloning new genes, revealing the pathogenesis of immune-related diseases, and exploring more effective anti-autoimmune disease, transplanting rejection and tumor immunotherapy will improve human health and promote multi-disciplinary development. Thus these studies are of great theoretical significance and social efficiency[18-20].

References

[1]   Xu CL. Food immunology. Beijing: Chemical Industry Press, 2007.11

[2]   Anonymous. A brief history of immunology. http://www. biox. cn/content/20060620/45378. htm. 2008-9-1

[3]   Riott I and others. Immunology. New York: Gower Medical Publishing Ltd. , 1989: 956

[4]   Shepherd P, Dean C. Monoclonal antibodies. New York: Oxford University Press, 2000: 297-300

[5]   Burmester, Gerd-Rüdiger, Pezzutto and others. Color atlas of immunology. Beijing: China Science and Technology Press, 2006

[6]   Lydyard PM, Whelan A. Instant notes in immunology. Beijing: Science Press, 2000

[7]   Qian CS. Combination of flow injection-time resolved fluorescence to analyze CEA. Modern Laboratory Medicine, 2005, 20(2): 38-39

[8]   Luo JX, Wu LP, Yang XC and others. Research progress for flow injection immunoassay. Journal of Pharmaceutical Analysis, 2001, 21(1): 64-66

[9]   Huang GX. Evelution of haemolysis assay in vivo & liposome immunoassay of complement (CH50). International Medicine & Health Guidance News, 2004, 10(2): 65-66

[10]   Wen ZL, Wang SP, Shen GL. A summary description of development in immunosensor. Journal of Biomedical Engineering, 2001, 18 (4): 642-646

[11]   Zhen XD, He D. The progress on the research of immunoassay for pesticide residues in food. Journal of Chinese institute of food science and technology. 2006, 2(6): 88-95

[12]   Li XX, Chen YL, Wang GL and others. The application of new immunoassay methods in food inspection. Science and Technology of Food Industry, 2006, 27 (3):170-174

[13]   Wang LL, Fang H, Zen QY and others. Detection of complement factors using piezoelectric immunosensor array. Journal of Analytical Science, 2001, 17 (5):358-363

[14]   Zhang JB, Wu YZ. MHC-peptide tetramer complexes technology and its application in T cell research. Chinese Journal of Immunology

[15]   Yu W, Wu JG. Research advance and clinical application of ELISpot assay. Chinese Journal of Laboratory Medicine, 2006, 24(6): 496-498

[16]   Wu XC, He L, Zhou KY. The research and clinical application of time-resolved fluoroimmunoassay. Medical Recapitulate, 2006, 12(7): 434-436

[17]   Yu W, Sun YK, Gu L and others. Protein and antibody microarrays and their biomedical applications. Progress in biochemistry and Biophysics, 2002, 19(3): 491-494

[18] Zhang Y, Su YH, Ba CF and others. Application of cytometric bead array capture technique in the medical field. Chinese Journal of Clinical Rehabilitation, 2006, 10(37): 129-131

[19] Xie MS, Hao JM, Liu GZ. Clinical value of multi-tumor markers liquid BioChip in detection to malignant tumor. Central China Medical Journal, 2006, 30(1): 25-26

[20] Wang DC, Wang LD. SELDI-TOF-MS protein chip technology and its application in cancer research. Tumour Journal of the World, 2004, 3(4): 301-304

Chapter 2　The Basic Principles of Antigen-antibody Reaction

2.1　The general principles of antigen-antibody binding

2.1.1　Antigen

2.1.1.1　Classification of antigen

Antigen can be divided into four kinds according to the source of antigens and their genetic relationship with body: ① xenoantigen: a kind of antigen from another species. A variety of animal serum (e. g. horse serum), various microorganisms and their metabolites (e. g. exotoxin) are xenoantigens to human being. ② alloangtigen: antigens from individuals of the same species but of different genotype (such as human erythrocyte antigens, leucocyte antigen). ③ autoantigen: the composition of body self which can cause autoimmune response, including the isolated components which have never been in contact with its own lymphocytes in embryonic period (lens protein, brain tissue, etc.), and the non-isolated components whose conformations change under the influence of infection, drugs, burns or ionizing radiation. ④ heterophile antigen: common antigens presented on the cell surface of animal, plant and microorganism of different species. They have broad cross-reactivity. One typical example is Forssmann antigen. Forssmann found that the antibody, prepared by immunizing rabbits with suspension made of a variety of guinea pigs organ, can not only react with corresponding organs antigen, but also agglutinate with sheep erythrocyte cells[1].

According to the immune response dependence on T cell, antigen can be divided into two kinds: ① thymus-dependent antigen (TD antigen), most of which require the aid of T cell to stimulate body to produce antibodies and cause memory response and TD antigens, such as blood cells, bacteria serum components, bacteria, can stimulate body to produce antibodies, mostly IgG, and also cellular immunity; ② thymus-independent antigen (TI antigen), which can stimulate body to produce antibody without or with less dependence on T cells, and does not cause memory response and most of them are polymers, and have repetitive epitopes. For example polysaccharides, which can stimulate B cells to produce IgM antibodies, but don't cause cellular immunity [2].

Other classifications are as follows: ①They can be divided into endogenous antigen and exogenous antigen according to their natural source. ②They can be divided into natural antigen and artificial antigen according to preparation sources, methods, chemically modified antigen and synthetic antigen. ③ According to performance they can be divided into complete antigen and incomplete antigen[3].

2.1.1.2 Immunogenicity of antigen

Immunogen is the antigen with immunogenicity and immunoreactivity. Immunogens are antigens, but not all antigens are immunogens. Antigens involve a relatively wider range, including immunogen, hapten and complex bodies composed by different antigen such as microorganisms, blood cells and tissue cells. However, these two nouns have long been used indiscriminately because of tradition. (The antigens commonly used for immunization are listed in Table 2-1).

Table 2-1   Antigens used commonly in immunology

Name	Molecular mass/kDa	Characters
Bovine serum albumin, BSA	66 (containing 585 amino acid residues)	Apt to be separated, available at a large number, moderate molecular weight, soluble in water, stable structure, and strong immunogenicity
Ovalbumin, OVA	About 40	The main proteins in egg white
Bovine γ-globulin, BGG	160	Most of immunoglobulin in normal bovine serum is IgG.
Keyhole limpet hemocyanin, KLH	450-13,000	Extracted from the keyhole limpet (a fairly ancient animal, unrelated to human beings and had never been in contact with human beings), the molecular mass is up to several million, one of the strongest immunogenic proteins as people have known
Flagellin	40(monomer)	The major protein monomer of organs (flagellum) of motivated bacteria, a weak TI antigen; polymer of flagellum is a TI antigen with high immunogenicity.
Sheep red blood cell, SRBC	—	One of the most widely used particulate antigen. The biggest advantage is that it is cheap, easy to obtain. Moreover, it can be used as target cells for hemolysis caused by antigen-antibody reaction.

Immunogenicity refers to the performance of causing immune response, including the induction of producing antibodies and effector T lymphocytes. It involves the interaction of antigen molecules and immune cells, which means that the antigen molecule must go through the processing, handling and presenting by antigen-presenting cells and be recognized by the antigen recognition receptors of T cells and B cells. The immunogenicity of antigens relates with not only the chemical properties of antigen, but also the characters of body's immune response.

The immunogenicity of antigens, firstly, is determined by the chemical properties of itself, and, secondly, by the species of animal. The same antigen will present very different immunogenic strength to different species of animal or animal individuals, so the success of an antigen inducing immune response of host depends on three factors: the nature of antigen, the host reactivity and the immune method. Here we will focus on thu factors and the nature of antigens[4].

(1) Foreignness

Normal mature immune system can distinguish the substances of themselves and of non-host substances, and generally do not take immune response to the substances of themselves, but only take response to the non-self-substances. As for human being, pathogenic microorganisms and some of its products, animal serum proteins and other foreign tissue cells, all are good antigens. The immune recognition is not determined by spatial location of material, but the recognition of lymphocytes, so sometimes the material of host can also be antigens of itself.

Usually antigen is non-self-material. The antigenicity of different materials depends on their chemical heterogeneity, which is the material basis of immune recognition. In general, the farther the distant of the genetic relationship of materials is, the greater the difference between their chemical structures is, the stronger the antigenicity will be; in contrast, the closer the genetic relationship is, the weaker the antigen will be. The best example is organ transplantation: heterograft will cause strong rejection and cannot survive, but homoplastic graft only leads to weak rejection and can survive for a certain period; the autologous graft will not cause rejection and can survive for a long periods. Another example is: the duck serum protein is a weak antigen for chickens and yet a strong antigen for rabbite; many homologous tissue proteins of mammalian, such as thyroglobulin, brain, testis and placenta, have the same organ-specificity because, in the germ line, the differentiation of these substances is low and the differences in the structure are small.

(2) Molecular size

Molecular size can affect the formation of immunogenicity. The molecular weight of an effective immunogen is mostly more than 10 kDa, and the greater the molecular, the stronger the immunogenicity will be. This may be because polymer materials are easy to form gel in aqueous solution and will stay for a long time in body, and have more opportunities to contact with immune cells, which would be propitious to boost the body's immune response. In addition, the chemical structure of macromolecules are more complex, and they contain relatively big amount and multi-type of effective antigen gene.

Generally the larger molecular weight proteins (more than 10 kDa) have excellent immunogenicity. The molecular weight of carbohydrates is always low, so most sugars don't have immunogenicity, but when aggregated into polysaccharide, it will be immunogenic. But the molecular weight of 10 kDa is not an absolute limitation, for example, molecular weight of gelatin is up to 100 kDa, but its immunogenicity is very weak, however, the molecular weight of insulin is only 5,734 Da, but it is immunogenic.

(3) Chemical structure

The formation of the immunogenicity also requires the complex chemical structure of molecule. The materials with straight-chain structure always lack immunogenicity, but the materials with multi-branched or ribbon structure are easy to become immunogens and spherical molecules always have stronger immunogenicity than the linear molecules. Synthetic linear polymers consisting of single amino acids, (e. g. poly-L-lysine and poly-L-glutamic acid) have no immunogenicity, but random linear copolymers consisting of a variety of amino acid are immunogenic; moreover, the immunogenicity of them will increase with the increasing of amino acid types in copolymer and it is more effective when containing aromatic amino acid. The gelatin molecule is of straight chain structure and no branches, and lacks cyclic groups, so its immunogenicity is weak, but the immunogenicity of gelatin will be significantly enhanced when connected with 2% of tyrosine.

2.1.1.3　Antigen specificity

One of the greatest characteristics of antigen reaction is its specificity. The specificity is very prominent in both their immunogenicity and reactogenicity. For example, the immune response induced by typhoid bacillus is only against typhoid bacillus, that is, shigella can't induce immunity against salmonella typhi, and will not react with anti-salmonella typhi antibodies. This is the fundamental basis of traditional immunization to carry out immunoprophylaxis and immunodiagnosis.

Antigen specificity relates with the types the arrangement, specific genes and space configurations, and even their charge properties and hydrophilic properties, i. e. amino acids in the protein molecules. But its specificity is not determined by the molecule, but the specific sequence and its spatial structure consisting of several amino acid residues in the molecular surface, and they are known as epitopes or antigenic determinant. When these epitopes are recognized by lymphocytes, it will induce immune response, and when recognized by antibody molecular, it will cause antigen-antibody reaction, which is the basis for study of antigen-specificity.

Responsiveness of different animal species and even different individuals of the same animal species to the same antigen varies widely and it is related with different genetic, physiological state, individual development and other factors. Immunization methods include antigen injection method, dose, frequency and interval of time, and the use of adjuvants etc. All these factors can affect the immune response. In short, only with a good organism, the host in a better physiological state and by a appropriate immune style, it can cause the immune response, then the antigen really has the immunogenicity[5].

(1) Epitope

Antigenic determinant, also known as epitope, refers to the special chemical group in the surface of antigen, which determines the antigen specificity and is the material basis of the specificity of the immune response and immune reaction. Epitope is the activity region of immunogen, the antigen specific receptor (TCR / BCR) or the basic unit of antibody binding on lymphocyte surface. Epitope is only a special structure of an antigen molecule composed of a few amino acid residues and can fully contact with lymphocyte or antibody in immune responses. Epitope is small and its size is equal to the corresponding antibody binding sites, about 3 nm $\times$ 1.5 nm $\times$ 0.7 nm, which is the size of 5 to 7 amino acids, monosaccharide or nucleotide residues, and at most no more than 20 amino acid residues.

The compositions of epitope have at least two types: ① continuous linear sequence consisting of certain amino acid residues arranged in a certain order, called sequential epitopes or linear epitopes. Sequential epitope is the primary structure of protein molecules, very stable and insusceptible to heat denaturation and protein conformational change. ②These epitopes, formed by the three-dimensional structure which comes from the folding and arrangement of 2-3 incontinuous amino acid residues in molecule, are called conformational epitopes. Sometimes, the continuous peptide sequences which presents α-spiral arrangement may also play the role of conformational epitopes. Antibodies of conformational epitopes can be used to study three-dimensional structure changes of proteins in the process of physiology and pathology. However, conformational epitope is the secondary or tertiary structure of protein, less stable, It will be completely destroyed and cannot be restored when the protein is denatured by heat or enzyme hydrolysis. Therefore, it's difficult to isolate and study it.

The number of antigen epitopes can be measured by calculating the number of antibody molecules which combined in saturated conditions. In general, the number of epitopes was positively correlated with the antigen molecular weight. For instances, chicken egg protein with molecular weight 42 kDa has 5 epitopes, and thyroglobulin with molecular weight 700 kDa has about 40 epitopes. An epitope can combine with an antigen binding site on antibody molecule, so the number of antigen epitopes is called antigenic valence. Antigenic valence is the total number of epitopes combining with antibody molecule. Each hapten can be regarded as a single epitopes. Natural antigen contains many different epitopes and a large number of various "natural" hapten molecules, for examples, chicken egg protein has 5 antigenic valence, and thyroglobulin has 40 antigenic valence.

Although a molecule has multiple antigen epitopes, only one or one kind of epitope plays a major role in inducting host immune response and stimulates host to produce the specific immune response mainly to this epitope. This phenomenon is known as immunodominance and the epitope which plays a key role is called dominant epitope. This principle also applies to different amino acid residues in the same epitope: an epitope can also has dominant group and the epitope specificity will be significantly changed if the groups are replaced. It may be related to the position of epitope in antigen molecule or the location of dominant groups in epitope. The hapten plays the role of dominant groups in epitope.

It has been approved that only epitope in the surface of antigen can contact with lymphocytes and antibody molecules and then bring immune effects. The research with poly-lysine as frame, with alanine (A), glutamate (G) and tyrosine (T) sequence as branch chain can clearly prove that: binding the G and T to the external frame can mainly induce the production of antibodies against G and T, but binding consequent A to the external frame will mainly induce the production of antibody against A.

The antigen immunogenicity firstly exhibits that lymphocyte recognizes the epitope. It has been considered that cell-mediated immunity and humoral immunity are against different parts of the same antigen. For example, immunizing the mice with glucagon can produce the antibody against its amino acid, but cell-mediated immunity is against the carboxyl end. Thus the conclusion is drawn that the T cell and B cell don't recognize the same type of epitope (Table 2-2). Then epitope can be divided into two types: B cells (recognition) epitope and T cell (recognized) epitope.

B cell epitope: B cell can recognize the epitope, which then will induce antibody responses, but the epitope can't be recognized by T cell. B cell recognizes soluble antigens which bind to the membrane surface of immunoglobulin, and the epitope recognized is on the accessible part of antigen surface. B cell antigen recognition receptors (BCR) or antibody molecules can react with the antigen which haven't been treated by antigen processing cell (APC), and the epitope recognized by B cell are mainly located in the molecular surface. There is a large quantity of research on such epitopes, and most of the specific antibodies currently applied are derived from the research about B cell epitopes.

Table 2-2　Comparison of B cell epitopes and T cell epitopes

Subject	B cell epitopes	T cell epitopes
The receptor distinguished	BCR (membrane surface Ig)	TCR
The property of the epitopes	Natural peptides, nucleic acids, polysaccharides, small molecule compounds	Denatured peptide fragments
The type of epitopes	Sequence epitope, conformation epitope	Sequence epitope
The size of epitope	5-15 amino acids, simple sugars, nucleotides	8-12 amino acids (CD^{8+} T) 12-17 amino acids(CD^{4+} T)
The location of epitope	Surface of antigen	Any part of antigen molecules
APC treatment	Need	Needn't
MHC molecule	Needn't	Need

B cell epitope includes continuous sequence and incontinuous sequence amino acid. The epitopes recognized by B cell must match the antigen binding site of antibody, and the antibody molecules combine with epitope through non-covalent bond. This binding reaction can only occur at close range. Antigen binding sites of antibody can enhance the binding to epitope only when they are relatively complemented in space. B cell epitope can be composed of sequent amino acids in polypeptide sequences (sequent determinant), and can also be composed of the amino acid residues which aren't connected in sequence and yet are connected in space structure (conformation determinant). After denaturation and degradation of protein antigen, the sequence determinant can maintain the ability of binding to the antibodies, but the conformation determinant will devitalize because of the disruption of protein spatial structure. B cell epitopes are often on the surface or at the folded and bent area of immune molecules, and the movability of these sites is propitious to present the best complement state of antibody and epitope. But the affinity between antibody and flexible epitope lower than when it is non-flexible epitope.

T cell epitope is used for T cell recognition and induction of cell-mediated immunity, whereas it can't be recognized by the B cell. T cells only recognize MHC molecule complexes which are antigenic peptide complexes presented on the surface of antigen-presenting cell. T cell epitopes of complete antigen can't be identified by TCR, even if the antigen only contains a few amino acid residues. Therefore, the antigen recognized by T cells must firstly go through certain treatment, for example the protein will be degraded to peptides and then connect with MHC molecules. Because only a few parts of T cell antigen receptor (TCR) are kept outside of the membrane, it can't combine with free antigen like antibody molecule and can only recognize the MHC binding epitopes which are presented by antigen presenting cells. T cell antigen receptor can only recognize small molecular peptides containing about 10 to about 20 amino acids and MHC molecules, but can't recognize natural soluble antigen. The peptide fragments connected with the peptide binding groove of MHC molecules usually binds with MHC groove wall through both ends residues anchoring, and the middle of the hump will be recognized by TCR directly.

T cell epitope normally presents in the molecule surface. The antigen must be processed, and become small peptides by the APC and conjugate with MHC molecule, and then it can be identified by TCR. T cells recognize peptide fragments of antigen which are processed by APC. Therefore, the conformation changing of the antigen molecules does not affect the main features of T cell epitopes. T cells, like the antibody molecules, can also lead to cross-reactivity by common antigens, but the combination is not as effective as the original induced antigen.

T cell epitopes can induce cellular immune responses. Meanwhile, as attacking target of cytotoxic T cells, T cell epitopes are necessary for inducing antibody response, because activation of B cell requires the assistance of activated T cells and T cell activation must be started by T cell epitopes. It can be deduced that each antigen must have at least one T cell epitope, and then the antigen will probably be immunogenic. Molecules only having B cell epitopes can be a target of antibody, but it can't induce antibody responses. Only a few molecules may be exceptions.

(2) Cross-reactivity

Specificity refers to the pertinence between materials match each other and is an important factor of immune response and immune reaction. The specificity of antigens is expressed not only in immunogenicity, but also in the immune response. Immunological identification, diagnosis and prevention can be widely used, which owes to the specificity of immune response and immune reaction. The specificity of antigen-antibody reaction can be proved by various experiments. Antigen can cause specific reaction not only with corresponding antibody but also with other antibodies, which is called cross-reaction. Antibodies induced by an antigen can also be combined with its common antigen, and this phenomenon is also called cross-reaction. Cross reaction can be used to explain some of the immune pathology and can also diagnose some diseases. However, the combination between antibody and cross-reactants isn't as strong as that between antibodies and its inducing antigen. When an antibody combines with cross-reactants, only some parts can match and cannot achieve the full match of entire space. The reasons that cause the cross-reaction include following three points.

1) Antigen heterogeneity

Antigenic heterogeneity refers to the material for animal immunization containing multiple antigen molecules. This immunogen will cause animal to produce corresponding specific antibodies for a variety of antigens. Any substance, containing one or more antigen molecules which are the same with the aforementioned material, will have cross-reaction with the multiple specific antiserum.

2) Common epitope

The antigen which only has one epitope is called a pure antigen; but a pure antigen is rarely found in nature, and the majority of antigens have a variety of epitopes. In general, different antigens have different epitopes, so they all are specific; but sometimes the same epitope can also appear in different antigens, and this epotope is called common epitope, and the antigens with common epitope are called common antigens. Common antigens in nature, especially in microorganisms, is a very common phenomenon. If common antigen is present in the same genus or closely related species, it is called group antigen, and if present in different and unrelated species, it is called heterophileantigen.

There are many examples of common epitope, for example, salmonella can be divided into more than 40 serogroups and contain more than 2,000 serotypes according to their O antigen, and the members of the same group contains common O-epitope which is determined by specific monosaccharides. In human beings, animals, plants and microbes, a common epitope also widely exists. The antigen is named by the name of discoverer, and called Forssmann antigen, and the common epitope is composed of N-acetyl galactosamine amine, galactose and glucose which are covalently crosslinked on cerasome.

Considering the cross-reactions more accurately, antigen Ag1 containing a number of determinants can cause the production of antiserum containing a variety of specific antibodies. If another antigen Ag2 has one determinant the same with Ag1, it can react with the antiserum of Ag1. The determinant which two antigens contain is called a common determinant.

3) Similar epitope

Some common epitopes are structurally similar and are known as similar epitopes. If the structures of two different determinants are approximately similar to each other, the antibody for one determinant may have cross reaction with the similar determinant. Obviously, this cross-reactivity is different from the cross-reaction that occurs between the antibody and corresponding antigen or epitopes. The binding force between antigen and antibody when their conformation are coincided is significantly higher than that of cross-reaction when the determinants are similar.

2.1.1.4 Hapten and its application

In the early 21st century, Land Steiner, a pioneer of immunology, found that some small molecules without immunogenicity also can combine with antibodies. If this small molecule is coupled with an immunogenic macromolecular protein, the antibody response against the small molecule can be induced. He named this small molecule as hapten which derived from Greek language "haptien" (means impose and grasp); and the protein coupled with hapten was called carrier[5].

Between the 1920s and 1940s, the research on the specificity in antigen-antibody reaction reached a climax. The common process will be implemented through conjugating hapten with a kind of animal serum protein, for example horse serum protein, and immunizing this complex to another kind of animal, for example rabbit, and then detecting the obtained antiserum. This antiserum induced by conjugated antigen has two kinds of antibodies. One is against the hapten, and another is against the carrier protein. So the reaction between original conjugated antigen and antiserum can't prove the specificity of hapten. To eliminate the influence of carrier, hapten can be conjugated to a third kind of animal protein, for example egg albumin, then if the reaction between this conjugated antigen and antiserum can be observed, it can prove that this is owing to the specific binding between the hapten of conjugated antigen and the antibody in serum against the hapten.

The specificity of hapten is usually investigated through the hapten-carrier complex. For instance, four haptens aniline, benzoic acid, sulfanilic acid and p-amino benzene arsenate whose structures are known as respectively conjugated to a macromolecular protein, and immunized to animals respectively. As a result, the induced antibody can only react with the corresponding hapten, and can precisely distinguish the four analogous haptens. Even if the analogues are slightly different, for instance three kinds of aminobenzoic acid, the ortho, meta, and para isomers, the antibodies can still distinguish these molecules. This method can be used to produce many antibodies against small molecules, even anti-metal ions antibody, which has greatly promoted the research of antigen properties. The carrier in the hapten-carrier complex owns an inherent specificity, but it will not interfere the specificity of hapten. Nevertheless, the specificity of carrier will significantly influence the effect of antibody response induced by hapten. The carrier not only endows the hapten with immunogenicity, but also is closely related to the memory of the hapten immune response. Further studies have shown that the hapten-specificity is identified by B cells, and carrier-specificity is recognized by T cells, and only the collaboration of T cell and B cell can launch the antibody response to hapten and result into the re-response effect. Hapten-carrier conjugate has both the carrier protein epitopes and the hapten epitopes, thus it can stimulate the body to produce antibody specific to the carrier and hapten, respectively.

Studies have shown that the anti-hapten antibody can only be produced after the hapten is bound to carrier, which is called carrier effect. The hapten is the key group in epitopes. After being bound to macromolecule carrier, hapten can change the original antigenic epitope of carrier and also form a new epitope.

2.1.1.5　Superantigen

Superantigen(SAg), a class of powerful immune activators, can activate T cells and antigen presenting cells, ranging from the induction of T cell proliferation, release of a number of cytokine, immune suppression to the immune tolerance and immune anergy, and are related to autoimmune and insufficiency [6].

Superantigen, produced by certain bacteria, viruses, mycoplasma, and parasites, can activate thousands times more T lymphocytes (CD^{4+} and CD^{8+} T cells) than ordinary antigen, and also can activate NK cells and single nuclear cells. Non-specific binding can occur between superantigen and the MHC II molecules of antigen presenting cells (APC) as well as T cell receptor V_β area in a way different from ordinary antigen. Without the treatment process in APC, Sag can directly bind to MHC II class molecules with complete protein molecules. The binding site is not in the peptide binding groove, but on the lateral grooves. The biological specificity of superantigen is closely related to the immunologic mechanism of microorganism-induced disease.

Common bacterial superantigen are listed as follows: ① Staphylococcus aureus enterotoxin (SE) which can cause food poisoning and shock; ② Streptococcus SPE2A, SPE2B, SPE2C, which can cause shock, PePM protein which can cause rheumatic fever, SSA which can induce toxic shock-like integrated signs; ③ Pseudomonas which can produce PEA superantigen; ④ Clostridium perfringens which can produce CPS and CPE superantigen; ⑤ Mycobacterium tuberculosis which can produce MTS superantigen.

The effect of endogenous retroviruses and bacterial SAg in vivo has been studied extensively. Since the endogenous expression level of the former can't be controlled, bacterial SAgs are studied most, in which SEB is one of the most important. Administrating SAg in vivo will induce T cell activation widely, followed by immune anergy, apoptosis and immune suppression.

2.1.1.6　Natural antigen epitope

The binding sites of antibody and small molecules, such as carbohydrates, small molecules oligonucleotides and peptides, are in the deeper parts of antigen binding groove.

Natural protein antigens may be composed of multiple B cell epitopes some of which are immune dominant epitopes. Some residues play a greater role than the others when they combine with the antibody, and thus they have immune advantage. The B cell epitopes in natural protein antigen are usually hydrophilic amino acid on the protein surface and are accessible to surface membrane Ig molecules or free antibodies. Unless protein is denatured, the natural protein antigen won't hide inside the protein. The amino acid sequence, mainly composed of hydrophobic amino acid residues can't be B cell epitopes.

The expression or not of antigen T cell epitopes, depends on whether there are MHC allelic molecules which can be bound to. Thus, individuals with different MHC alleles have a different ability to provide epitopes. This may result in different immune responses of individuals to the same antigen. Since most of natural antigen-induced B cell immune responses depend on helper cells of T cells, T cell epitope are necessary for most of B cells to undergo immune response. Hapten has only B cell epitopes but not T cell epitopes, so it has only immune response but not the ability to induce a body to produce antibody.

2.1.2　Antibody

Antibody is the main effectors of humoral immune response, and exists in blood and tissue fluid, can recognize and bind to invading pathogenic microorganisms, and therefore, is an important component of host defense system. Immune globulin (immunoglobulin, Ig) usually refers to a group of immunoglobulin with antibody activity and (or) antibody-like structure. Thus, the concept of immunoglobulin is different from that of antibody. The latter is a kind of immunoglobulin with antigen-specific binding function to corresponding antigen, produced by plasma cells from the B lymphocyte under effective antigen stimulation. Globulins with antibody activity or chemical structure similar with antibody are collectively called immunoglobulins.

Antibody is a protein and has the general properties of protein: thermolabile, can be hydrolyzed by a variety of protease, can be precipitaed by ethanol, neutral salts, or trichloroacetic acid. Immunoglobulin antibodies can be extracted from the serum by 50% saturated ammonium sulphate or sodium sulphate, and protein denaturing agents also can inactivate antibodies.

Since immunoglobulin is mainly distributed in the γ zone in serum electrophoresis, it is also called γ-globulin antibody. But eventually globulins with antibody activity are not only in the γ zone, they can also be extended to the β zone and even $\alpha2$ area, which reflects that antibodies produced by different cell clons have heterogeneity and structural diversity[7].

2.1.2.1　The structure of immunoglobulin

(1)Basic structure

Ig molecule consists of four peptide chains, in which two long chains are known as heavy chain (H) and consist of about 440 amino acids, and molecular mass of each long chain is about 50-70 kDa, two short chains are called light chain (L) and consist of about 220 amino acids, molecular mass of each short chain is about 22.5 kDa. Light chain has two types: kappa (κ) chains and lambda (λ) chain. An antibody molecule has only one of the two types of light chain, and the proportion of two types of light chain varies in different species. The κ/λ ratio in mouse antibody is 20/1, while it is 2/1 in humans.

Heavy chain consists of the μ, δ, γ, α and ϵ five components. Five kinds of heavy chain determine the antibody categories: IgG, IgA, IgD, IgM and IgE. Some categories can be divided subgroups. Human IgG is divided into four sub-categories: IgG1, IgG2, IgG3, IgG4, and IgA is divided into IgA1 and IgA2. Four peptide chains are connected together by disulfide bonds (-S · S-) among the chains. The structural model is shown in Figure 2-1.

At the position of 1/2 L-chain and 1/4 H-chain (before about 110th amino acid) if regarding the N-terminal of Ig peptide molecule as start-point, the sequence and the type of amino acid varies from one IgG molecule to another one, the sequence is known as variable region (V area). The other amino acids of the peptide chain alongside the C terminal is almost constant between the types and the order, which is known as stable region or constant region (C area). V area is located on the N terminal and has three HVR (hypervariable region) in H chain and L chain region. In the three HVR, the sequence and type of amino acid residues are particularly volatile, which is directly related with the antigen recognition. The three HVR eventually are antigen binding sites of Ig molecule, and they are also referred to as complementarity determining regions (CDR). The other amino acid residues in variable region are known as framework region (FR), and account for nearly 75% of the V area, and the sequence hardly changes (about 5%). The feature of FR is to support CDR and maintain the stability of three-dimensional structure of V area. FR of H and L chain has the same amino acid residues in some locations. According to the variation of V_H/V_L amino acid sequence homology, Ig can be divided into groups and subgroups.

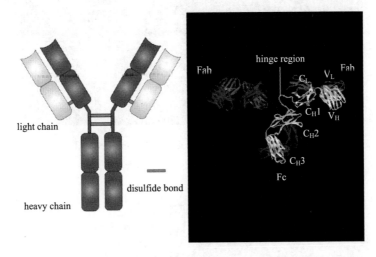

Figure 2-1   Structure model of immunoglobulin (IgG) (See Colored Figures)

(2) Three-dimensional structure

Each peptide chain of Ig molecules comes into being through folding adjacent secondary structural elements into spherical localized regions (Figure 2-2-A) by disulphide bond between chains, and each spherical area has about 110 amino acid residues. In different Ig molecules, amino acid residues sequences of corresponding spherical regions is similar to a high degree, which is called homology.

IgG, IgA and IgD molecules have total of 12 homologous regions, and Lκ or Lλ of them have 2 homologous regions (V_L and C_L), and H chain have 4 homologous regions: V_H, C_H1, C_H2 and C_H3. IgM and IgE molecules have 14 homologous regions respectively, because they have one more C_H4 in H chain than IgG, IgA and IgD. Although different C_H of H-chain are homologous and they are also homologous with the C_L of L chain, their amino acid sequences are rarely the same with the V region, indicating that V region and C region are coded by different genes (*V* genes and *C* genes). Each homology region is responsible for a certain immune function, and it is also known as functional region.

The amino acid sequence in variable region is of a high variability, the high variable regions of the corresponding V_L and V_H form a pocket, and have different shapes according to the different amino acid residues to work with a wide variety of suitable epitopes, which constitutes the molecular basis of antibody specificity (Figure 2-2-B). Meanwhile, the amino acid sequences in C_H and C_L are relatively stable, which attributes to many specific biological effects of them, such as fixation of complement and regulation of Ig catabolism rate in the C_H2, and affinity function to the cell in the C_H3 or C_H4.

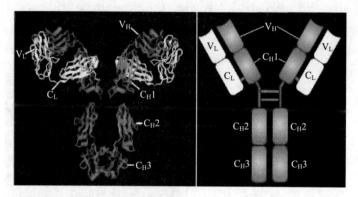

Figure 2-2-A　A model of immunoglobulin ribbon (See Colored Figures)

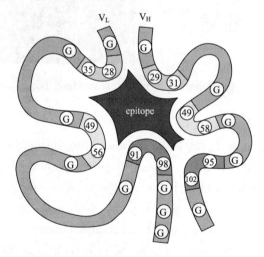

Figure 2-2-B　A model of immunoglobulin hypervariable region (See Colored Figures)

The area between C_H1 and C_H2 domains of heavy chain is rich in proline and cysteine. These residues which have less free groups, and can hardly form immobile secondary or tertiary structure with the neighboring regions. This free soft peptide is called hinge region. The flexibility of this structure allows the antigen binding sites of antibody molecule to change direction freely, which greatly enhances the binding ability of antibody to antigen molecules. Meanwhile, the complement fixation point can be exposed due to Ig allosteric.

(3) The Ig fragments in hydrolysis

Ig molecules can be hydrolyzed by many protease, resulting in different fragments. The papain and pepsin are commonly used enzymes in immunological studies. At physiological pH, papain can split IgG molecule at the 219th residue of N-terminal of H-chain, and generate two identical Fab fragments and an Fc fragment (Figure 2-3). Fab fragment (antigen binding fragment) contains a complete L chain and part of the (Fd) segment of H chain, and the molecular weight of it is 45 kDa; Fab segment still has the antigen binding activity, but the activity is weak due to mono valence. Fc fragment (crystallizable fragment) is the remaining part of the two H-chain C terminal, and the molecular mass is 55 kDa, and can form crystal under certain conditions. Fc segment can't combine with antigen, but has many other biological activities, such as fixing complement, affinity to cells (macrophages, NK cells and granulocytes, etc.), going through placenta, mediating the combination with bacterial proteins (e. g. protein A and G) and reacting with rheumatoid factor.

Pepsin can split IgG molecule at low pH on the 232th positions of the H-chain C terminal, and generate a F (ab')2 fragment which contains two Fab segments and a smaller pFc' fragment. F (ab') 2 fragment is antibody activity section with bivalence, and can produce two Fab' frangments. The molecular mass of Fab'is slightly larger than that of the Fab, but their biological activity is the same. The molecular mass of pFc' is smaller than that of Fc. Although it still maintains affinity to some macrophages and binding ability with rheumatoid factor, the original fixing complement activity of Fc fragment is lost.

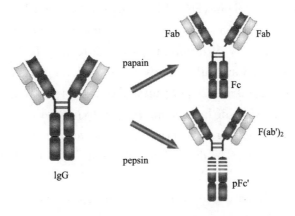

Figure 2-3　Diagram of IgG molecules hydrolysis fragments (See Colored Figures)

(4) Ig assistant component

Joining chain. Beside H chain and L chain, the polymer forms of Ig molecules such as IgA and IgM molecules still contain a chain (joining chain, J chain); but IgA mono-mer or monomer IgM doesn't. Connecting Ig monomers into polymer by J-chain is not necessary, but J-chain may be related to maintaining the stability of polymers. The J chain of a human being is of about 15 kDa molecular weight and is highly homologous with other species. J chain gene is not part of the gene cluster of Ig molecule. It is loca-ted at the 15th chromosome and produced not only in plasma cells IgA and IgM, but also in immature plasma cells synthesizing IgG, But it will not combine with IgG molecule.

Secretory piece: The secretory IgA molecule also contains a secretory component (SC) which is also known as secretory piece (SP). It is a part of poly-immunoglobulin receptor (poly-IgR) on epithelial cells and poly-IgR is a member of immunoglobulin su-per family (Ig-super-family). This receptor was produced by epithelial cells, and then connected with polymer IgA. In the process of outputting IgA-poly-IgR complex from epithelial cells, the receptor molecules are split by protease and the rest which is still at-tached to Ig is secretory piece. The molecular mass of free secretory is 80 kDa, and it covalently combines with SigA by disulphide bonds. The function of Secretory piece is to protect SigA molecules from being split by proteinase in secretion, then SigA in the mucosal surface can maintain stablity and favours its biological activity.

2.1.2.2　The types of antibody

Immune proteins are macromolecular proteins with various properties of antigens. They are good antigens for xenogenesis, allograft, or even the host itself and also anti-gen complexes with multiple epitopes. The heterogeneity of Ig molecules reflects the ge-netic differences of antibody-forming cells and represents genetic variability of antibody molecules at different levels. It usually can be detected by serological methods and ac-cordingly Ig and its peptide chain (H and L chain) can be divided into different serologi-cal types. The site and duration of synthesis, serum concentrations, distribution and half-life and biological activity of different Igs are different. Ig is produced by plasma cells and is present in blood and other body fluids (including tissue fluid and the outer secretion) and accounts for about 20% of total plasma protein; it can also be distributed in the surface of B cell. Most of Ig have antibody activity and can specifically recognize and bind antigens, and then trigger a series of biological effects. The structure of Ig is inhomogenous and thus Ig can be divided into different types by structure.

(1)IgG

IgG is mainly synthesized and secreted by the spleen and plasma cells in lymph nodes and exists in monomer form. During the development of body, IgG is synthesized later than IgM. The synthesis of IgG is begins in the third months after the birth of in-dividual. The IgG level of individual 3 to 5 years old will be close to the adult. IgG is a major component of antibodies in serum, accounting for about 75% of total serum Ig.

Human IgG can be divided into four subclasses, IgG1, IgG2, IgG3 and IgG4 (4 subclasses of mouse are IgG1, IgG 2a, IgG 2b and IgG3), according to antigenic differences of γ chain in IgG molecules. The hinge region of IgG3γ3, containing 62 amino acid residues, has 4 replicates of γ1 hinge region (15 amino acid residues) and about 10 to 15 disulphide bonds between heavy chains, so it can be cleaved susceptibly by proteinase and the half-life is also short. The biological activity of different subclasses of IgG is different. The half-life of IgG is relatively long, about 20 to 30 days. IgG activates complement through the classical pathway. The order of its ability to fix complement is IgG3> IgG1 > IgG2, and IgG 2b>IgG2a>IgG3 for mice. IgG4 in human and IgG1 in mice haven't the ability to fix complement. IgG is the only Ig that can pass through the placenta and plays an important role in the natural passive immunity. IgG also has the function of opsonophagocytosis, cytotoxicity (ADCC) and combining SPA, etc. Because the characteristics of IgG, it plays a major role in the immune protection, and the majority of antibacterial, anti-virus, anti-toxin antibodies belong to the IgG class. Injecting the gamma or the placenta globulin of a normal human into the body of puerpera without immune to measles or hepatitis A can realize artificial passive immunity and prevent corresponding infectious diseases. Many auto-antibodies, such as anti-thyroglobulin antibodies, systemic lupus erythematosus LE factor (anti-nuclear antibodies), and antibodies of immune complexes causing the allergic reaction type III, are IgG; therefore, IgG is known as multifunctional immunoglobulin.

(2) IgM

IgM is pentamer (Figure 2-4), the Ig of the largest molecular weight 970 kDa is called macroglobulin. The ability of IgM to activate complement is stronger than that of IgG. Antibody of natural blood group is IgM. The IgM in serum is pentamer which is composed by five monomers through disulphide bonds and a J chain. The sedimentation coefficient is 19S. IgM haven't hinge region in the molecular structure; $C_\mu 2$ may replace the functions of the hinge region. IgM is the

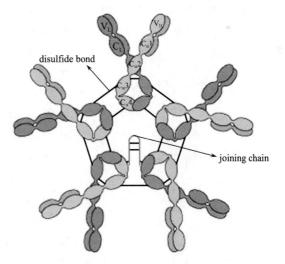

Figure 2-4   IgM pentamer (See Colored Figures)

earliest immunoglobulin arising in the process of biological evolution, for example, the lamprey can produce IgM. In the process of individual development, for either the Ig in the surface of B cell (SmIg) or synthesized and secreted into serum, IgM is the earliest Ig. Fetus in the late of embryonic development has the ability to produce IgM.

In general, IgM is produced firstly in the humoral immune response which is caused by antigen stimulation. IgM accounts for 5% to 10% of total serum Ig and is produced in early immune response. With the participation of complement, its hemolysis is 500 times stronger than of IgG, and after the activation of complement it plays the role of opsonization by C3B, C4b and other clips; therefore, IgM plays an important role in the body in the early immune protection. Natural blood group antibody (agglutinin) is IgM, and incompatible blood transfusion will cause high incidence of severe hemolytic reaction. IgM can't pass through placenta. If IgM for some pathogenic microorganisms appears in umbilical cord blood, it indicates a certain embryo infection of corresponding pathogenic microorganisms such as Treponema pallidum, rubella virus or cytomegalovirus, namely embryonic infection or vertical infection. Serum of normal a human also contains a small amount of monomer IgM.

As for B cell, membrane surface IgM is a major SmIg identifying antigen receptor. Mature B cell contains SmIgD, and SmIgM$^+$B cells account for about 80% in B cell repository of normal human. In memory B cells, SmIgM fades away and is replaced by SmIgG, SmIgA or SmIgE.

(3) IgA

IgA are produced mainly by mucosa-associated lymphoid tissues. Most of them are synthesized by the gastrointestinal lymphoid tissue, and a small part by the respiratory tract, salivary glands and reproductive tract tissue. Glandular tissue of lactating mothers contains a large number of IgA producing cells which mainly come from stomach and intestines. In human beings, there are a small amount of IgA coming from bone marrow. IgA begins to be synthesized from 4 to 6 months after a human being is born and reaches the serum level of adults at 4 to 12 years old. Serum-type IgA account for about 10% of the total Ig and their half-lives are about 5 to 6 days. IgA have two subclasses, IgA1 and IgA2. IgA1 is mainly present in serum and accounts for about 85% of total IgA, and the molecular weight of α1 chain is 56 kDa. IgA2 is mainly present in the outer secretion and small part in the form of serum-type IgA. IgA2 accounts for about 15% of serum IgA, α2 chain of IgA2 is short of hinge region and is of 52 kDa molecular weight. In serum, IgA are present in monomer form and also in dimer or trimer form covalently linked by J chain (Figure 2-5).

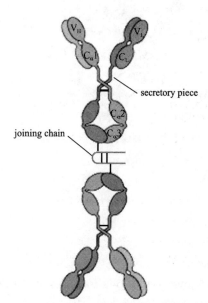

Figure 2-5   Structure of immunoglobulin Ig A dimer (See Colored Figures)

Secretory IgA are composed by double body and secretory component which are connected by J chain. They are mainly present in colostrum, saliva, tears, gastrointestinal fluid, and bronchial secretions, etc. and the most important factor for the mucosa local immune. Secretory IgA can bind to the corresponding pathogenic microorganisms (for example polio virus) to inhibit its adsorption to susceptible cells and also neutralize toxins such as cholera toxin and *E. coli* toxins. Newborn baby is susceptible to infect respiratory tract and gastrointestinal tract infections may be related to insufficient synthesis of IgA. Onset of chronic bronchitis also has a certain relationship with the reduction of secretory IgA. Puerpera can transmit IgA to baby by colostrum, which is an important natural passive immunity. Eosinophils, neutrophils and macrophages can express $Fc\alpha R$, and serum IgA monomer can mediate regulation, phagocytosis and ADCC effect. In addition, secretory IgA have the function of immune exclusion. Secretory IgA combine with a large number of soluble antigens in diet, normal intestinal flora and pyrogen materiald which are released by pathogenic microorganisms to prevent them entering bloodstream.

(4) IgD

IgD is of 175 kDa molecular weight. It was found from human myeloma protein in 1995 and mainly produced by cells of the tonsils, spleen and plasma etc. It is synthesized in the later stages of individual development. its concentration is 3-40 $\mu g/mL$ in human serum and its amount is less than 1% of total serum Ig. The hinge region of IgD is very long and sensitive to protease, so the half-life of IgD is very short, only 2.8 days. The exact immune function of IgD in serum is not clear until now. In the stage when B cell differentiates into mature B cell, B cell can express SmIgD and immune tolerance after stimulated by antigens. When mature B cell is activated or becomes memory B cell, SmIgD disappears gradually.

(5) IgE

IgE, a class of Ig, found in 1966, is of 188 kDa molecular weight. Its content is very low in serum and it only accounts for 0.002% of total serum Ig. It is synthesized in the later stages of individual development. The subype of heavy chain of IgE belong to ε chain, which includes four CH regions ($C\varepsilon 1$-$C\varepsilon 4$) and rich in cysteine and methionine, but has not hinge region. It is sensitive to heat and will lose biological activity at 56℃ for 30 min. IgE is mainly produced by plasma cells of the nasopharynx, tonsil, bronchial, gastrointestinal and others in mucosal lamina propria, through which allergens often invade and in which I-type allergic reactions often take place. IgE is pro-cells antibody, $C\varepsilon 2$ and $C\varepsilon 3$ functional areas can combine with high affinity $Fc\varepsilon R \ I$ in membranes of basophils and mast cell. When allergens enter body and combine with IgE which has been fixed onto basophils and mast cells, type-I allergic reaction will take place. When parasitic infection or allergic reactions take place, local IgE levels in exocrine fluid and serum will significantly increase.

2.1.2.3　The main biological activity of antibody

The important biological activity of immunoglobulin is its specific binding to antigen and mediating a series of biological effects by heavy chain C including activating complement and combining cell to cause phagocytosis, extracellular killing and immune inflammation, and to lead eventually to the purpose of removing foreign antigen[8].

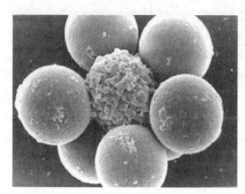

Figure 2-6　Specific binding of membrane
cell surface antigen and antibody

(1) The combination of antigen

The materials that can stimulate body to produce antibodies is antigen. That antibody molecules only combine with their corresponding antigens is called specific binding (Figure 2-6). For example, diphtheria antitoxin can only neutralize diphtheria exotoxin, but can't neutralize tetanus toxin, and vice versa. When binding with antigen, the V region of antibody's Fab fragment must be consistent with three-dimensional structure of epitope (conformation), which is especially related directly with amino acid residues of high variable region, therfore, the antigen-antibody binding is of high specificity. Although some amino acid residues are very far away in the amino acid sequence of peptide chain, they still can close tightly and form a double row of concave or bursiform surrounding the active site of antigen since the peptide chain can go back and forth for several fold along the parallel direction of long axis in functional areas. There are many hydrophobic sulphur amino acid side chains between layers. The interaction of antibody and antigen molecules is kept by a variety of non-covalent forces, such as hydrogen bonding, electrostatic and Van der Waals forces and reversible. Antibody has the activation function only after connecting to antigen, whereas natural Ig molecules can't play this role. But when antigen is absence, some physical treatment (such as heating and cooling) can also simulate the conformational changes of Ig to be activated.

(2) The activation of complement

Under certain conditions, the antibody can combine with complement molecules in serum, activate it and result in a variety of biological effects, which is known as the complement fixation phenomenon of antibody. This phenomenon reveals the interaction between antibody molecule and complement molecules.

The complement system is composed by a group of soluble proteins in serum and tissue fluid of human beings or vertebrate and a group of membrane-binding proteins and complement receptors in the blood cells and other cells' surface. It is responsible for fighting infection and is immunomodulatory in immune system and participates in immune pathological response.

The complement receptor (C1q) only has very weak combination with free Ig molecules, but strong with IgG or IgM (classical pathway) or agglutination Ig (alternative pathway) in immune complexes. C1q can react with C_H2 functional areas of Fc segments in IgG, and the binding site is in the side chains of three amino acids. All single Fc fragments of IgG subclasses have the same affinity to C1q, but in intact protein it's mainly IgG1 and IgG3 affecting the binding with C1q.

The ability of activating complement is strongest. At least two IgG molecules in close parallel are needed to activate C1q effectively, but single IgM molecule can activate complement after binding to antigen. Circulating IgM shows that single binding site with low-affinity to C1q and is similar with the efficiency of IgG. But when IgM combines with multiple determinants of antigen, the conformation will change into ring-like and hook-like, resulting in exposing C1q binding sites which are hidden by similar subunit and enhancing the ability of activating complement.

(3) Effect of parent cell

Cells can induce a series of biological effects after binding to the appropriate Ig through surface Fc receptor and different effects in different cells. In vitro experiments, adding immune serum into neutrophil suspension can enhance phagocytosis effects to corresponding bacterial. This phenomenon is known as antibody opsonization. It reveals the interactions between antibody molecules and immune cells. For example, antibody can promote phagocytosis of monocytes-macrophages and neutrophils, mediate antibody-dependent cellular cytotoxicity (ADCC) in antibody-dependent cells such as NK cells and macrophages, and induce type I allergy in mast cell and basophils, etc.

IgG can bind to Fc receptors which are anchored on cell surface. All of these receptors are members of Ig superfamily, including FcγRI (CD64), FcγRII (CD32) and FcγRIII (CD16). FcγRI is rich in the surface of monocytes. Neutrophils can also express this receptor after regulated by appropriate cytokines. FcR is a receptor of high affinity and can bind strongly to IgG1 and IgG3, and also can bind to IgG4, but can't combine with IgG2. FcγRII and FcγRIII receptors are present in many cells such as neutrophils, eosinophils and platelets, and have low affinity interaction with IgG1 and IgG3.

Activated B cell surface bears a IgM binding protein (FcμR), which is absent in T cell, monocytes or granulocytes. There are FcαRs on the surface of monocytes and neutrophils, so IgA also has the opsonization. Recently, it has been reported that there are IgD receptors on T cell, but its significance remains unclear. FcεRI receptor appears in mast cell and basophils; FcεRII are located on the surface of B cell, macrophages, eosinophils and platelets. Their interactions are related to IgE response regulating.

(4) Other biological activity

Fc segment of IgG1, IgG2 and IgG4 of human beings can combine with staphylococcal protein A (SPA), and its binding site is between C_H2-C_H3 of IgG. The IgG3 of xanthoderm can also combine with protein A. This phenomenon may be because of the replacement of histidine by arginine. Streptococcal protein G (SPG) can bind to all subunits of IgG of human beings and also almost all mammalian. The combining capacity is higher than that of SPA. But these two proteins have no affinity to other class of Ig.

Transmembrane IgG of human beings can be transported into blood circulation of fetus through the placenta. This transportation is not a passive diffusion, but implemented through selective binding between Fc segment of IgG and fetus capillary vessel and subsequent penetrating. Only γ chain of Ig has this capacity, whereas other types of Ig haven't. In addition, IgA can be transported to exocrine from the mucosa by combination with secretory component, such as transportations to intestines and latex.

2.1.3　The general principles of antigen-antibody binding

The fundamentals of antigen-antibody reaction involve three aspects: the interaction between determinants of antigen and antibody (be consistent with each other and complementary); specific, reversible and weak binding on molecular surface; action in ultra short distance. The basic conditions are binding force, affinity and that hydrophilic colloid transform into lyophobic one.

2.1.3.1　Structural basis

There is complementarity between determinants of antigen and hypervariable regions of antibody Fab. The variable region near the N end of antibody can form a trench, the size of which is about 3 nm×1.5 nm×0.7 nm. The trench is variable due to the variability of amino-acid residues of hypervariable regions. Antigen, wedge-shaped, can embed into this trench. However, the Ag can embed only when it is complemental with Ag determinant, therefore, the relationship between antibody and Ag is just like that of key and clock [9].

2.1.3.2　Chemical change

(1) Binding force of antigen-antibody

The combination of antigen-antibody, a specific complemental binding, acts through non covalent bond without forming a covalent bond. This weak adhesion involves four intermolecular forces (Figure 2-7) [10]. ①Electrostatic attraction: the attraction power between the amino and carboxyl groups, which bear opposite electric charge, of antigen and antibody. The gravitation is inversely proportional to the square of the distance between interaction groups, and the average bond energy is 20.9 kJ/mol.

②van de Waals force: the attraction due to molecular polarization when antigen and antibody close each other. The combination is proportional to the product of the polarization degrees of two interaction-groups, and inversely proportional to their distance to the seventh power. The bond energy is about 4. 2-12. 5 kJ/mol. This power of van de Waals force is less than the electrostatic attraction, but more specific. ③ Hydrogen bond: the attraction between hydrogen atom of hydrogen donor and hydrogen acceptor. The hydrogen donors include carboxyl, amino and hydroxyl groups; and the hydrogen acceptors include carboxyl oxygen, carboxyl carbon and peptic oxygen, etc. The power depends on the direction of hydrogen bonds (hydrogen bonds are highly directional). Hydrogen binding is inversely proportional to the distance to the sixth power between hydrogen donor and hydrogen acceptor, and the bond energy is about 20. 9 kJ/mol. ④Hydrophobic interaction. This force is mainly involved in the non-polar amino acid of the molecular side chain in Ag and Ab. When these two hydrophobic groups are in contact with each other in aqueous solution, they tend to gather for exclusion of water molecules (the force between polar atom and molecule). When antigenic determinant is close to the binding site of antibody, the polarity between them will disappear, and then hydrophilic layer will be lost immediately. This force is very important in binding of antigen-antibody reaction, most powerful and accounts for about 50% of all the binding force.

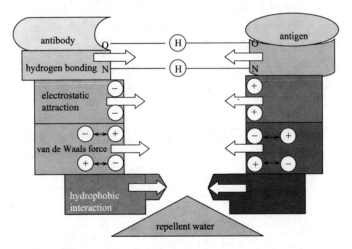

Figure 2-7　The schematic diagram of antigen-antibody binding

(2) Antigen-antibody affinity

Avidity is the binding capacity of complex antigen and corresponding antibody in reaction system. The binding can maintain the mutual adaptability between antigenic combining site and antigenic determinant, and is inherent adhesion between antigen and antibody. Avidity is related to affinity, antibody binding valence and the number of effective antigenic determinant. The greater the avidity is, the firmly antigen-antibody will combine with each other. The binding between antigen and antibody is reversible, and the affinity constant is the equilibrium constant when the action is balanced. The formula is:

$$\text{Affinity constant } k = \frac{\text{concentration of antigen—antibody complex}}{\text{concentration of free antigen} \times \text{concentration of free antibody}}$$

Ab of a higher k value can bind firmly to Ag and the AG–Ab complex is not easy to dissociate, which indicates this antibody has a high avidity.

(3) Changes in physico-chemical property: hydrophilic colloid transform into hydrophobic

Visible reactions between antigen and antibody can occur, such as precipitation and agglutination, which is because there is not only specific binding phase, but also non-specific promotion of agglutination in antigen-antibody reaction. In other words, antigen-antibody reaction is also a process in which hydrophilic colloid transform into hydrophobic one and agglutinate to large complex then.

Antibodies are globulins, and many antigens are also protein. They are common negative charged in serologic-reaction condition, which lead polar water molecules to hydration shell surrounding them, and become hydrophilic colloid. This is why they never self-aggregate and self-precipitate. When specific binding between antigen and antibody occurs, it can induce reduction or disappearing of electric charges of antibody and antigen and sequent destruction of hydration shell. Eventually the surface area exposed to water decreases and complexes form, which enables hydrophilic colloid to transform into hydrophobic colloid. At this point, in the effect of electrolytes (such as NaCl), it is easier to aggregate hydrophobic colloid and form visible complex. The changed process of colloidal state in antigen-antibody reaction is shown in Figure 2-8.

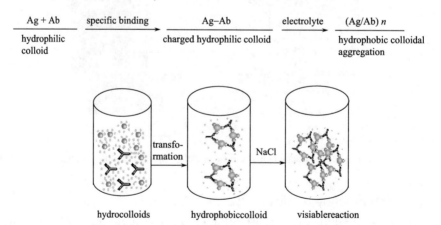

Figure 2-8　The schematic diagram of hydrophilic
colloid transforming into hydrophobic colloid

2.1.3.3　Reaction

The antigen-antibody reaction can be divided into two stages: ① Between antigen and antibody, a specific and invisible binding reaction will occur firstly. This stage is completed very fast, only a few seconds to several minutes. ②Visible stage: Under the effect of surrounding (certain pH, electrolytes, temperature, complements), antigen-antibody complex can undergo visible reaction, such as agglutination, precipitation, solution, complement fixation, etc. In this stage, the visible reaction is very slow and usually requires several minutes to hours. Of course, in the serologic reaction, these two stages usually can't be separated strictly.

2.1.4　The characteristics of antigen-antibody reaction

2.1.4.1　Specificity and cross-reaction

Antigen is of specificity, which means it can only bind to the antibody produced under its stimulus. This is determined by the complementarity between antigen determinant and antibody hyper variable region. Variable region near the N end of antibody can form a trench of 3 nm×1.5 nm×0.7 nm size; and only the determinant of antigen which is complemental to the hyper variable region can embed the trench (Figure 2-9). This is determined by the structural basis, just like the relationship between key and clock. For example, the antitoxin can only combine with diphtheria toxin, but can't bind to tetanus toxin. Since this high specificity is the basis of all the serological reaction,

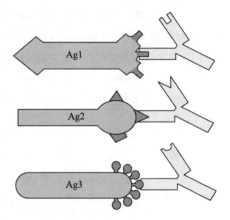

Figure 2-9　The schematic diagram of antigen-antibody specific reaction

we can diagnose or help diagnosing diseases by clinical detection of Ag or Ab.

Since two different antigens have similar antigen determinants or structure, they can act with both the antibodies for them, which is expressed as a cross-reaction. The response of antibody to antigen is actually to antigen determinant. If two different antigen molecules have the same antigen determinant; they can cross-react with the antiserum of each other. Therefore, Ag or Ab used in clinical serologic reaction usually need identification and purification to ensure the accuracy of experiment [11].

2.1.4.2　Proportionality

The optimal ratio or equivalence point is the concentration or quantity ratio of antigen/antibody at which precipitation occur rapidly. Antigen-antibody binding can exhibit a visible reaction only when they are at certain concentration or quantity ratio range. For example, precipitation, by adding a fixed volume of antigen with progressive concentration into a row of test tube, in which a certain quantity of antibody are added, precipitation will appear soon as the antigen concentration increases. But beyond the certain ratio range, the speed and quantity of precipitation will decrease rapidly and even disappear as antigen concentration increases. The rate of precipitation reflects the suitable ratio of antigen/antibody participating in the reaction. When the ratio is right, they will react fast, and contrarily slowly.

Experimental results showed that, in the same antigen/antibody reactive system, the optimal ratio of concentration of antigen and antibody for precipitation was kept unchangeable.

In the optimal reaction conditions, almost all the antigen and antibody will bind to each other to precipitate, and free antigen or antibody will be hardly found in the supernatant. Actually, the formation of precipitate will be up to the maximum when there is a little excess of antigen. When the ratio of antigen/antibody exceeds this range, both the precipitate quantity and the rate will decrease, which is called zone phenomenon. Thus, we can divide the antigenantibody reaction into three parts, equivalence zone, pro-zone and post zone, according to quantitative precipitation curve (Figure 2-10). The equivalence zone is the condition at which the antigen/antibody ratio is right for precipitatation.

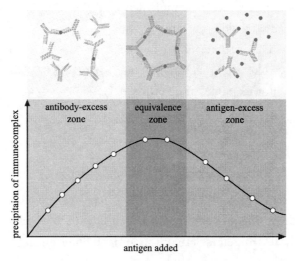

Figure 2-10　The schematic diagram of antigen-antibody optimal ratio

Grid theory can be used to illustrate the formation mechanism of antigen-antibody complex. When antigen and antibody combine together in equivalence zone, they can form grid and the visible precipitate since most of the antigens are multivalent while antibodies are two valents at least. When antigen or antibody is overmuch, they can only form small-grid complex because of unsaturated bind valence, and many free antigen or antibody molecules exist. This theory has been supported strongly by observation through electron microscope. When antigen or antibody is monovalent, they can't form visible precipitate whether the reaction ratio is right or not.

2.1.4.3 Reversibility

This means the property that antigen and antibody can be dissociated into free antigen and antibody under a certain conditions after they form a complex. This dissociation conditions include low pH and high-concentrations salt, etc. The substances commonly used for the dissociation of antigen-antibody complex include 3 mol/L potassium thiocyanate, 0. 1 mol/L glycine (pH 2. 4) and 7 mol/L urea, etc.

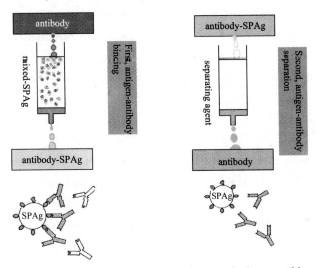

Figure 2-11   The schematic diagram of antigen-antibody reversible reaction

Figure 2-11 shows that antigen-antibody binding is an interaction on the molecular surface. This action complies with the law of mass action and keeps in dynamic balance, and is reversible under certain conditions. Beside environmental factor, the dissociation degree of Ag-Ab complex also depends on the Ag-Ab affinity. The greater the affinity is, the more stable the combination between Ag-Ab will be.

After Ag-Ab complex is dissociated, Ag and Ab can still keep their original physicochemical and biological properties. Therefore, Ag and Ab can be purified based on this feature, which is the basic principle of affinity chromatography.

2.2   The factors affecting Ag-Ab binding

2.2.1   Antibody

Antibody comes from animal immune serum and antibodies from different animals are different in reactivity. Immune serum of rabbits and other animals have a wide equivalence zone, and can produce a soluble immune complex only when the Ag is excess. This kind of Ab is known as R-type Ab. The immune sera coming from horses, humans or many other big animals has a narrow equivalence zone and is called H-type Ab. Generally, monoclonal antibodies are not applied in precipitation or agglutination reaction since they are monovalent and can only bind to one determinant of Ag, and form small molecular immune complexes.

Ab concentration: Ag-Ab reaction needs an appropriate concentration to produce a visible reaction.

Specificity and affinity: Specificity and affinity are the two key factors affecting serological reactions. They can affect the accuracy of experimental results together. Diagnostic reagents should choose Ag or Ab of high specificity and affinity to ensure and improve the reliability of experiments. The immune serum obtained in the early stage of immunization usually is of lower affinity, but higher specificity; the serum obtained in the late period is of higher affinity, but the Ab types and reactivity in the immune sera will become more complex after immunization for a long term.

2.2.2　Antigen

The serological reaction results will be affected by the physico-chemical properties, molecular weight, the types and the number of determinant of antigen, etc. For example, precipitation won't appear if monoclonal Ab was used in reaction; it can't appear agglutination when erythrocytes react with IgG Ab; rough-type bacteria will self-agglutinate in physiological saline.

2.2.3　Electrolytes

Apparent precipitation or agglutination will appear after Ag-Ab binding in an appropriate amount of electrolytes. If the electrolytes concentration is too high, there will be salting-out phenomenon. The solution containing 8.5 g/L NaCl solution (physiological saline) is commonly used as the diluent agent of Ag or Ab.

In complement fixation and dissolution reaction, beside isotonic NaCl, appropriate Mg^{2+} and Ca^{2+} are also needed in order to achieve a better response.

2.2.4　pH

Ag-Ab reaction must be in a suitable pH condition. The reaction results will be directly affected if the pH is too high or too low. When the pH is near the isoelectric point of bacteria, it may induce acid agglutination and lead to a false positive. The serologic reaction generally is carried out in pH 6-8. The most suitable pH of complement fixation and dissolution reaction is at pH 7.3-7.4, and the complement enzyme reaction activity will be reduced to some extent if the pH is beyond this range.

2.2.5   Temperature

Usually, temperature can affect the rate of Ag-Ab reaction, but rarely the results. With temperature increasing, Ag and Ab will move faster, and thus the collision probability of Ag-Ab will increase, and eventually accelerate the binding reaction; however, high temperature can also induce Ag-Ab complex dissociation easily. If the temperature is low, although the reaction is slow, Ag-Ab will combine together more stably, which enables the observation of Ag-Ab reaction more easily.

The common temperature for Ag-Ab binding reaction is at 15-40 ℃, and the optimum temperature is at 37 ℃. Some serologic reactions need special optimum temperature, for example, the combine of cold agglutinin and erythrocytes is better at 4 ℃, but the complex will be dissociated at above 20 ℃.

Many factors can effect Ag-Ab reaction, therefore, great attention must be paid to the choice and stability of experimental conditions and the selection of a strict control (such as the positive control, negative control and standard control), which is the only way to keep the accuracy of the results[12].

2.3   The main application of Ag-Ab binding reaction

2.3.1   The detection of complement

Complement not only has an effect to keep homeostasis, but also can induce immunity pathological injury. The detection methods include 50% complements hemolytic activity (CH50) detection and complement component (C3, C4 and C19, etc.) detection. Recently, the clinical applications of complement detection include the issues such as epidemic hemorrhagic fever (EHF), systemic lupus erythematosus(SLE), pulmonary disease, hepatic disease, renal disease, cardiovascular disease, cerebrovascular disease, and diabetes mellitus, etc [13-15].

2.3.2   The detection of immune globulin

Immune globulins are a group of proteins with antibody activity and mainly distributed in organism blood, interstitial fluid and exocrine secretion. It is an important indicator of humoral immunity functions of organism. Human beings have five kinds of immune globulins including IgG, IgA, IgM, IgD and IgE. Among them, IgD and IgE are in low levels, thus the commonly detected Ig are IgG, IgA and IgM[16,17]. Serum immune globulin abnormalities can be divided into three categories as the following.

(1) The level increasing of several different Igs

This is mainly seen in infection, tumor, auto-immune disease, chronic active hepatitis, cirrhosis of liver and lymphoma, etc. For example auto-immune diseases, the simultaneous increasing of IgG, IgA and IgM are commonly seen in SLE, while the simultaneous increasing of IgG and IgM are commonly seen in polyarthritis destruens.

(2) The increase of a single Ig

This is also known as "M" protein disease, and mainly seen in: ①Multiple myeloma, which is often shown as only one kind Ig abnormally increase, while others decrease notably or keep at normal level. Among them, the most common one is IgG, which can reach 70 g/L in serum, then IgA as the second, IgD as the less common, and IgE as the most rare. ②Macroglobulinemia, Plasma cells which can produce IgM express malignant proliferation, and in serum IgM can reach 20 g/L or even higher.

(3) The decrease of one or several Ig

This type can be divided into primary and secondary, the former are genetic, such as Swiss gammaglobulinemia, and selective IgA, IgM deficiency diseases, etc. Secondary cases defects are often seen in malignant disease of reticulate lymphatic system, chronic lymphocytic leukaemia, nephritic syndrome, extensive burns and ambustion, long-time usage of immunosuppressive agents or radiographic exposure.

The increase of IgD is often seen in IgD-type MM. It often increases in epidemic hemorrhagic fever (EHF), allergic asthma, and atopic dermatitis patients and it also physiologically increases among pregnancy trimesters and smokers.

Increase of IgE is often seen in the hypersensitivity disease, such as allergic coryza, extrinsic asthma, bostock catarrh, chronic urticaria parasitic infection, acute/chronic hepatitis, drugs interaction-induced interstitial pneumonia, allergic bronchopulmonary aspergillosis (AbPA), polyarthritis destruens, and IgE-type MM, etc.

2.3.3 The detection of immune complexes

There are two types of immune complexes in vivo; one is circulating immune complex (CIC) present in blood, and the other one is the immune complex which is fixed in tissue. The detection technologies of immune complexes can be divided into Ag-specific and non- Ag-specific methods. In most cases, the characters of Ag in immune complexes are not clear or very complex, thus the Ag-specific approach is not commonly used.

There are three evidences for detecting whether the immune complexes are pathogenesis. ①There are IC precipitate in local lesion; ② CIC levels increased significantly; ③ Identified Ag characters of IC. The third evidence is difficult to find, but at least the first two evidences must be contained, whereas only CIC detection is not abundant. Human beings in healthy states also have a few CIC (about 10-20 μg/mL), its cut-off value between physiology and pathology is difficult to distinguish. In addition, there are many methods based on different mechanisms for CIC detection. A positive result is obtained in one method whereas a negative one maybe be obtained in another method. It will be more significant if the results are agreed with immunohistochemistry tests.

Recently, it has been proved that some diseases, including SLE, polyarthritis destruens, some glomerulonephritis and vasculitis, etc. , are related with immune complexes. CIC is an auxiliary diagnosis index of these diseases, although it is significant for disease diagnosis and therapy. It should consider the potentiality of immune complex diseases and detect CIC and tissue deposition IC when symptoms, such as purpura, arthralgia, proteinuria, vasculitis and serositis, appear. In addition, when there is a tumor, the detectable rate of CIC is higher, but the injury of type III allergic reaction does not appear, which is called clinical hiding IC disease. However, this state is usually associated with tumor conditions and prognosis.

2.3.4   The detection of cytokines

Cytokines, a type of protein antigen, can combine with specific monoclonal antibody. The detecting methods based on antigen-antibody reaction have been developing fast and include enzyme-linked immunosorbent assay (ELISA), radioimmunoassay (RIA), immunofluorescence analysis, etc. [18,19].

Recently, the cytokines have been widely used in clinic and involve EPO, IFN-α, G-CSF, GM-CSF and interleukin (on trial), etc. To study the effects of these cytokines in physiological system and the role of cytokine products used in clinic, the detection of them is necessary. Moreover, the method should be suitable for routine operation. The cytokines detection yet have not been applied extensively in clinical diagnosis, since some methods developed are not satisfactory and the results from different methods vary widely, which brings some difficulties to its clinical therapy. Therefore, it is necessary to understand the characteristics and the factors of these methods.

References

[1]   Griffiths AD, Duncan AR. Strategies for selection of antibodies by phage display. Curropin Biotechnol, 1998, 9(1): 102-106

[2]   Hoogenboom HR, Debeuine AP, Hufton SE and others. Antibody phage display technology and its applications. Immunotechnology, 1998, 3(4): 169-176

[3]   Yin TB and others. Poultry immunology. Beijing: China Agricultural Scientech Press, 1999: 33

[4]   Meng PF and others. Fundamental immunology. Zhengzhou: Zhengzhou University Press, 2004: 2

[5]   Larrick JW, Fty KE. PCR amplification of antibody genes. Methods: A companion to methods in enzymology, 1990, 2(2): 106-114

[6]   Hong XH. Clinical microbiology examination. Beijing: Science and Technology Document Press, 2005: 44

[7]   Zhang JB and others. Medical molecular cell biology. Beijing: Peking Union Medical College Press, 2002: 398

[8]   Chen R and others. Medical immunology and microbiology. Beijing: Chemical Industry Press, 1989: 13

[9]   Pereira S, VaN Belle P, Elder D and others. Combinatorial antibodies against human malignant melanoma. Hybridoma, 1997, 16(1): 11-15

[10]   Yang TB. Immunology and immunologic detection. Beijing: People's Medical Publishing House, 1998: 75

[11] Chamow SM, Ashkenazi A. Immunoadhisin: principles and applications. TIBECH, 1996, 14: 52

[12] Jespers LS, Roberts A, Mahler SM and others. Guiding the selection of human antibodies from phage display repertoires to a single epitope of an antigen. Biotechnol, 1994, 12: 899

[13] Schier R, McCall A, Adams GP and others. Isolation of picomolar affinity anti-c-erbB-2 single-chain Fv by molecular evolution of the complementarity determining region in the center of the antibody binding site. J Mol Biol, 1996, 263: 551

[14] Song YP and others. The detection of serum humoral of cardiovascular diseases in immunity levels. Shanghai Journal of Immunology, 1987, 7(1): 57-58

[15] Li PC and others. Determination of plasma C_9 and its preliminary clinical application. Shanghai Medical Laboratory Science, 1991, 4: 235

[16] Vaughan TJ, Williams AJ, Finnern R and others. Human antibodies with sub-nanomolar affinities isolated from a large non-immunized phage display library. Nat biotechnol, 1996, 14: 309

[17] Li M, Xu W, Huang J and others. The clinical application value of albumin and Immunoglobulin assay in diagnosis of nervous system diseases. INTERNATIONAL MEDICINE & HEALTH GUIDANCE NEWS, 2008, 14(6): 18-21

[18] He YL. Detection and clinical significance of cellular factor. Journal of Youjiang Medical College for Nationalities, 2001, 23(6): 974-975

[19] Ju TW, Cao XT. Detection and application of cellular factor. Chinese Journal of Laboratory Diagnosis, 1997, 1(3): 45-48

Chapter 3 Fabrication and Purification of Antibodies

3.1 Polyclonal antibodies

3.1.1 Preparation principle

The process that immune system produces antibodies on the stimulus of immunogen, is called immune response against immunogen. Specific immune responses include humoral immunity and cellular immunity, which are implemented with B cell and T cell involvement, respectively. In addition, macrophage plays an important role in processing antigen, immune response and immune regulation [1].

The production of antibody can be divided into three stages. The first stage is sensitization in which immunogen will be processed (activated) and recognized. The second one is reaction stage, in which activated lymphocytes will experience series of proliferation and differentiation. The third one is the effective stage, in which the sensitized lymphocytes produce lymphokine and function cellular immune. Plasma cell will produce specific antibody which enter lymph, blood, tissue fluids or mucosa surface and function humoral immune.

Every time animal organisms will produce antibody through the above-mentioned three stages after meeting antigen. But body only produces less amount of antibody after a latent period when it meets a kind of antigen for the first time. Analyzing the essence and type of antibody, it is found that generally IgM is produced after immunization for the first time. IgM appears first, disappears most rapidly and maintains only for weeks to months. IgG appears followed by IgM, reaches the highest level, when IgM almost disappears, and maintains for a long time.

In order to improve the titer of antibody, the animal should be immunized again by the same antigen. The content of antibody in the body will decrease after the body meets the antigen again because part of the existing antibody will bind to the antigen injected. However, the antigen injected again will activate memory cell produced in primary immunization and prompt plasma cell to produce a large number of IgG and less amount of IgM. The content of IgG in body will increase in secondary immunization followed by primary immunization. After secondary immunization, the content of IgG in body will fall to a momentary low period and experience a peak again on the effect of memory cell, which was produced in earlier immunization. It should be noted that the content of antibody in last low point is higher than in the former although the content of antibody in blood will fall to a low period after immunization every time. That is, the content of antibody in body will present a wave-like uplift [2].

3. 1. 2　The choice of species

It is very important to select an appropriate species for immunization. The following aspects should be considered: ① The difference between antigen and species for immunization. Big is better when it comes to species difference because it is difficult to generate immune response between two genera with close relationship, such as between rabbit and rat and between chicken and duck. ②The amount of antiserum needed. Big animals such as horse and mule can produce a large amount of antiserum (an adult horse can produce 10,000 mL antiserum). However, rabbit or guinea pig also can meet the requirement but for not too much antiserum. ③ The type of anti-serum. Anti-serum can be divided into R (rabbit) and H (horse) types. It is hard to carry out precipitation reaction with H type of antiserum. ④ Choice of antigen. Protein antigen is suitable for most of animals such as goat and rabbit, the most common species. But some proteins have very low immunogenicity against these animals and result in the problem that no antibody is produced, such as IgE against sheep and enzymes (for example, precursor to pepsin) against goat. This is because similar materials occur in these animals or because of other causes. Some species, such as guinea pig, turkey, dog, pig and cat, are alternatives for immunization with these proteins. ⑤ Rabbit is suitable for immunization with steroidal hormones and guinea pig suitable is for immunization with enzymes. It should be noticed that the animals must be healthy, without infectious diseases and at right age. And male animal is better for immunization. In addition, attention should be also paid to animal feeding in order to avoid death of animals and eliminate individual differences. As for rabbit, a group of 2-3 kg weight animals, consists of 3 pure new Zealand rabbit, is better for immunization [3,4].

3. 1. 3　Processing the antigen

Firstly we should learn some knowledge about the type, character and concentration of an antigen in order to process the antigen properly. Polypeptide antigen can be divided into three kinds namely polypeptide-KLH conjugate, polypeptide-BSA (or GGG)conjugate and bare polypeptide. Generally polypeptide-KLH conjugate is applied to immunization, and polypeptide-BSA conjugate and bare polypeptide are applied to detection. The antigen obtained should be stored in aliquots and freezing and thawing should be avoided. For example, *Escherichia coli* recombinant protein probably contains insoluble precipitation and should be broken up by ultrasonic cell disruptor. Note that the processing should be implemented in ice-bath and the strength of ultrasonic wave should be moderate to avoid the creation of bubble. This is because excessively strong ultrasonic wave will denature proteins and too weak ultrasonic wave can't do the work. Ultrasonic wave processing should be done for 2-3 seconds at a time and repeated by 3-4 times.

Tube for aliquots shouldn't adhere proteins of sample. Two labels with different colours should be utilized to distinguish the antigens for immunization and for detection. In addition, date, name, concentration and dose of the antigen all should be labelled on the aliquots [5].

3.1.4　Immunization of animals

There are individual differences between the same species animals. In addition, different antigens have different immunogenicity. Thus, some immunogenic materials, called immune adjuvant, usually will be added into the antigen before animal immunization, which can enhance the immunogenicity of the antigen and stimulate strong immune reaction in animal body.

Most common immune adjuvant is Freund adjuvant, which consisting of a piece of lanolin and five pieces of paraffin is called incomplete Freund adjuvant. It is worth noting, the ratio of lanolin and paraffin can be adjusted ranging from 1 : 2 to 1 : 9 (V/V) to meet the demand. If incomplete Freund adjuvant is added with 1-20 mg BCG vaccine per mL, it becomes complete adjuvant.

About 300-500 μg dose of antigen is needed in primary immunization and about 1/4 of the first dose is enough for booster immunizations. A booster immunization usually is carried out per 2-3 weeks. In primary immunization animal can be given 0.5 mL pertussis vaccine as adjuvant through subcutaneous injection, and yet in booster immunization it needn't be given adjuvant again.

Two weeks after second booster immunization, 2-3 mL of blood are taken from the vein at the ear edge of the animal, prepared into serum and detected about the titer of the serum. If the titer of the serum is lower than what is expected, new booster immunization is needed again until the titer reaches the value expected. At the moment, the animal can be bled and serum can be extracted [6].

3.2　Monoclonal antibodies

3.2.1　Principle

Antibody is majorly synthesized by B lymphocytes. Every B lymphocyte has a hereditary gene which can synthesize an antibody. Every animal's spleen has millions of lymphoid cell series, which have different hereditary genes and can synthesize different antibodies.

If body is stimulated by an antigen, many determinants of antigen molecule can activate different B lymphocytes with different hereditary genes. The cell activated will divide and proliferate into daughter cells. That is, a clone divides, proliferates and forms multiple clones, which can synthesize multiple antibodies, from a single activated B cell. If a cell also called a monoclone, which only secretes a specific antibody, is selected and then cultured, groups of cell can be obtained from the single cell division and proliferation. Monoclonal cell can synthesize an antibody with a determinant, which is called monoclonal antibody (McAb)[7].

    Monoclonal B lymphocyte excreting specific antibody must be selected firstly if wanting to fabricate monoclonal antibody. One problem, that B lymphocyte can't grow in vitro, must be fixed out. However, myeloma cell can grow in vitro and myeloma cell can be fused with B lymphocyte through cell hybridization, depending on which hibridoma cell can be created. This hybrid cell has not only the advantage of B lymphocyte, excreting specific antibody, but also of myeloma cells, growing and reproducing forever in vitro culture, which comes from the parental cells, Monoclonal antibody, recognizing only a single epitope on the antigen, can be fabricated by culturing and proliferating single fused cell into groups of cell (Figure 3-1).

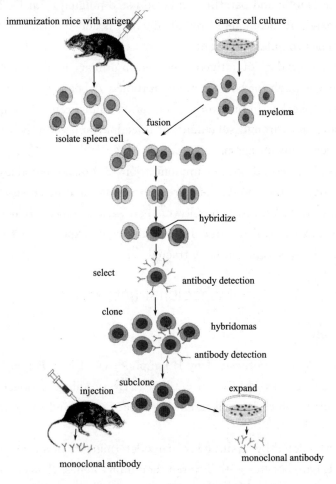

Figure 3-1   Fabrication of McAb

3. 2. 2   The choice of animal and immunization

Pure Balb/C mouse, whose nest is always close-by, is compliant and weak, has a small scope of actives, with poor appetite and less excretion amount. Generally it can survive in a clean laboratory and is selected for hibridoma cell technique by most of the laboratories. Appropriate immunization schedule is a key point for cell hybridization and obtaining excellent McAb. Generally about two months before hybridization, immunization schedule should be listed according to the character of immunogen and primary immunization to be carried out.

Soluble antigen is of poor immunogenicity and should blend adjuvant. Hapten should be transferred into immunogen and then blend adjuvant. The common adjuvants include complete Frenud's adjuvant and incomplete Frenud's adjuvant.

Nowadays immunization strategy for soluble antigen, particularly for poor immunogenic antigen, has been innovated endlessly. Such as preparing granule of soluble antigen and immobilization of the antigen, they can not only enhance the immunogenicity of the antigen, but also decrease the immunization dose. Granules of antigen are of strong immunogenicity and can do the deed very well. Only 1×10^7 to 2×10^7 cells are needed in immunization when it comes to cell antigen. Changing administration way is also effective. For example, primary immunization can be given directly through injection in spleen. Utilizing cytokines as adjuvant also can improve immune response in body and enhance the response of immune cells against antigen.

3. 2. 3   Cell fusion

Myeloma cell should be the same strain with the animal immunized in order to improve fusion rate and inoculate the hibridoma cells into the abdominal cavity of the same strain mouse. This inoculation can produce a large amount of McAb. Myeloma cell can be cultured in common culture medium such as RPMI 1640 and DMEM. 10%-20% of fetal bovine serum, in which generally the concentration of the cells is 1×10^4 to 5×10^5/mL and not beyond 1×10^6/mL, is suitable for culturing hibridoma cells. When cell is in mid-log growth phase, it can be cultured at the passage ratio of 1 : 3 to 1 : 10. The passage time of the hybrid cells is 3 to 5 days. In passage, part of cells will occur atavism, thus they should be treated with 8-azaguanine which can keep the survival cells sensitized with HAT culture medium [8].

In tissue culture, single cell or a few dispersed cells can't grow well, but some living cells, called feeder cells, can help them grow and proliferate well when added into them. In fabrication of McAb, feeder cells are needed thoroughly in many processes, such as screening hibridoma cell, cloning and mass production. Some common feeder cells include abdominal cavity macrophages, spleen cells and thymocytes from mouse.

Sometimes 3T3 cell line, a kind of fibroblast of mouse, after projected by radioactive rays, is also an alternative. The feeder cells are usually added at the concentration of 2×10^4 or 1×10^5 cells per well [9].

3. 2. 4　Screening of hybridoma cell and detecting antibody

After treated with PEG, spleen cells and myeloma cells will form several kinds of hybrid cell. However, only the hybridization of spleen cell and myeloma cell is valuable. In the selective culture medium, HAT culture medium, myeloma cells and spleen cells can't grow since they lack thymidine kinase and hypoxanthine guanine phosphoribosyl transferase (hgprt), respectively. On the contrary, the hibridoma cells possess the two enzymes simultaneously, which makes them grow well in HAT culture medium.

After cultured in HAT for 1-2 days, a large number of myeloma cells will die. After 3-4 days, myeloma cells will disappear and hibridoma cells will grow into small clones of cell. After 7-10 days, HAT culture medium should be replaced with HT culture medium, which will be utilized for two weeks and then replaced with general culture medium. Usually half volume of the culture medium for hibridoma cells will be replaced every 2 to 3 days. It is the time to detect specific antibody and select the hibridoma cells needed when the hibridoma cells are covered 1/10 area of well bottom in the selective culture.

Screening method should be chosen according to the character of antigen and the type of antibody. Generally rapid, facile, specific and sensitive method is preferred.

Some common methods are listed as follow: ① radioimmunoassay (RIA) which is suitable for detecting soluble antigen and cell; ② enzyme-linked sorbent immunoassay (EISA), suitable for detecting soluble antigen, cell and virus; ③ immunofluorescent test, suitable for detecting surface antigen of cells; ④ other methods such as indirect hemagglutination test and cytotoxicity assay[10].

3. 2. 5　Subcloning of hybridoma cell

Generally subcloning of hybridoma cell means cloning of antibody-positive well. After selection through HAT medium, clone of hybridoma cell in a well does not necessarily contain only one clone. Maybe there are several or more clones including antibody-secreting cells, antibody-nonsecreting cells and irrelevant antibody-secreting cells, etc. To separate these cells, they must be subcloned. Principle of subcloning is that antibody-positive hybridoma cell should be subcloned as early as possible lest antibody-secreting cells possible are inhibited by antibody-nonsecreting cells. Since antibody-secreting cells usually grow faster than the antibody-nonsecreting, competition between them will result in losing the later. Even hybridoma cell subcloned has to be recloned at intervals so as to prevent from losing the ability of antibody-secreting because of mutation or chromosome loss. There are many methods of subcloning of hybridoma cell, in which limiting dilution and soft agar plate are most common methods [11].

3. 2. 6　Hybridoma cell freezing and thawing

Hybridoma cells are prone to get contaminated and lose the ability of secreting antibody in cultrue at any time until they are cultured into a stable cell strain with antibody-secreting capacity. If the original hybridoma cell hasn't been frozen, the aforementioned accident, once happening, will waste all the previous efforts. Thus, it is important freezing subclone recoloned from original hybridoma cell in time.

Cryoconservation of hybidoma cell is the same as other cells. In principle, every ampoule should contain more than 1×10^6 cells. Nevertheless, the requirements for original hybridoma cells are changeable according to their culture condition. As for the hybridoma cells cultured in 24-wells culture plate, a well of cells can be loaded into an ampoule when they grow and till with the bottom of a well.

Frozen stock solution is best precooled and manipulation should be gentle and rapid. Cells can be cooled from room temperature to 0 ℃ at once, stored in ultra-low freezer and then transferred into liquid nitrogen the next day. Instrument for cell sub-conversion is another alternative. The frozen cell should be recovered at intervals and checked for their activity and stability of antibody-secreting. Cells can be stored in liquid nitrogen for several years or even longer time.

The cell recovery method is as followed: take out the ampoule from liquid nitrogen, place it at 37 ℃ water-bath, thaw the cells in 1 minute and wash them with complete culture solution twice, then transfer into culture bottle, culture in incubator at 37 ℃ and 5% CO_2, and detect antibody activity when cell colony forms.

3. 2. 7　Mass production of monoclonal antibody

There are two methods for mass production of monoclonal antibody [12]: ①Rotate tube culture in vitro can culture mass hybridoma cells and separate monoclonal antibody from the cell supernatant. However, this method can only provide a few antibodies. Generally, the cell culture solution only contains 10-60 μg/mL antibody, in this case the mass antibody production is high-cost. ②Introducing hybridoma cell into vivo and provide ascite or serum. It can be divided into two methods. One is solid tumor method, details of which are as follows: mouse is subcutaneously injected with hybridoma cell in logarithm stage, the injection dose is 0. 2 mL at each point and totally 2-4 points are injected with the cell concentration of 1×10^7-3×10^7 cell /mL. Collect blood usually after 10-20 days until solid tumor grows up. The antibody concentration is as high as 10 mg/mL, but the amount of serum is limited. Another method is ascite technology. Usually Balb/C mouse is intraperitoneally injected with 0. 5 mL pristane at first. After 1-2 weeks it is intraperitoneally injected with 1×10^6 hybridoma cells. Ascite will appear after 7-10 days cell injection. Closely monitor the health of mouse and the change of ascite. When the ascite is enough and the mouse is on the edge of death, the animal should be executed. Extract the ascite with dropper. Generally a mouse can produce 5-10 mL of ascite. Syringe is also an alternative for ascite extraction, with which the ascite can be collected for several times. The monoclonal antibody concentration can be as high as 5-10 mg/mL. The aforementioned method is the most common nowadays. Moreover, ascite can be cryoconserved, recovered and intraperitoneally injected into another mouse when needed. It will bring up ascite tumor and produce mass of antibody.

3. 2. 8　Identification of monoclonal antibody

Systematical identification of McAb prepared is very important. McAb should be identificated from the following aspects [13].

(1) Specificity of antibody

Besides immunogen, the cross-activity of other antigens with related component should also be tested. ELISA and IFA methods are suitable for the test. For example, when testing monoclonal antibody against melanoma cell, the activity between the antibody and mela noma cell should be analyzed. Beyond that tumor cells from other organs and normal cells also should be tested about their cross-activities in order to select specific monoclonal antibody against melanoma cell and related antigens. In preparing monoclonal antibody against recombinant cytokine, first consideration is whether the antibody has cross-activity with proteins of expression strain and then is cross-activity with other cytokines.

(2) Identification of types and subtypes of McAb

Generally when an antibody is screened with enzyme label or fluorescence label, the type of antibody has been confirmed. If enzyme or fluorescence labels of rabbit against mouse IgG or IgM are adopted in antibody identification, that indicates the antibody identified is IgG or IgM. As for subtype of antibody, it must be confirmed through standard anti-serum against subtype antibody by double immunodiffusion or sandwich ELISA method. In double immnodiffusion, appropriate PEG (3%) can benefit forming precipitate.

(3) Identifying the neutralization activity of McAb

Biological activity of McAb can be identified by animal or cell protection experiment. For example, the neutralization activity of anti-virus McAb can be confirmed by administrating the antibody and virus simultaneously to susceptible animal or sensitive cell, meanwhile observing whether the animal or cell is protected by the antibody.

(4) Identifying antigenic epitope of McAb

ELISA additive assay can analyze the epitope of McAb and confirm whether several

McAb recognize different epitopes.

(5) Identifying the affinity of McAb

The affinity of McAb against corresponding antigen can be tested by ELISA or RIA.

3.3 Purification of antibody

3.3.1 Principle

Antibodies prepared through various methods usually blend many impurities which must be purified in order to obtain antibody of a relatively single component or realize special applications. Antibody, a kind of protein, is of a certain isoelectronic point, solubility, charge and hydrophobicity as same as other proteins. These characters can be utilized to separate and purify antibody through electrophoresis, salting-out or other chromatographic technology. Purification of antibody is implemented according to the following two principles: ① Firstly the application of antibody should be confirmed before purifying it because there are different requirements for antibody in different applications. ② Various kinds of antibodies from different sources, including polyclonal antibody, monoclonal antibody and genetic antibody, contain different impurities. In addition, type of heavy chain, subtype and molecule weight of antibody are also important factors in selecting purification method [14-16].

3.3.2 Purification methods

3.3.2.1 Salting out precipitation

Salt has double effect on the solubility of proteins. Less salt can increase the solubility of proteins through the effect of salt and water molecules on polar group of protein molecule. On the other hand, a large amount of salt will decrease water activity since the ions of salt can destroy hydrate shell around protein and neutralize the charges on the surface of proteins, resulting in protein congregating and precipitating. The salt concentration at which a certain protein precipitates depends on its solution behaviour. Hydrophobic proteins are precipitated at clearly lower salt concentrations than hydrophilic proteins. Generally a certain concentration of salt will be selected at which most of the proteins precipitate except antibody. The most common salt for salting-out is ammonium sulphate which is highly soluble, insensitive to temperature, low in price and can grade well. Moreover, it can stabilize protein structure and deal with a large amount of samples.

3.3.2.2　Gel filtration

Gel filtration, also called exclusion chromatography, can separate proteins with similar structure and but different molecule weight. The separation medium consists of porous beads in different sizes. Smaller molecules can diffuse into the pores of beads, while larger molecules enter less or not at all. That leads to the fact that smaller molecules must travel a longer way than larger molecules. Because of that, the former will be eluted slowly, while the latter will be eluted rapidly. Generally, the elution consequence of proteins is from the higher to the lower according to molecule weight in gel filtration.

Most of gel matrix are chemical crosslinking polymers such as dextran, agarose, polyacrylamide and vinyl-type polymer. The average pore size of gel depends on the crosslinking degree. The higher the crosslinking degree of gel is the smaller its average pore size is and the stronger its rigidity. The pore size of gel selected depends on the molecule weight of target antibody and major protein impurities.

3.3.2.3　Ion-exchange chromatography

Ion-exchange chromatography (IAC), as a purifying technology, can implement the separation of ions or polar molecules by utilizing the affinity difference between exchangeable ions on the chromatography medium and the components to be separated. IAC, the most common method of chromatography, has advantages of rapid separation, good selection and repetition, etc. It can be applied to separate various biological molecules such as protein, amino acid, polypeptide and nucleic acid.

Ion exchange resins, which are synthetic polymers with many ionization groups, can be divided into cation and anion exchange sorbent according to the ionization group of them. If the ions to be separated flow through the chromatography column, the speed of the ions depends on the affinity between them and the medium, and the character and concentration of ions in solution.

Ion exchange sorbent, which is insoluble in water and composed of matrix, electronic group and counterion, can release counterion and bind to ions and ionic compounds in solution. However, the binding won't change the physicochemical properties of sorbent and ionic compounds.

3.3.2.4　Immunoaffinity chromatography

Immunoaffinity chromatography (IC) is a separation method implemented by utilizing the specific affinity between antigen and antibody. After antigen bonds to the medium of IC, the IC can separate antibody from antiserum. In the fermentation liquor of protein engineering bacteria, the protein to be separated usually is at low concentration and hard to be separated by ion exchange chromatography and gel filtration. Nevertheless, IC is an effective way for separating this target protein. The immunoaffinity sorbent can be obtained by conjugating the antibody, which was prepared through animal immunization of the target protein, with chromatography matrix. Because of the strong interaction between antigen and antibody, high eluant strength is needed in this IC separation. Reducing the affinity between antigen and antibody can facilitate elution by choosing different type of them or analogs. In addition, protein A from staphylococcus aureus also can bond IgG and be applied to separate IgG.

References

[1] Qian HL. Microbiology. Beijing: China Medical Science Press, 1993: 284

[2] Drew MP, Suzanne LT. The role of CD^{4+} T cell responses in antitumor immunity. Current Opinion in Immunology, 1998, 10(5): 588-594

[3] Wu XW, Liang ZH. Practical immunological experimental technology. Wuhan: Hubei Science and Technology Press, 2002: 1

[4] Ma Y, Liu HZ. Postgraduate textbook of immunology. Dalian: Dalian Press, 1998: 47

[5] Bundesen PG, Drake RG, Kelly K and others. Radioimmunoassay for human growth hormone using monoclonal antibodies. J Clin Endocrinol Metab, 1980, 51(6):1472-1474

[6] Barnard R, Quirk P, Waters MJ. Characterization of the growth of hormone-binding protein of human serum using a panel of monoclonal antibodies. Journal of Endocrinology, 1989, 123: 327-332

[7] Li ZG, Fan MY. Doagnosis Technology of infection. Beijing: Science Press, 1990: 73

[8] Luo LX. Cell fusion technology and its application. Beijing: Chemical Industry Press, 2004: 4

[9] Tijssen P, Kurstak E. Highly efficient and simple methods for the preparation of peroxidase and active peroxidase-antibody conjugates for enzyme immunoassays. Analytical Biochemistry, 1984, 136(2): 451-457

[10] Sun JF. Methods of animal experiments. Beijing: People's Medical Publishing House, 2001: 389

[11] Ma JX, Liu XJ. Practical clinical nuclear medicine. Beijing: Atomic Energy Press, 1990: 92

[12] Luttmann W, Bratke K, Küpper M, Myrtek D and others. Heidelberg: Elsevier Inc. , 1987: 265

[13] O'Sullivan MJ, Gnemmi E, Chieregatti G, Morris D and others. The influence of antigen properties on the conditions required to elute antibodies from immunoadsorbents. J Immunol Methods, 1979, 30(2): 127-37

[14] Carter RJ, Boyd ND. A comparison of methods for obtaining high yields of pure immunoglobulin from severely haemolysed plasma. J Immunol Methods, 1979, 26(3): 213-232

[15] Chen YF, Hu JL, He ZT. Advancement of antibody preparation technology. Chemistry of Life, 2001, 20(3): 250-253

[16] Zhu XT, Xie Q, Ma RJ. Advancement of monoclonal antibody preparation technology. Gansu Science and Technology, 2005, 21(3): 108-109, 98

Chapter 4　Antibody Engineering

4.1　Cell engineering antibody

In 1975, British scientists Kohler and Milstein fused the antibody producing lymphocytes with tumor cells and firstly developed monoclonal antibody of homogeneity by using B lymphocyte (B cell) hybridoma technique. They won the Nobel Prize in Physiology or Medicine 1984. The hybridoma monoclonal antibody is also called cell engineering antibody [1], which is defined as hybridoma that can not only secrete antibodies but also have unlimited production capability. This kind of cell is obtained by fusion of myeloma cell, which has unlimited production capability but cannot secrete antibodies, with B cells, which can secrete antibodies but has limited production capability, under specific condition. Cell colony proliferated from a single cell (a cell clone) can be acquired after mass screening work. Hybridoma cells from the same clone can only synthesize and secrete uniform antibodies which are identical in physical and chemical properties, molecular structure, genetic marker and biological properties. This kind of antibody is called monoclonal antibody.

Hybridoma technique is regarded as the first leap of antibody engineering as well as a milestone in modern biotechnology development. The monoclonal antibodies produced using this technique have wide applications in disease diagnosis, treatment and scientific research. The hybridoma technique is based on three principles: animal immunizing, cell fusion and screening of hybridoma cells[2].

4.1.1　Animal immunizing

Under stimulation by specific exogenous antigen, B cells in animals will massively proliferate and become plasmocytes to secrete antibodies against the antigen. This kind of antibody is specific. Animal immunizing, utilizing specific exogenous antigen to immunize animals once or more, can stimulate the animal proliferating large quantity of B cells having specific response to the antigen.

4.1.2　Cell fusion

Stimulated by exogenous antigen, B cells will secrete antibodies. Yet these B cells can only survive for a very short time in vitro (apoptosis occurs after two weeks at most), while the myeloma cells do not secrete any immunoglobulin and can survive for a long time in vitro. If these two kinds of properties can be integrated, we can obtain cells not only secreting antibodies but also surviving for a long period of time in vitro. Since it is inevitable to produce multiple cells in the hybrid process, further screening of the useful hybridoma cells is necessary.

4. 1. 3   Screening of hybridoma cell

HAT medium (H-hypoxanthine, A-aminopterin, T-thymine) is commonly used to screen hybridoma cell. There are two pathways, the endogenous and the exogenous, for DNA synthesis. The endogenous pathway utilizes Gln (glutamine) or uridine monophosphate to synthesize DNA under the catalysis of dihydrofolate reductase (DHFR); while exogenous pathway uses hypoxanthine or thymine to synthesize DNA remedially under the catalysis of HGPRT (hypoxanthine-guanine phosphoribosyltransferase) or TK (thymine kinase). Aminopterin in HAT medium is the inhibitor of DHFR which can effectively restrain the DNA synthesis by endogenous pathway. The B cells contain the two enzymes HGPRT and TK, so that DNA synthesis can still be completed by using hypoxanthine and thymine in HAT medium, even when endogenous pathway is restrained. However, B cells are normal cells which cannot survive for a long time, while hybridoma cells inherit the properties from both B cells and myeloma cells and they can survive in HAT medium for a long time by producing HGPRT and TK. After two weeks the hybridoma cells can be obtained and served as a resource for producing monoclonal antibodies.

Once the hybridoma cells are identified and cultivated to enough quantity, the monoclonal antibodies can be produced massively. The primary methods include in-vitro culture, in-vivo culture and microencapsule technique. The major defect of in-vitro culture is that the amount of monoclonal antibodies produced is much less, so that the apoptosis of cells might occur because of over proliferation and growth in the massive preparation of antibodies. In contrast, the amount of antibodies obtained from animal serum or ascites by in-vivo culture is about 1,000 times high than that from in-vitro culture. Therefore in-vivo culture is the most suitable method for mass monoclonal antibodies preparation. According to some reports, microencapsule technique can raise efficiency for about 50% and cost less compared with traditional methods. Similar to previous methods, once hybridoma cells are obtained and identified, antibodies can be produced and collected in microsphere, which is made up by carbohydrate porous membrane. This method increases the output of monoclonal antibodies significantly.

4.2   Genetic engineering antibody

In early 1980s of the 20th century when the genetic engineering antibody technique was born, by combining the research achievements in the gene structure and function of antibody with recombinant DNA technique, scientists began to produce antibodies through genetic engineering to reduce the immunogenicity of mouse antibodies. With the help of new biotechnology such as genetic engineering, the coding genes for antibodies can be modified and reassembled according to different needs and be expressed in proper recipient cells.

As an exogenous protein, mouse monoclonal antibody can induce the production of human anti-mouse antibody (HAMA) in human body and there is no breakthrough in producing human monoclonal antibodies through cell fusion, hybridoma method. Therefore, the application in human body treatment obviously lags although monoclonal antibody have wide applications in many areas. Since early 1980s, mouse monoclonal antibodies have been developed from humanization of constant region to humanization of variable region, from retaining some mouse residues for guarantee of the affinity to obtaining fully humanized sequence by using antibody library technique.

Since the first genetic engineering antibody (human-mouse chimeric antibody) was developed in 1984, new genetic engineering antibodies continuously appear, such as humanized antibody, monovalent small molecular antibodies (Fab, single-chain antibody, single-domain antibody, hypervariable region polypeptide, etc.), polyvalent small molecular antibodies (double-chains antibody, three-chains antibody, microantibody), some special forms of antibodies (bispecific antibody, antigenized antibody, intracellular antibody, catalytic antibody, immunoliposome) and antibody fusion proteins (immunotoxin, immunoadhesin).

Compared with monoclonal antibodies, genetic engineering antibodies have the following advantages: the rejection response of human body to the antibodies can be reduced or even eliminated by modification of the antibodies with genetic engineering techniques; the genetically engineered antibodies have smaller molecular weight, and can partially reduce the murine origin of antibodies; it is easier for them to penetrate through vascular wall and enter the core part of nidus; novel antibodies can be developed according to therapeutic needs; the antibodies can be massively expressed by prokaryotic, eukaryotic cells and plants so that the production costs can be reduced significantly.

The most important value of them is the reduction of immune response induced by body treatment, which makes it possible to repetitively administrate immunotoxin in clinical application. The difference between types of immune globulins reflected by constant region of antibody heavy chain and its subtypes may influence the function of antibody in human body. Therefore, modification of the types and subtypes of antibodies may improve therapeutic effect in human body when constructing vectors for antibody expression. Since the deficiency of constant region, single-chain antibody cannot combine with cells whose constant region receptors are positive, so that nonspecific destructions can be avoided in body treatment. Because of its low immunogenicity, smaller molecular weight, better penetrability, it is easier to penetrate into tumor and has better therapeutic effects than complete antibodies[3,4].

There is a broader prospect for application of genetically engineered immunotoxin. After years of research and development, there are some progresses in application of immunotoxin in the treatment of breast cancer, melanoma, leukemia and in-vitro bone marrow transplantation. Many agents used in tumor treatment are applied in Ⅰ and Ⅱ phase clinical trial. But until now, application of immunotoxin in body is not so effective as people expect, since there is a big difference between clinical treatment and the cell test in vitro or animal models. Concentration of antibodies in tumor cells depends on the synergism of both immunological and non-immunological factors. Therefore, it needs further scientific researches before immunotoxin serve as the main force of targeting therapy. Cooperating with other therapeutic methods, its defects can be remedied to contribute to the cancer treatment[5-8].

4. 2. 1   Chimeric antibody

There is a wide clinical application of mouse monoclonal antibodies. However, such antibodies still have limitations. For example, mouse antibodies will be removed from human circulatory system quickly after administration. Generally, mouse antibodies cannot effectively activate the immune defense system of the host. On the contrary, in most cases, mouse antibodies will cause the immune responses against mouse antibodies themselves and induce the production of HAMA, which is also the human antimouse response. To solve this problem, chimeric antibody was invented to replace part of mouse monoclonal antibody with part of humanized antibody. Chimeric antibody is the first successfully produced genetic engineering antibody[9]. It is produced through joining mouse antibody genes to human antibody genes to form chimeric genes and then inserting the chimeric genes into vector and transfecting the vector into myeloma tissues for expression. Thus, the negative effects caused by mouse monoclonal antibody components is reduced and the curative effects is promoted.

The genes for construction of genetic engineering antibodies were firstly acquired from hybridoma cells, so the construction also must undergo a long and complicated process of animal immune, cell fusion, cloning and screening. Besides, humanized antibodies and the antibodies against private antigens still cannot be produced by genetic engineering and neither the affinity of those antibodies can be improved. These disadvantages limit the wide applications of genetic engineering antibodies.

4.2.1.1 Humanization of the constant region

At first, Fc region in mouse antibody was considered to be replaced with human gene because this region activates immune response in human body and was most likely to cause human antimouse response. This method fuses the variable region of mouse antibody to the constant region of human antibody so as to form the human-mouse chimeric antibody. Compared with other humanization methods, it is easier to operate and the affinity and specificity of antibody are guaranteed since complete variable region of mouse antibody is maintained. However, the variable regions of exogenous mouse antibody are maintained and HAMA still will take place. Since the homology between the coding sequences for variable region of different mouse antibody and the coding sequences for variable region of human antibody are not entirely the same, there may be significant differences between HAMAs produced in human body by different mouse antibody. Many human-mouse chimeric antibodies have been developed and applied in clinical therapy, mainly for malignant tumor. In August 1998, FDA formally authorized the application of primary antibody (TNF-α human-mouse chimeric antibody) in inflammatory bowel disease treatment. It was the first case that human-mouse chimeric antibody was used as chronic diseases drug, which showed more optimistic prospect of applying human-mouse chimeric antibody in human body. In the mid 80s of the 20th century, China began to research in this area. Since then many successful constructions of human-mouse chimeric antibodies have been reported and relative techniques became gradually matured.

4.2.1.2 Humanization of variable region

Since human-mouse chimeric antibody retains mouse monoclonal antibody variable region, HAMA still will be produced in human body. It is necessary to humanize the variable region too. The modified antibody was firstly constructed through CDR transplant.

The antigen combining site in the antibody molecule is formed by 6 CDR of VH and VL. Other region (framework region) is responsible for maintaining specific spatial conformation of the molecule. Therefore, modified antibodies produced by replacing CDR from human variable region with that from mouse monoclonal antibody have specificity of parent mouse monoclonal antibody. However, it is not as easy as CDR transplant for modified antibody construction. Since amino acid residues of framework region may influence the conformation of CDR plane, construction of an ideal modified antibody usually involves a series of analysis, design and reconstruction work. For example, the affinity of 27 modified antibodies have been measured and 63% of them showed affinity decline from 10% to 87%.

Some researchers proposed a method to humanize surface residues from another concept. Since antigen-antibody reaction only involves molecule on the surface of antibody, they changed surface residues of mouse variable region to make them similar to human variable region, which was expected to eliminate HAMA. Now there are only reports about humanized antibodies that still have specificity and affinity of parent mouse monoclonal antibody by using this method, but no effects of heterology elimination has been shown so far.

4.2.2   Humanized antibody

In the two afore-mentioned techniques, the former maximally reduces the mouse sequence, which may influence the binding ability of antibodies, while the later maximally retains the mouse sequence, which may maintain the binding ability of the parent mouse monoclonal antibody, but the effects of reducing heterology may be influenced. Replacing the CDR in human antibody with the CDR from mouse monoclonal antibody, we can obtain antibody with minimum murine origin which is called humanized antibody[10].

Nowadays, the humanization method based on CDR transplant integrated the above two methods. By using database retrieval system and computer-aided molecule modelling methods to find the human antibody variable region template with maximum homology and taking a comprehensive consideration of the surface residues, residues which have interaction with CDR or have significant influence on spatial structure, the key residues which need to reserve or change can be selected. And then through molecule modelling, gene synthesis and expression as well as necessary correction, the actual effect will be confirmed. This is really a complex system. Until now, a large quantity of modified antibodies have been constructed and over 20 humanized antibodies have been applied in clinical treatment with good results. The disadvantage of this technology lies in the high operational difficulties and uncertainty of the result such as the reduced affinity and the retention of mouse sequences.

4.2.3   Fully humanized antibody

With the rapid development of genetic engineering technique, the therapeutic monoclonal antibodies developed from 100% of the mouse monoclonal antibody (Mab) to chimeric antibody, humanized Mab and most recently fully humanized Mab (Figure 4-1). The immunogenic problems are eliminated gradually while the high affinity to antigen is retained and the pharmacokinetic of antibody is also improved. By using gene knockout to delete mouse Ig genes and replacing them with human Ig genes, followed by the im-

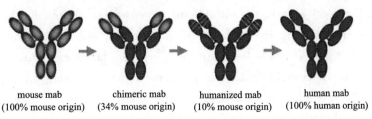

mouse mab   chimeric mab   humanized mab   human mab
(100% mouse origin) (34% mouse origin) (10% mouse origin) (100% human origin)

Figure 4-1   Humanization progress of mouse mab (See Colored Figures)

munization of mouse with antigen, large quantity of the fully humanized antibodies can be produced with hybridoma technique.

With the appearance of phage antibody library technique, Jespers reported the humanization of mouse Mab through epitope-oriented screening [11]. The main procedures were as follows. Using the mouse Mab as template, the heterozygous antibody library was created by fusion of the coding genes for mouse Mab light chain or heavy chain with the coding genes for human Mab heavy chain or light chain. Screening of the antibody was implemented by using corresponding antigens so as to pick out the coding genes for human heavy or light chain which matched another mouse origin chain, followed by matching these genes with the gene library coding for the variable region of the other human origin chain. Thus the full variable region of human Mab can be obtained. The advantage of this method is that we can obtain fully humanized antibody, but there is no more similar report appeared after that. Recently, Rader reported the result by using a method similar to aforementioned method. In this method, all of the CDR3 in the gene library coding for human heavy chain variable region and light chain variable region were replaced with the CDR3 from mouse Mab, and then the humanized antibody with high quality was obtained after screening. The practicability of this method has been approved and it should have a bright application prospect.

A large quantity of mouse Mabs have been produced until now and some of them have clinical prospect. Thus, the humanization of mouse Mabs is of great significance. Remarkable progresses have been made in this area, while more and more humanized Mabs are being used in clinical treatment. However, there is still a lack of a simple, effective and satisfying method for humanization of mouse Mabs [12].

4. 2. 4　Small molecule antibodies and antibody fusion protein

The small molecule antibodies were noticed because of their small molecular weight, strong penetrability, low antigenicity, easy to be genetically engineered and can be expressed in prokaryotic system. There are many kinds of small molecule antibodies, moreover, different forms of such antibodies continuously appear. The small molecule antibodies fully studied or having clear application prospects include Fab fragment, Fv fragment, ScFv, Fv fragment fixed by disulfide bond, diabody and minibody, etc. Fab fragment is a heterodimer which is suitable for secretory expression, but not suitable for massive expression by inclusions. Single-chain antibody, also called single chain fragment of variable region (ScFv), is the antibody which is produced by fusion of the coding genes for V region of H and L chains in Ig, transfection to *E. coli* and expression.

ScFv has been studied most among all the small molecule antibodies. With the properties of small molecular weight (about 1/6 of complete antibody) and simple structure, ScFv is the smallest antibody fragment with complete antigen binding site[13].

Compared with complete antibodies, ScFv has features such as retaining short time in the non-target tissues, being removed quickly from blood, good tissue penetrability. In addition, ScFv can be expressed in bacteria and is easy for genetic manipulation and mass production by genetic engineering. It can be reconstructed and connected with other effector molecules to produce anti-tumor fusion proteins by chemical coupling or genetic engineering methods. Therefore, ScFv is a valuable molecule for tumor treatment and targeting diagnosis.

With strong penetrability ScFv can easily penetrate into local tissues to work. Based on ScFv, some better small molecule antibodies have been developed recently. DsFv, a Fv fragment fixed by disulfide bond, is produced after inserting a cysteine in the proper position of each variable region of light chain and heavy chain. It has been found that both that the binding ability and the stability of DsFv are better than those of ScFv. Moreover, the immunotoxins constructed by DsFv have been applied in pre-clinical trial. By reducing the length of the linker between two variable regions of ScFv, the V_H and V_L are forced to pair and form bivalent small molecule antibody, which has better binding ability than univalent antibody. If the coding genes for two different variable regions were matched and cross-linked, the bispecific diabody could be obtained. Compared with the bispecific antibodies produced by chemical crosslinking, tri-hybridoma and tetra-hybridoma, diabody is easier to be produced. Moreover, it has smaller molecular weight and is more stable and effective. Thus, it has a broad application prospect. A typical example is the recently developed anti-CD19/anti-CD3 bispecific diabody, which can mediate T cells to kill tumor cells.

Holliger[14] and Kortt [15] had reported that ScFv monomer is hard to form when reducing the linker of V_H and V_L in ScFv molecules to less than 12 amino acid residues (normally 15). Since the intramolecular fragments of V_H and V_L can't match, the V_H and V_L are promoted to match intermolecularly, which forms the 50 kDa double-chain antibody with non-covalent bonds. Experimental data showed that double-chain antibody was removed more slowly from kidney and had higher affinity. So it could stay for a longer time in tumor cells and had higher tumor: blood ratio compared with ScFv monomer. After analysis of the structures with different valents of the same antibody and evaluation the relationship between the structure and the functional affinity, it was found that the ratio of functional affinity from tetravalent, divalent and monovalent molecules was 140 : 20 : 1. This showed the necessity of constructing multivalent antibody molecules. The relationship between molecular weight and tumor-targeting, tumor penetrability and pharmacokinetics was studied and the results showed that the optimal molecular weight of antibody for tumor-targeting was 60,000 to 120,000 Da. The smallest molecules (such as CDRs, single domain antibody) will be removed from human body quickly and accumulate very small amount in tumor. Some bigger antibody molecules like ScFv will accumulate more in tumor, but still not enough for treatment. Complete antibody molecules have defects such as poor penetrability and poor tumor-targeting accuracy. Double-chain antibodies have shown the best balance of tumor-targeting, tumor tissue penetrability and blood clearance so far. Thus, double-chain antibodies might be the most promising carrier for immunotherapy and more and more reports about them have appeared.

Besides, many alternative strategies have been designed to construct divalent or bispecific small molecule antibodies. The basic idea is to add a dimerized structure on one end of ScFv. In the modified molecules, minibody is the better one with effect in human body treatment. It is produced by joining the CH3 group of Ig to the carboxyl terminal of ScFv to form a divalent molecule. Comparing the immune imaging effects and the therapeutic effects of Ig, F(ab')2, ScFv, diabody and minibody on tumor-bearing mice, the results showed that minibody and diabody had the best imaging effects whereas complete Ig and minibody had the best therapeutic effects.

Fusion proteins with several biological functions can be obtained by fusion of the antibody fragments to other proteins. Such fusion proteins can be divided into two categories. One is to combine Fv fragments with other biologically active proteins and the resulted fusion proteins can deliver bioactivities to specific tissues with the help of the specific identification function of antibodies.

Targeting therapy (especially in tumor treatment) is the main application area of these fusion proteins. Until now, targeting therapy has achieved good effects in vitro tests and animal experiments, but its applications in human body is insufficient and need further assessment. Nowadays, antibody fusion proteins made up of the exotoxin from *Pseudomonas aeruginosa* and ricin have entered the Phase II clinical trial. The other one is the antibody fusion protein which contains Fc fragment. The fusion protein made up of Fc fragment and some proteins with adherence or combining functions is also called immunoadhesin. The Fc fragment can give the immunoadhesin the following functions: ①Application in detection or purification by combining with anti-Ig molecule or protein A. ② Effector functions of antibody mediated by Fc fragment such as antibody-dependent cell-mediated cytotoxicity (ADCC), complement fixation, and opsonic action. ③ Increasing the half-life of this protein in blood.

In China, many researches have been conducted on small molecule antibodies, with an emphasis on ScFv and Fab, and there are also some reports on the success of diabody developing. Some laboratories are developing antibody fusion proteins for targeting therapy. Since there is groundwork in mouse Mab research in China, the development of small molecule antibodies and antibody fusion proteins have clinical application prospects.

4. 2. 5　Bispecific diabody

Bispecific diabody is defined as the combination of two Mabs by some methods like chemical crosslinking, dihybridoma or genetic engineering. Such antibodies contain different antigen identification units and can bind different antigens. The antibodies which identify the effector cells and the target cells can be combined so as to promote the antitumor effects of immune effector cells. For example, the antibody which identify tumor antigen and the antibody (CD3 antibody or CD16 antibody) which identify cytotoxicity immune effector cells(CTL cell, NK cell, LAK cell)can be combined to produce bispecific antibody.

In 1985, Milstein and others [16] further developed hybridoma technique based on Mab and produced bispecific monoclonal antibody (BsMAb) by cell engineering method. He fused two hybridoma cells which can secrete different specific Mabs to obtain a quadroma and two bispecific Ab that have two parental specificities. This method is time-consuming and difficult, whereas the yield is low and the products are unstable and of low activity. And these Mabs are usually secreted by hybridoma cells fused with mouse B cells and mouse myeloma cells so as to have murine origin and cause rejection in human body. In addition, such complete antibody molecules have the disadvantage of high molecular weight, poor vascular penetration ability as well as high production cost and are not suitable for large-scale industrial production.

The appearance of recombinant technique leads to the reform of screening, humanization and production of antibody. It replaces hybridoma technique and makes it possible to design antibody-based pharmaceuticals. Recombinant antibodies are becoming smaller and smaller and can be reconstructed as polyvalent molecules or fused with other molecules such as radionuclides, toxins, enzymes, liposomes and viruses.

Holliger constructed recombinant Bispecific Single-chain antibody by the genetic engineering method for double-chain antibody construction. He fused V_H and V_L from two different antibodies (A and B) with 5 amino acid residues linker (Gly4 Ser)$_3$ to form two different single chain, $V_H A$-$V_L B$ and $V_H B$-$V_L A$. Their coding sequences shared the same promoter, but had their own initiation codons and signal peptide coding sequences. They were coexpressed in the same cell and secreted into periplasmic space. Since the linker is too short, VH and VL on the same chain cannot match each other, but can match the V region of the other chain and fold to form a dimer molecule which has two antigen combining sites.

Crystal diffraction shows that the two antigen combining sites of bispecific antibody formed as dimer locate in the opposite direction of antibody molecule so that they can connect two cells. In human body, the antitumor activity of bispecific antibody formed as dimer is similar to that from the quadroma, because the bispecific antibody from quadroma has much higher molecular weight and stays longer in blood. Nevertheless, bispecfic antibody prepared as dimer has lower immunogenicity than that prepared from quadroma technique, and is easy be massively expressed in bacteria. Meanwhile, such bispecific antibodies have better effects for killing tumor cells under mediating of T cells and NK cells in vitro than those produced from quadroma.

4.3　Antibody library technology

The antibody genes for construction of genetic engineering antibodies were initially derived from hybridoma cells. So a long and complicated process of animal immunity, cell fusion and clone screening was also needed. In addition, we still can't prepare antibodies for the antigens from rare species and humanized antibodies, and the affinity of antibodies were also hard to improve. These drawbacks restrains more wide application of genetic engineering antibody. Antibody library technology, developed from the application of combinatorial chemistry method in antibody engineering, has enabled us to find effective way to solve these problems. In antibody library technology, all genes of the antibody variable regions can be amplified through DNA recombination and then the functional active antibody fragments (the library) expressed in prokaryotic systems, followed by screening of the genes coding for the specific antibodies from the library.

The phage antibody library is the most commonly used antibody library technology. Using phage antibody library technology to screen antibody, the animal immunity process is not necessary, so it is easy to prepare antibodies for the antigens especially from the rare species and screen the fully humanized antibodies or the antibodies with high affinity. The technology of phage antibody library is one of the breakthroughs in life science, and pushes the research and study of antibody engineering to a new era. On the basis of phage antibody library, the technology of ribosome display has been developed in recent years. The process of screening an antibody through ribosome display technology is entirely done in vitro, therefore, the high-capacity and high-quality libraries can be built without the steps of transformation in *Escherichia coli*, which makes it easier to screen antibodies with high-affinity and modify antibody molecules in vitro evolution. The technology of ribosome display represents the future development trends of antibody engineering.

The technology of phage display library is regarded as the third revolution of antibody technology. The key of the technology is to construct and develop the humanized library of large capacity, high specificity and high sensitivity. The phage antibody is prepared by expression of the antibody molecule, Fab segment or ScFv, on the surface of membrane, which is completed through the combination of the coat protein of the single-chain phage with the Fab segment or the ScFv fragment. The feature of this technology is that it can not only identify the corresponding antigens and combine with them, but also infect the host bacterium for amplification. The genes coding for the full variable regions of B-cell was cloned and assembled in groups of the phage antibodies, namely the technology of phage antibody display.

The emergence of the phage library technology relies on the developments of three experimental techniques. The first one is the development of PCR technology. This technology makes it possible to clone the genes coding for the complete set of the variable regions through RT-PCR, using the total RNA as templates. The second one is the successful accomplishment of the secretion of immunoglobulin fragments with binding affinity in *Escherichia coli* of. The third one is the development of the phage display technology, in which target molecule fuses with the coat proteins of the single-strand phage, and then is displayed on the surface of phage. With the reproducible feature of the phage, the target molecules can be solidified on the surface. The ligand peptides to the target molecules can be screened and obtained through the process of affinity adsorption-elution-amplification, which can be implemented on the surface of the phage. The methods of mutation and strand displacement can be used to improve the affinity of the selected antibody, thus the specific antibodies with high affinity can be obtained.

4.3.1   Application of antibody library

The conception of antibody library technology has been proposed in the early of 1990s. In short, the antibody library technology is to express the library with bacteria cloning instead of B-cell cloning. The phage display antibody library is the most important progress in the field of antibody engineering. Before that, genetically engineered antibody mainly included recombinant monoclonal antibody obtained through DNA recombination technology. Antibody library technology further uses the genetic engineering technology to clone novel antibodies, and, so the development of antibody engineering entered a new period of time.

4.3.1.1 Preparation of human antibodies

Because of the inefficiency of human hybridoma monoclonal antibody system, the development of human monoclonal antibody is very slow. Antibody library technology provides an effective way to solve this problem. As for the immunogen which can immunize *in vivo*, such as microbial infection and vaccination, it is easy to harvest human antibodies with high affinity from the lymphocytes of the host immunized. A variety of the anti-virus human antibodies have been developed, including antibodies against RSV, HIV, HCV and HBV. In China, several human antibodies against hepatitis viruses have also been reported. With the maturity of the antibody library technology and the construction of large-capacity antibody libraries, it is possible to screen human antibodies with high affinity from the master library without immunization[17]. The wide spread of this technology will greatly promote the development of human antibody. Nowadays, two strains of human antibodies have been applied in the phase I/II clinical trials as treatment of Rheumatoid Arthritis (anti-TNF-α) and ocular Fibrosis Syndrome (anti-TGF-β). These progresses indicate the great development of the application of human antibodies in clinical trials.

4.3.1.2　Improvement of antibody

Before antibody library technology appeared, the modification of antibodies was confined to the known structures, such as the humanization of antibodies, and the construction of small molecule antibodies and antibody fusion proteins. The antibody library technology pushed the capability of antibody modification to a new level. Aiming at some specific performance, the unknown structure can be screened. It is particularly effective in the improvement of the antigen binding capacity, for example, the improved mutants can be screened and obtained by introducing mutations in corresponding sites and affinity adsorption. Turner randomized the joint sequences which connected V_H to V_L in ScFv, and then built a library. With the library, he obtained joint sequences that had higher expression levels and were easier for renaturation. Hennecke used similar methods to screen the joints that were more stable and easier for genetic manipulation. Coia introduced random mutations to gene coding for ScFv and selected mutants with higher expression level. Saviranta modified the specificity of anti-estrodiol Fab that had cross reaction to testosterone. By introducing mutations to V_H with the mismatch PCR method, the mutant without cross reaction to testosterone was selected. Ftdji randomized the sequence of V_HCDR3 and harvested antibody with enhanced catalytic capability.

In particular, the antibody library technology can imitate the maturation process of the antibody affinity *in vitro* to improve the affinity. There are many reports on the success of this kind of works. A typical example was reported by Schier and others. In that work, the affinity of the primary antibody c-erbB-2ScFv was increased to $0.77 \times 10^{11}/M$, 1,230 times higher than that of the original antibody; Yang [18] increased the affinity of the anti- HIV antibody by 420 times, reaching $0.67 \times 10^{11}/M$. These reports showed that antibody library technology may be used to improve various properties of the antibodies, so that we can obtain more valuable antibodies.

4.3.1.3　Preparation of antibodies without immunization

The antibody library technology can imitate the generation of antibody *in vitro*. It opens a new era of antibody preparation without immunity. There are three key points in the generation of antibody in vivo: the variable B-cell repertiore, the clonal selection, and the affinity maturation. The powerful filtering abilities of the antibody library technology have basically solved the problems in the last two points. Therefore, the key element for preparation of antibodies without immunization lies on the construction of a good phage antibody library. T issues need to be considered: the large capacity and the diversity of the library.

(1) Capacity of the antibody library

So far, more than ten antibody libraries built without immunization have been reported. The capacity of these libraries ranges from 10^7 to 6.5×10^{10}. Analyzing these libraries, it was found that the antibody affinity varied in proportion with the library capacity. The affinity of the antibodies obtained from the library with a capacity of 10^7-10^8 were 10^6-10^7/M, while those from the library with a capacity of 10^8-10^9 were 10^7-10^8/M. So it is vital to construct a library with large capacity in order to clone antibodies of high affinity without immunization.

The capacity of the library is mainly confined by the efficiency of *Escherichia coli* transformation. One microgram DNA ligations can obtain about 10^7 of library capacity with the most effective method, electroporation. Therefore, the construction of a library with the capacity of more than 10^9 needs repetitive construction works for many times. Then a combined infection method was put forward, in which the genes coding for one chain of the antibody was transformed and the other chain was prepared for phage infection with *Escherichia coli*. Since the efficiency of phage infected is about 100%, the capacity obtained is close to the colony number of the bacterial transformants. In order to ensure the genes coding for light and heavy chains can recombine into the same vector and form the proper expression unit in the host, oriental recombination sequences should be designed on the two vectors which carry two parts of the coding sequences. For instance, high-efficiency oriental recombination will take place in the LoxP locus of phage P1 when protein Cre exists. Griffiths and others built the largest antibody library so far with this method. But successful construction of the antibody library in similar ways hasn't been reported since then, so the general feasibility of this technology is yet to be proved.

(2) Diversity of antibody library

Nowadays we have three methods to construct antibody library with diversity: the natural antibody library, the semi-synthetic phage antibody library and the synthetic antibody library. Natural antibody library is constructed by amplifying the coding genes for antibodies from lymphocyte in body. Semi-synthetic phage antibody library is built by imitation of the V(D)J rearrangement *in vitro*, that is to combine the artificial synthesis of randomized CDR3 with the V segment (containing CDR1 and CDR2) in genome. Synthetic antibody library has all variable sequences artificially synthesized.

So far, five natural antibody libraries have been reported. The first attempt to clone antibody without immunization used a natural antibody library, but the result was not very encouraging because of the small capacity. Recently, two natural antibody libraries with large capacity were reported and more than 20 antigens with high affinity has been selected from them.

More semi-synthetic phage antibody libraries have been reported, quality of which are directly related to the library capacity. Griffiths and others constructed a library with the capacity of 6.5×10^{10}. More than 30 antigens were tested and the antibodies with high affinity had been selected from his library. Both the two kind of libraries have advantages and disadvantages. As for the natural antibody library, the body has effects of self-tolerance and it contacts with multiple antigens, which cause bias on diversity of the result antibodies. For this reason, some scientists prefer the semi-synthetic phage antibody libraries. But the V genes recombined *in vivo* by V(D)J all are active and the length of V_H CDR3 ranges from 5-30 amino acid; whereas the CDR3 produced by semi-synthetic phage antibody library may result in inactive clones and the length of artificially synthesized CDR3 is also restrained. In spite of the advantages and disadvantages of these two kinds of libraries, both have achieved success as reported.

There is no formal report on the construction of synthetic antibody library, except some reports in academic conferences. With analysis on human antibody sequences and their spatial structures, Pack synthesized 7 V_H coding genes and 7 V_L coding genes, which can represent all the antibody types and structures. These V_H and VL genes can be reconstructed into 49 sub-libraries, in which the CDRs were of cassette structure and can be easily replaced, and the codons which *Escherichia coli* host preferred were used for expression. Pack reported the result of a small sample. He used a V_H and a V_L fragments and the randomized V_HCDR3 for construction of an antibody sub-library. The capacity of the library was just 3×10^7. Antibodies against 8 antigens were successfully selected from the library and the affinity reached 0.2×10^8/M. This result was positive, yet the reports on the synthetic phage antibody library is insufficient and the comprehensive evaluation of such libraries need more data.

There are preliminary studies on semi-synthetic phage antibody library in China. In view of its great potentiality, and the fact that the feasibility of producing antibodies using "Master Library" without immunization has been confirmed, we should develop such master libraries for ourselves. In general, the development of the genetic engineering antibodies, especially the progress of phage antibody library, has pushed forward the antibody preparation technology. It successfully solved the problems in the preparation of humanized antibody and promoted the development of various types of antibodies. It can be predicted that the development of genetic engineering antibodies will reach a new peak.

4.3.2　Phage display human antibody library

Phage display human antibody library is a new technique invented in recent years.

For its potential to produce human antibodies, the technology has attracted more and more researchers. The phage display human antibody library created a simple and rapid way to produce genetic engineering antibody. Phage display human antibody is a milestone in the history of the development of monoclonal antibodies[19-21].

According to the sources of human antibody genes and different compositions, phage display human library can be divided into four categories.

(1) Natural antibody gene bank

In this library, the antibody gene fragment comes from unimmunized donors. The library mainly stands for the antibody spectrum of natural antibodies in the body, but has a low affinity.

(2) Immunization antibody library

Antigen antibody gene comes from immunized donors. The library is characterized by its strong antigen specificity and high affinity, but the diversity of antibody is lower.

(3) Semi-synthetic phage antibody library

In this library, CDR1 and CDR2 in the variable region of the heavy chain come from 49 V_H gene fragments in human embryo, while CDR3 is the one of Fd in antibody Fab which is replaced by artificial synthesized random primers of coding 5-8 or 6-15 amino acids. It is characterized by the large capacity that can reach 10^{12}, but without selection of the immune system *in vivo* the affinity is generally lower.

(4) Transgenic mouse antibody library

Human antibody genes were transformed into rats to get fully human antibodies. It has solved the problem of mouse antibody, and will be the trend of research and application of human antibody.

4.3.3   The construction of phage display human antibody

4.3.3.1   Amplification of genetic fragments Fab in human antibody

The lymphocytes from peripheral blood, spleen, tonsils, lymph node tissue or bone marrow of individuals who have been vaccinated or been immunized with antigen are taken for extraction of total RNA or genomic DNA. In principle, each static Ig mRNA of B cells can produce 100 copies, and plasma cells can produce 300 copies during proliferate phase. Therefore, the immunization can significantly increase the number of Ig mRNA and genes encoding for antigen binding site V. RT-PCR amplification can be used to get a complete set of antibody genes. The primer for 5′ end-of antibody gene is normally based on the conservative sequence in FR1 or leader; while that of 3′ end is based mainly on antibody conserved sequence of hinge J. A set of primers were designed according to the family sequence of antibody genes, DNA or cDNA of antibody is amplified, respectively, and then the amplified products are mixed. Sometimes amplification of degenerate primer will get the largest complementarities between primers and templates. To prepare genes coding for antibody ScFv, the linkers should be designed, for instance the oligonucleotide encoding the $[Gly_4\text{-}Ser]_3$ unit can be utilized to prepare V_H-linker-V_L.

4.3.3.2　Cloning of genes for human antibody

Vectors used in phage antibody library are reconstructed on the basis of the existing cosmid. These vectors contain LacZ promoter, ribosomal binding sites, pelB leading sequence and MCS. Besides, these vectors contain the coat protein coding gene of filamentous phage M13. Plasmid pComb$_3$, a common vector for construction of phage library, was constructed by Lerner's group in 1991, through double digestion (*Sac* I-*Eco*R II) of the fragments in vectors λHC$_2$ and λLC$_2$ constructed by Huse in 1989 and combination of them. Vector pComb$_3$ retained the pelB leading sequence and multiple cloning sites for cloning of heterologous genes, meanwhile, this vector introduced LacZ promoter from M_{13} Mp$_{18}$ and the CAP binding sites used to control light chain genes, fused the gene *gIII* coding for the coat protein of M_{13} phage, and inserted a fragment from the plasmid pBluescript. Thus the 4,029 bp vector pComb$_3$ was constructed. When using these vectors for construction of antibody gene library, the whole genes coding for the heavy chain and light chain were digested with proper restriction enzymes and inserted into the corresponding sites. In experiments, a kind of antibody genes were cloned previously, and the other kind of the genes were then cloned. After random rearrangement, these genes were then inserted into the MCS of phage or cosmid expression vectors. The combination of the light and heavy chain coding genes were rather random, thus increasing the antibody diversity.

4.3.3.3　Expression of human antibody genes

In the phage antibody display library, the coat protein of phage M_{13} was introduced into the vector for antibody genes, which led to that the genes coding for antibodies were fused to the coat protein coding gene and the antibodies were expressed and displayed on the surface of phage M_{13}. The fusion of Fab coding gene and the gene *gIII* was under control of the LacZ promoter and operator, so that the fusion protein was secreted into the periplasma of the bacterial host by mediation of the pelB signal, followed by refolding of the heavy and light chain into the antibody Fab. The pelB signal sequence was cut down before the whole chain entered the periplasma, while the antibody Fab was located on the inner membrane of the bacterial host with the help of the fusion molecule between heavy chain and cpIII. By infection of helper phage (such as VCSM13), Fab was packaged in the phage and displayed on the surface with the help of cpIII. Since the phage contained natural cpIII protein, it infected other host bacteria again. So the phage colonies resistant to antibiotics were recombinant phages. Thus the phage antibody library for human antibodies was constructed, in which every phage particle represented one kind of monoclonal antibody expressed by B cell, while the combination of multiple phages formed the polyclonal antibody. Using the antigen-antibody reactions of the Fab on phage surface and the infection activity of the phage, the phage display library can be easily screened by the solid phase antigen.

Besides the display on the surface of phage, antibody Fab can also be expressed in a soluble form. The Lerner's group introduced two restriction sites, SpeI and NheI, on the two termini of gene $gIII$ in the vector pComb₃. The coat protein coding sequence can be cut off by the restriction enzymes SpeI and NheI, and the digestion can self-link into a circle vector. After transformation it with bacterial host and inducement of the host with IPTG, the antibody Fab can be harvested from the host periplasma or the intracellular medium. The Winter's group used another way to realized the same goal. They introduced an amber stop codon in the antibody coding gene and the gene $gIII$. If the fusion gene was transformed the amber suppressor host such as XL-$Blue$, the antibody coding gene and the gene g III formed an open reading frame and the antibody was displayed on the surface of the phage. If the fusion gene is transformed a host without amber suppressor, $HB2151$, the stop codon will terminate the expression of gene $gIII$ and the soluble Fab was obtained.

4.3.3.4　Screening of specific human antibody fragment

In the effect of antigen presentation, B lymphocyte proliferates and differentiates into plasma cells that secrete antibodies, and mutates into antibodies with different affinity under antigen's selection pressure. Phage display antibody library simulates the choice of the immune system. The antibody displayed on the surface of the phage can be screened in $vitro$ with solid phase antigen. Since the phage infects $Escherichia$ $coli$, phage antibody library technology will provide a simple and efficient operation system for rapid selection of specific antibodies in the way of scouring screening. The detailed steps are as follows: ① Phage absorb target antigens. ② Wash repetitively to remove nonspecific binding. ③ Elute and collect phage combined of antigen. ④ Infect $E.$ $coli$ again and screen the specific phage antibodies. After such a screening circle, "adsorption-eluting-expansion", specific antibody phage will be enriched by 20-1,000 times. If the enrichment rate of each circle is measured by only 50 times, after 4 screening circles the enrichment rate of phage antibodies will achieve more than 10^8, which can accomplish the screening of phage antibodies accounting for $1/10^8$ of the capacity. With the help of the high efficient screening system, phage antibody library can easily screen antibody with capacity of more than 10^8 and leave the early phage antibody library technology far behind. In addition, antibodies can be also successfully selected by methods such as solid screening, captured screening, complete cells screening, organization screening and organs screening. All of these methods have their advantages and add new ways for the screening of phage antibodies.

Since the transformation efficiency of the bacterial host limits the capacity of phage antibody library, building an antibody library with adequate capacity is difficult for phage antibody library technique. Waterhouse designed a method to combine the heavy and light chains of antibodies together *in vivo*: the plasmid carrying the heavy chain coding gene was firstly transformed *E. coli*, and then the phage carrying the light chain coding gene was used to infect the *E. coli*. As bacteria in the log phase are easily infected, the light chain gene entering the host can recombine with the heavy chain gene. This method of recombination is expected to build library with capacity of 10^{11}, which is close to the number of bacteria after culture and proliferation. But in fact, cosmids which are respectively cloned with heavy and light chain gene are not easy to package into a phage at the same time. As last, Waterhouse constructed a phage P1 recombinant containing specific Lox-cre points system, which can integrate heavy chain gene and light chain gene located in two different carriers in the infected *E. coli* to realize restructure of V_H and V_L genes.

Waterhouse adopted plasmid pUC_{19}-210x containing the mouse anti-phox antibody V_H gene as a donor, while the phage containing the mouse anti-phox antibody V_L gene and mouse anti-TNFα V_H gene were adopted as receptor. Thus the recombinant antibody displayed on the surface of the phage only reacted with the antigen phox through ELISA. Waterhouse indicated that the establishment of the restructuring system of phage P1 provided an effective way to increase the capacity of phage library and make the capacity of phage antibody libraries closer to natural antibody library. Meanwhile, reorganization of the heavy and light chain gene in the restructuring system of phage P1 will greatly promote antibody affinity mutation, facilitating the selection of high affinity antibody.

Antibodies expressed in the surface of antibodies can be monovalent or polyvalent. There are only 3-5 copies of the *g*Ⅲ in a phage gene, therefore, the surface antibody fused with *g*Ⅲ protein is of low valence. After the gene *g*Ⅲ fuse with the antibody coding gene, the phage lose the infection ability and need helper phage to provide the cpⅢ protein for infection of host again. So the protein cpⅢ and the antibody-cpⅢ fusion protein enter the phage competitively and thus the antibody displayed on the surface is monovalent. CPⅧ is the major coat protein for M_{13} phage and its gene copies are about 3,000. Just as gene *g*Ⅲ, gene *g*Ⅷ will fuse with antibody gene in phage. But the antibodies are often polyvalent, and can express 24 different antibody fragments in phage. Generally speaking, the monovalent phage antibodies are used to select antibody with high affinity, while polyvalent phage antibodies are for antibodies with low affinity. Therefore, in the construction of the phage antibody library technology, the selective infection of phage cp Ⅲ is preferred for display of antibody fragment.

4.3.3.5 The features and advantages of the phage display human antibody libraries

The phage display antibody library can imitate the natural antibody library and doesn't need immunization of human and animals. On the one hand, the coding genes for the heavy chain and the light chain are obtained by amplification templated with the Ig cDNA in human peripheral blood lymphocyte, bone marrow cells and spleen cells. On the other hand, the recombination of the genes coding for heavy chain and light chain in vitro leads to random pairing of these two kinds of chains, so this combination put forward the diversity of the antigen-antibody recognition. In general, 10^7 of the specific antibody molecules can recognize 99% of the epitopes. Therefore, a combinatorial phage library with capacity of 10^8 is theoretically thought to include all the antibody molecules.

Antibody engineering strains are more stable and easier for preservation than hybridoma cells and DNA operation is suitable for large-scale industrial production. It takes at least a few months to prepare monoclonal antibody, from spleen cells to the stable sub-lines, while phage antibody library technique needs less than a few weeks. Therefore, phage antibody library technique is simpler and faster than hybridoma technique. Phage antibodies can be more easily screened from the phage antibody library with the capacity of over 10^8. Especially after the enrichment of screening, a wide range of specific antibodies with different antigen-binding activity can be screened in a combinatorial phage antibody library of a complete set of heavy and light chains. Production of these antibodies requires neither cell fusion, nor extraction of RNA or DNA from immunized individuals. Marks reported the amplification of human V_H and V_L coding genes from the peripheral blood lymphocyte in individuals without immunization, and then link them into single chain antibody gene, followed by construction of a "natural" human antibody library with the help of phage vectors. More than ten different specific single chain antibodies are screened from the library, including anti-thymoglobulin, protamine, TNF, CEA, CD4 and blood cell nuclear antigen and autoantibody molecules, etc. As estimated, 10%-30% of human B cells will produce autoantibody. However, they won't be activated and proliferated because of immuo-tolerance of body. Phage antibody library technology makes it possible to make antibodies freely.

It is generally believed that the antibodies produced by the body in first immunization are of low affinity. With repeat stimulation of antigen, immune responses again and again cause mutations of antibody genes of immune cell and produce high affinity antibody. In the construction of phage antibody libraries, the recombinant between heavy and light chains stimulates the affinity maturation process of antibody in vivo. In the phage antibody library, there is great arbitrariness in matches of heavy and light chains, which can often change the matches of the existing antibodies in B cells, producing antibodies with different affinity. Using two sets of carriers with gene $g\text{III}$ or gene $g\text{VIII}$, these antibodies with different affinity can be screened. In general, "natural" antibody library constructed from individuals without immunization has lower antibody affinity, mostly $10^{-4}/M$. Phage antibody library technology can cause the original antibody genes mutation and reconstruction to improve the affinity by using methods of molecular biology.

Affinity can be changed through chain displacement, which does not exist in the body but may cause matching between heavy and light antibody chain genes again in the library to improve the affinity of antibodies. This is a common way by which phage antibody library technique obtains high affinity antibodies. Marks screened specific single chain antibody for phox with affinity of $3\times 10^{16}/M$ from phage antibody library. He used different lightchain to recombine with the heavy-chain gene of original antibody, changing the heavy chain and light chain pairing mode. After the light chain exchanges, specific antibodies with the affinity of $6\times10^7/M$ can be screened in the antibody libraries. The antibodies obtained are of 20 times higher affinity than before. Maeks further replaced genes in CDR1 and CDR2 area of heavy chains. After pairing with the light chain, he constructed a library and screened antibodies of affinity of $10^9/M$, which was 320 times higher than the original library and equal to antibodies produced from mouse hybridoma after several times immunization.

Utilizing PCR mismatch or random mutagenesis technology, phage antibody library technique can change antibody affinity. This is the replacement of amino acid on purpose, which applies to the sequence of known antibody. It also imitates the organism somatic mutation process, resulting in antibody gene mutation and improvement of antibody affinity. The recombinant phage will naturally random mutate in proliferation. Recently the natural mutation rate of each base in a division is estimated as 1/10,000-1/30,000. With the artificial mutation *in vitro* using topoisomerase, the antibody with high affinity can be obtained by spontaneous mutation of the coding genes for heavy and light chains. PCR mismatch will cause random mutations of the base sequence and raise the affinity of antibodies by 100 to 1,000 times. The unique and efficient screening system plays a vital role in construction of phage antibody library. The gene coding for the antibody with low affinity, which was screened from the natural antibody library, was subjected to PCR mismatch to obtain the mutant and the shuffled heavy chain and light chain. These mutated or shuffled genes were used for construction of sub library. With the process of "adsorption-eluting-expansion", high affinity antibodies could be screened from it. Gram and Hanks obtained mutated antibodies with affinity over 10 times higher than the original antibody by this method. At present, after reorganization of light and heavy chain genes, the chains of the antibodies are randomly combined. The antibody diversity can exceed the diversity of immune system itself. As the phage antibody genes are primarily from antibody variable region, the mutation happening in antigen binding sites CDR areas will has more direct influence on antibody affinity. It is thus clear that phage antibody library technology has great potential to enhance antibody affinity.

4.3.3.6 Phage display human antibodies constructed in mouse

The purpose to construct the phage display human antibody library is to obtain specific antibodies with high affinity. The existing technologies like strand displacement, the mismatch of PCR and random mutagenesis technology, all imitate the organism somatic mutation process, and cause antibody gene mutations to improve the affinity of antibodies. The difficulty of these methods restrained the development of human antibody. In recent years, much attention is paid on these methods: ①Severe combined immune deficient disease (SCID). It is now popular to produce human antibodies by phage display antibody library in mouse. Through transplanting the human peripheral blood lymphocytes or lymph nodes cells and spleen cells into SCID mice abdomen, human immune system in SCID mice will be established. After immunization with antigen, the peripheral blood lymphocyte and the spleen cells were taken for production of human antibody with the humanized phage display antibody technology. ②Transgenic mouse for human antibodies. It is a very complex biological engineering, combining genome engineering, transgenic animals and hybridoma technique. Firstly, the mouse immunoglobulin genes are replaced with human immunoglobulin genes and then SCID mouse with mice containing light, heavy chain genes of human antibody are crossbred. Then double transgenic or double gene deficient homozygote mice are selected. Since this transgenic mouse contains human antibody genes profile, it is like a human antibody factory. When the mouse is simulated with antigen, the B cells in vivo of mouse will cause immune response and secret human antibody.

Currently, biology engineering companies, Cell Genesys and Genpharm, have obtained human antibodies from human immunoglobulin transgenic mice by different ways, and have analyzed specificity, affinity and biological effect of antibodies. The results show that human antibody transgenic mice can have immune response to human and other antigens, indicating that the produced human antibodies have high affinity and diversity to a certain extent. The problems of murine origin and the low affinity were solved and these have become the most useful human antibodies, some of which have entered the clinical trial phase.

Phage display antibody library technology has displayed its application prospects. Through this technology, not only human ScFv, Fab bispecific antibody and miniature antibodies with the function of recognition can be produced, but also human antibody enzyme with function of catalysis and cutting are selected. These human antibodies of different patterns and functions bring hope to diagnosis and treatment of human diseases. However, phage display of human antibody library technology is still in the stage of mature development. There are many issues that need attention and resolution, such as low affinity, immunogenicity and toxicity of antibody. In short, with the completion of the human genome project and the improvement and maturity of the transgenic animals, it will become possible to produce human antibody for clinical use, and brings hope for some diseases seriously harmful to human health.

Antibody Library technology has been invented for 10 years and was considered as a revolutionary progress of antibody engineering fields once it came up. Its development and application has brought about great changes in antibody technology. The problem of producing human antibody has been generally resolved. A lot of human antibodies have been made. Several human antibodies from antibody library have come into clinical application, and some have entered the clinical trial phase III. The performance of antibody can be more easily designed and improved according to need; all kinds of antibodies, including antibody for their own or for weak antigen, can also be prepared without immunization; the affinity maturation in vitro has broken through the maximum of that in vivo. All these prophesied of a new era when biological scientists and researchers will be capable of controlling the performance of antibodies and producing "super" engineering antibodies of high value.

4.4　Applications of antibody

Antibody molecules are the proteins most widely used in the field of biology and medicine. Polyclonal antibodies, monoclonal antibody and genetic engineering antibody have been produced against the targets such as tumor specific antigens or cancer relevance antigens, idiotypic determinants of antibody, cytokines and their receptors, hormones and some cancer gene products. etc. , by the traditional method of immunization, cell engineering or genetic engineering technology. These antibodies have been widely used in diagnosis, treatment and research of diseases. The intact murine antibody easily causes rejection of body, while genetic engineering antibodies of the small molecular weight can reduce the murine origin through reconstruction. Therefore, genetic engineering antibodies have become the focus and hotspot of biotechnology and pharmaceutical fields [22,23].

According to a report of PhRMA, the developing antibodies and developed antibodies drugs in the market mainly have functions as follows: reversal of organ transplantation rejection; tumor immune diagnosis, tumor immunity imaging and tumor targeting therapy; asthma, psoriasis, rheumatoid arthritis, erythematosus and acute myocardial infarction, sepsis and multiple sclerosis and other autoimmune diseases; anti- idiotypic antibody as a molecular tumor vaccine for cancer treatment. The multiple functional antibodies (bi-specific antibody, tri-specific antibody, antibody-cytokine fusion protein and antibody enzyme, etc.) have other special purposes.

Cancers are a kind of the major diseases threatening human health. Prevention and treatment of cancer are one of the major objectives of research and development of anti-body drugs. At the beginning, antibody was primarily used to tumors in diagnosis in vitro and immune imaging in vivo. With the continuous progress of antibody engineering in recent years, people pay more attention on development of antibody drugs for tumors treatment. First engineering antibody for tumor treatment of human approved by United States, Rituxanr, was initially used for non-hodgkin's lymphoma, and total efficacious rate of 60% was achieved. Now scientists are exploring for the treatment of AIDS-related lymphoma and central nervous system lymphoma.

Since the 1980s, Chinese scientists have closely followed the world's latest research trend of antibody engineering research, thus laying a solid foundation for the development of antibody engineering. In 1987, the "863" project (Biotechnology Field) made antibody-oriented drugs as a special theme. The development of antibodies against liver cancer, lung cancer, gastric cancer and leukemia were carried out. Meanwhile, the human-human monoclonal antibody and human-mouse monoclonal antibody were also studied. After a dozen years of hard work, remarkable progress has been made in the research of guiding drugs and genetic engineering antibodies.

In short, the research and development of antibody engineering in China have already achieved some success, but there are also some shortcomings. For example, though a number of research projects in China have moved towards or reached the world advanced level, they are just laboratory products, and still are considerable distance away from application.

References

[1]   He ZX. Introduction of modern biology. Beijing: Beijing Normal University Press. 1999: 262

[2]   Wang PH. Transfusion technology. Beijing: People Hygiene Press,1998: 231

[3]   Zheng MJ,Yang CZ. Design and application of genetic engineering antibodies. Chinese Pharmaceutical Journal, 2002, 37(4): 245-248

[4]   Walter G, Konthur Z. High-throughput screening of surface displayed gene products. Comb. Chem. High Throughput Screening, 2001, 4(2): 193-205

[5]   Todorovska A, Roovers RC, Dolezal O and others. Design and application of diabodies, triabodies and tetrabodies for cancer targeting. Immunol Methods, 2001, 248(1/2): 47-66

[6]   Zhang J, Zhang FC. Studies and application of genetic engineering antibodies, 2006, 16(10): 163-164

[7]   Zhu BQ. Some application advances of genetic engineering technology in pharmaceutical industry. Chinese Journal of Pharmaceuticals, 1997, 28(2): 56-58

[8]   Xie P, Song Hp. Development of gene engineering and medical implication: Medicine and Society, 2000, 13(4): 3-5

[9]　Cao Y. Guidance of experimental technique of molecular biology. Beijing: People Hygiene Press, 2003: 277

[10]　Wang JZ. Development and quality control of biotechnological medicine. Beijing: Science Press, 2002: 640

[11]　Jespers LS, Roberts A, Mahler SM and others. Guiding the selection of human antibodies from phage display repertoires to a single epitope of an antigen. Bio/technol, 1994, 12: 899-903

[12]　Wang Y. Progress in genetic engineering antibodies. Chinese Journal of Immunology, 1999, 5(15): 19-195

[13]　Liang GD. New experimental technique of molecular biology. Beijing: Science Press, 2001: 84

[14]　Holliger P, Prospero T, Winter G. Diabodies: small bivalent and bispecific antibody fragments. Proc. Natl. Acad. Sci. , 1993, 90: 6444-6448

[15]　Kortt AA, Lah M, Oddie GW, Gruen CL, Burns JE, PearceL A, Atwell JL, McCoy AJ, Howlett GJ, Metzger DW, Webster RG, Hudson PJ. Single-chain Fv fragments of anti-neuraminidase antibody NC10 containing five-and ten-residue linkers form dimers and with zero-residue linker a trimer Protein Engineering, 1997, 10(4) 423-433.

[16]　Berek C, Griffiths GM, Milstein C. Molecular events during maturation of the immune response to oxazolone. Nature, 1985, 316: 412-418

[17]　Liu XX. Principle and technology of protein engineering. Jinan: Shandong University Press, 2002: 80

[18]　Yang WP, Green K, Pinz-Sweeney S and others. CDR walking mutagenesis for the affinity maturation of a potent human anti-HIV-1 antibody into the pi-comolar range. J Mol Biol, 1995, 254(3): 392-403

[19]　Ho M, Kreitman RJ, Onda M and others. In vitro antibody evolution targeting geimline hot spots to Increase activity of an anti-CD22 immunotoxin. J Biol Chem, 2004, 279: 1-53

[20]　Moriki T, Kuwabara I, Liu FT and others. Protein domain mapping by lambda phage display: the minimal lactose-binding domain of galectin-3. Bio-chem Biophy Res Comm, 1999, 265(2): 291-296

[21]　Holt IJ, Enever C, de Wildt RM and others. The use of recombinant antibodies in proteomics. Curr Opin Biotechnol, 2000, 11(5): 445-449

[22]　Lin FQ, Zhang GS. Progress in celluar immunology. Beijing: People Hygiene Press, 1980: 348

[23]　Liu LX, Liu SF, Chang NB and others. Cloning, expression and preparation of polyclonal antibody for human Dcp2 gene. Journal of the Graduate School of the Chinese Academy of Sciences, 2007, 24(3): 362-367

Chapter 5　Labelled Immunoassay Technique

During the past two decades, immunological detection methods developed rapidly, especially after the utilization of labelled antigen and antibody which enhance the sensitivity and specificity of detection greatly. In the early 70s of the 20th century, enzyme-labelled antigen or antibody analysis technology was established after immunofluorescence (IFA) and radioimmunoassay (RIA) analysis were created in the 50s and 60s of the 20th century respectively.

In labelled immunoassay, a tag (such as radionuclide, fluorescein, enzyme, chemiluminescent agent, etc.), which can be micro-detected or ultramicro-detected, is labelled to antigen (antibody) to form a label. The label is then added to the antigen-antibody system and reacts with the corresponding antibody (antigen) to examine the presence or not of label, which can indirectly reflect the amount of the antigen or antibody in sample.

After conjugating with the antibody or antigen, the labels do not alter immune properties of the later, which endows this method with some advantages, such as sensitive, rapid, qualitative, quantitative and positioning. These immunoassay methods, which are combinations of these labeling technologies and antigen-antibody reaction, have being gradually replacing the coagulation, sedimentation and other classic immunologic detect technology due to their advantages, such as high sensitivity, good accuracy, to be handled facilely, to be commercialized easily, and automatability. These methods have been widely used to determine antigens, antibodies and other immune-related substances, and are the most widely used immunological detection techniques at present.

Currently, labelling technique of immunoassay mainly includes enzyme immunoassay, IFA, RIA, gold immunoassay and chemiluminescence immunoassay, etc. Enzyme immunoassay, utilizing enzyme-labeled antibody (antigen) as the main reagent, combines the specificity of antigen-antibody reaction with the efficiency of enzyme catalysis. As one of the three classic labelling techniques, enzyme immunoassay is widely used and continuously updated in various fields.

5.1　Radioimmunoassay

Radioimmunoassay combines sensitive radioisotope determination and specific antigen-antibody reaction to quantitatively determine immuno reactive substances in vitro. According to whether antigen or antibody is labelled with radioisotope, the technique are classified into two categories. One is radioimmunoassay (RIA) which utilizes radioisotope-labelled antigen to determinate unknown antigens. The other is immunoradiometric assay (IRMA) which utilizes radioisotope-labelled antibody to determinate the corresponding antigen.

Currently RIA, known as radioreceptor assay (RRA), has been developed to study cell receptor molecule. It utilizes radioisotope to label ligand for examination of the corresponding receptor molecule and is the most sensitive and reliable technology for quantitative and positioning determination of receptor molecule. Additionally, RIA demonstrates wide application prospect in design, mechanism and biological effects of drug, as well as diagnosis and treatment of diseases.

RIA is the most sensitive technology in serology and immunology. It has many advantages such as higher specificity and sensitivity, and the detection limits can reach nanogram or even picogram [1,2].

5.1.1　Technical evolvement

RIA was found by the American scholar, Yalow and Berson, in 1960, and was first utilized to determinate the plasma insulin of diabetic patients, which was a major breakthrough in medicine and biology methodology and opened a new era in the history of medical testing. It made it possible to analyze those traces substances quantitatively and precisely which were ever recognized as could not be measured and yet of important biologically significance. This accordingly opened a new path to further uncover the mystery of life, and made it possible to understand again the biochemical basis of biological phenomena at the molecular level. Since then, the rapid development of endocrinology in 30 years proved the tremendous impetus of this submicro analysis technology. In 1977, the inventor of this technology was awarded the Nobel Prize in medicine biology. After that, this new technology penetrated quickly into other areas of medical science, such as virology, pharmacology, hematology, immunology, forensic medicine, oncology, as well as biology and medicine-related science, such as agricultural science, ecology and environmental science. Analysis targets of RIA were expanded from hormone to almost all the biologically active substances. In China, RIA research was started in 1962 and developed and spread rapidly, which played a great role in promoting the progress of biomedical field.

For high sensitivity, specificity, high precision, and capable of detecting small molecule and macromolecule compounds, RIA was widely used in laboratory medicine, and usually in the determination of various hormones (such as thyroid hormones, sex hormones, insulin), trace proteins, tumor markers (such as AFP, CEA, CA-125, CA-199) and drugs (such as phenobarbital, chlorpromazine, gentamicin). There are kits supplied in various test items, and the relative equipments are not expensive, making it

widely used. However, due to radioactive nuclides are harmful, and so protection measures must be taken. Radionuclide laboratory construction should be supervised by epidemic prevention departments and the operators have to be specially trained. In addition, nuclides are of half-life time, and the expiry date of kits is not so long, and, therefore, there can be a lot of inconvenience during application of RIA. In particular, enzyme immunoassay and luminescence immunoassay developed rapidly in recent years. In
the long run, there is a trend that RIA might be replaced. But at present, when
considering the cost of equipment and testing, the RIA has some advantages and will be
adopted in medical laboratories for a period of time.

5. 1. 2   Principle

The antigen-antibody complex $(Ag+Ab \leftrightarrow Ag-Ab)$ will form when an antigen reacts with a corresponding antibody. A radioisotope-labelled antigen and antibody complex will form when labelled antigen reacts with a corresponding antibody, as follows:
$^*Ag+Ab \leftrightarrow {}^*Ag-Ab$. The labeled and unlabelled antigen will compete with each
other, and form labelled and unlabelled antigen-antibody compounds, when they are
added into a corresponding antibody. The content of labelled antigen-antibody complex
is inversely proportional to the unlabelled antigen's content in certain scope, and this
principle can be used to determine the unknown antigen or antibody [3].

$$^*Ag \qquad\qquad AgAb+{}^*Ag(F)$$

$$Ab$$

$$Ag \qquad\qquad {}^*AgAb(B)+Ag$$

Competitive assayes are the most common and can be performed in two ways, such
as the analyte and the tracer (labelled molecule) competing for a limited number of
binding sites (direct), or the anlyted and the immobilized ligand (antigen) competing for
a limited number of binding sites

If there are no anaytes in samples, the reaction equation is :

$4(Ag^*)+2(Ab) \rightarrow 2(Ag^*Ab)+2(Ag^*)$

Whereas,

$4(Ag^*)+4(Ag)+2(Ab) \rightarrow 1(Ag^*Ab)+3(Ag^*)+1(AgAb)+3(Ag)$

In this system, the content of labelled antigen and antibody is optimized . The
amount of antibody usally is set as to bind 40%-50% labelled antigen. Due to the
amount of objective antigen varied in samples, different amount of labelled antigen can
be captured by antibody and the amount of radioactivity measured id indicative of the
amount of analyte present.

When the labelled-antigen, unlabelled-antigen and specific antibody existed in a reaction system simultaneously, the two antigens will compete each other for combining specific antibody as the binding force with specific antibodies of the labelled antigen is same with the unlabelled antigen. The amount of labelled-antigen and specific antibody is fixed, so the content of labelled-antigen-antibody complex will change along with the unlabelled-antigen. As unlabelled-antigen increases, it will combine more antibody molecules, and thus inhibit the binding of antibody molecules with labelled-antigen, which make the labelled antigen-antibody complex decrease and correspondingly increase of free labelled antigen. In another words, radioactivity strength of antigen-antibody complex is inversely proportional to the concentration of antigen in samples.

If antigen-antibody complex and free labelled antigen are separated and determined respectively, the molar ratio (B/F) of combined labelled antigen (B) and free labelled antigen (F) or the binding rate $[B/(B + F)]$ can be calculated. When a series of standard antigen with different dose is used to test, the corresponding B/F values can be calculated to draw out a dose-reaction curve. Specimens are measured under the same conditions. A according to the calculated B/F value, the amount of antigen can be found from the dose-response curve.

5.1.2.1　Labels

Radioactive isotope labels can be categorized into two kinds, γ-rays and β-rays. The former contains ^{131}I, ^{125}I, ^{57}Cr, and ^{60}Co; the latter includes ^{14}C, ^{3}H and ^{32}P. The specific activity is firstly considered in the choice of isotope lable. For example, the theoretical specific activity of ^{125}I is 64. 38×10 ^{4}GBq / g(1. 74×10 ^{4}Ci/g). The maximum specific activity of ^{14}C with longer half-life is 166. 5 GBq/g (4. 5 Ci/g). Therefore the sensitivity of ^{125}I is about 3,900 times higher than ^{14}C when 1 mol ^{125}I or ^{14}C used as tracers. Because ^{125}I is of a suitable half-life and easy to label antigen or antibody, it is most commonly used in RIA [4].

5.1.2.2　Labelling method

The methods for labelling ^{125}I can be divided into two major categories, namely, direct and indirect labelling method[5].

Direct labelling method is binding ^{125}I to tyrosine residue on protein side chains directly. this method is simple and a single step binding reaction, and many ^{125}I atoms can bind to protein molecule, so the labell has highly specific activity. Nevertheless, this method can only be used to label compounds containing tyrosine. In addition, if tyrosine is important for the specificity and biological activity of protein, the activity will be easily damaged after labelling.

In indirect labelling method (also known as the connection method), ^{125}I is firstly labelled on a carrier, and then the labelled carrier is purified, and at last the carrier is combined with protein. As the operation is complicated, radioactivity of labelled proteins are significantly lower than that prepared thorough direct method. However, this method can label peptides and proteins which lack tyrosine. If the immune activity or biological activity is damaged for the protein tyrosine structure changes caused by direct labelling method, the indirect method can be applied. The indirect method is a moderate labelling reaction which can avoid the loss of biological activity caused by directly adding protein into ^{125}I liquid. The following is chloramine T method which is the most commonly used as direct labelling method.

Chloramine T is a sodium salt from toluene sulfonic amide N-chloro derivative. As an oxidant, it decomposes gradually and forms hypochlorous acid in aqueous solution. In weak alkaline solution (pH 7. 5), chloramine T oxidizes the I^- of ^{125}I into I^+, then I^+ replaces the hydrogen on benzene ring of tyrosine and forms diiodo tyrosine.

The radioactive iodine labelling efficiency is related to the exposure degree and the number of tyrosine on antigen (protein or peptide) molecule. If the molecule contains more tyrosine residues which are exposed, the labelling efficiency is higher.

Labelling method: the purified antigen and ^{125}I is put in the bottom of a small tube, and then freshly prepared chloramine-T is poured into the tube. After mixing and oscillating for several ten seconds to 2 min, the sodium metasulphite is added to terminate the reaction. Then KI solution is added to dilute them. The solution is separated on a dextran column and the fractions are collected by tubes. Radioactivity (number of pulses/min, cpm) is determined by well-type scintillation counter. The front part is the labelled antigen peak; the back is the peak of free ^{125}I. 1% albumin is added as stabilizer to the labeled antigen tube, and the solution is preserved as a fluid for labelling antigen.

5.1.2.3 Identification of the label

All proteins are precipitated with trichloroacetic acid (bovine serum albumin was added into the sample beforehand), then the cpm value of the sediment and supernatant are measured respectively. It is generally required that free iodine is below 5% of total radioactive iodine. The iodine will shed from the label after the labelled antigen has been stored for a long time. If the free iodine exceeds 5%, this part of free iodine should be removed.

Part of antigen activeness is always lost during immune activity labelling, which should be avoided as much as possible. The antigen activity testing is carried out as follows: excessive antibody are added to the few labelled antigen, the B and the F are separated after reaction, and then the radioactive materials are determined separately and the BT% is figured out. This value should be above 80%. The higher the value is, the less is the activeness of the antigen lost.

Specific radioactivity means the emission intensity of per unit of weight antigen. Specific radioactivity of labelled antigen is represented with mCi/mg (or mCi/mmol). The higher the value is, the more sensitive the determination is. The specific radioactivity of the labelled antigen is calculated according to the utilization rate (or labelling rate) of radioactive iodine.

5.1.2.4   Evaluating of antiserum

Antiserum containing specific antibody is the main reagent in RIA, and polyclonal antibodies are usually obtained through stimulating small animals with antigen. The quality of antiserum directly affects the sensitivity and specificity of analysis. The main quality indicators of antiserum are affinity constant, cross-reactivity and titer.

(1) Affinity constant

It is usually represented with Ka and reflects the binding capability of an antibody with the corresponding antigen. The unit of Ka value is mol/L, which means that 1mol antibody are diluted to several liters and the binding rate of the antibody with the corresponding antigen is 50%. The higher the Ka value is, the better the sensitivity, precision and accuracy of RIA is. Only the antibody with Ka value reaching 10^9- 10^{12} mol/L is suitable for radioimmunoassay.

(2) Cross-reactivity

Some compounds to be tested in RIA have very similar structures, such as thyroid hormone T3, T4, estrogen hormone estradiol, estriol, and so on. Antiserum against an antigen often has cross-reactivity with the analogues of the antigen. Therefore, cross-reactivity reflects the specificity of the antiserum. The accuracy of analysis will be affected by the higher cross-reactivity. The cross-reactivity is determined by the same method with analysis of the corresponding antigen of the antiserum: observe the amount of the analogue replace 50% of the zero standard tube. For example the antiserum to T3, 1ng T3 can replaces 50% of the zero standard, and its analogue, T4 needs 200 ng to achieve this result, so the cross-reactivity is 1/200 (0.5%).

(3) Titer

It means the maximum dilution of an antibody under which the antibody is still in effect an assay. It reflects the effective concentration of antibody in antiserum. In RIA, titer is the antiserum dilution when 50% of labelled antigen is bound to the antibody without free antigen.·

5.1.3   Basic reagents and technical methods

5.1.3.1   Preparation of antigen for RIA

(1) Complete antigen

In order to ensure the specificity of the immune response, the purify of antigen for

RIA is required above 90%. Antigen of high purity is usually obtained through electro-phoresis, gel filtration and ion exchange chromatography techniques. Purified antigen must be identified before application. If the antigen is separated in electrophoresis or disc electrophoresis, only one precipitation line appears, which can approve its purity.

(2) Hapten

To obtain better immunogenicity, hapten must be coupled to the carriers to form a complete antigen. The carriers usually include serum albumin, γ globulin, fibrin, thy-roglobulin and ovalbumin, etc. Coupling methods are determined by the functional groups of hapten and protein carrier. The commonly used coupling agents are diimine, diisocyanate and metaperiodate.

5.1.3.2　Antibody preparation and purification (see Chapter 3)

5.1.3.3　Labelling of antigen

(1) The choice of radioisotope

Principles of selecting isotope: ① Method is simple, economic and easy to apply. ② Easy to be protected. ③ Isotope can combine well with antigen and does not shed from the tag. ④ Isotope does not cause damage and denaturation of protein due to radia-tion. ⑤ Isotope has higher counting efficiency.

The commonly used isotopes are ^3H, ^{125}I and others include ^{14}C, ^{35}S and ^{32}P, etc.

1) ^3H. As all organic compounds contain hydrogen, replacing hydrogen with ^3H will not affect the chemical properties of the compounds. ^3H is of long half-life, of a weak β decay, and of low radioaction energy, and easy to be protected. It can be used for a longer period after being labelled once, and the measuring effectiveness can be up to 60% by scintillation measurement. ^3H labeling requires high working conditions, and is generally undertaken by specialized agency, thus it is difficult to be popularized.

2) The labelling of ^{125}I iodine is simple, and the label can be directly measured by well-type counter with sodium iodide crystals. ^{125}I can emit γ-ray, and the proteins and poly peptides containing tyrosine can be labelled with radioactive I. At present, the cou-pling technology is considerably developed, and many haptens without protein and pep-tides can also be labelled by iodine. Additionally, the radioactive energy of iodine is great and the half-life is short.

Although the radioactive specific activity of ^{125}I is only 13% of ^{131}I, but the half-life of ^{125}I is longer than ^{131}I, and also of smaller radiation damage and higher counting effi-ciency, Moreover, the isotopic abundance is larger. Thus ^{125}I is used more commonly than ^{131}I. Table 5-1 lists the common radioactive isotopes in RIA.

Table 5-1　Common radioactive isotopes

Isotope	Toxicity	Decay mode	Half-life
^3H	Low toxicity	β-	12. 26a
^{14}C	Low toxicity	β-	5730a
^{131}I	Highly toxic	γ · β-	8. 07d
^{125}I	Low toxicity	EC · γ	60. 20d
^{35}S	Neutral toxicity	β-	67. 48d
^{32}P	Neutral toxicity	β-	14. 26d

(2) The labelling methods

Nowadays iodine labelling is commonly used. There are many methods for labelling iodide, such as iodine chloride method, lactoperoxidase method, perchloric acid method and connection labelling, etc. But the chloramine T method is most convenient, effective, and easy to be utilized.

Principle of chloramine T method: Chloramine T is an oxidizer and can slowly release hypochlorous acid in aqueous solution, which can form an intermediate with mild oxidation effect. It can oxidize the radioactive iodine ion into an active iodine ion, which will replace one or two hydrogen atoms adjacent to tyrosine benzene hydroxyl of antigen molecule, and consequently transfer it into peptide chains containing iodinated tyrosine. The process is as follows.

1) Dilute the protein antigen with 0. 5 mol/L pH 7. 5 PB solution to the concentration 20 μg/μL. 5 -10 μL of diluted antigen is taken out.

2) 5 μL of $Na^{125}I$ (containing 2-3 mci) is taken out.

3) 1 mg chloramine T is dissolved into 0. 1 mL PB solutions 90. 5 mol/L pH 7. 5).

4) $Na^{125}I$ and the chloramine T solution are slowly added respectively and stirred into the protein fluid in the ice-bath. After the addition, they are stirred for 5 min.

5) 0. 1mL of metabisulphite (containing 3 mg, prepared in PB solution) is added.

6) 1 drop of 1% potassium iodide is added. If the solution is completely transparent, the next step will be done. If it still appears brown, metabisulphite should be continuously added.

7) The solution is separated with the Sephadex G50 column, the first peak should be collected and that is iodine labelled protein antigen.

(3) The condition for labelling

1) Radioactive iodine specific activity is high.

2) The reaction volume is small.

3) The labelling reaction is carried out at pH 7. 5. If the pH is more than 8. 5, iodine will replace other group beside tyrosine.

4) The amount of chloramine T must be determined in advance. The determination

is: the maximum amount of radioactive iodine can be precipitated with 10% trichloroace-tic acid when labelled protein is in presence, under which the minimum amount of chlo-ramine T is required, The dosage of chloramine T should to be appropriate. If it is less, although it can meet the requirements of the reaction, it is of low yield. When the dosage is too much, it is easy to denature a large amount of protein.

5) Iodination efficiency is determined by the concentration of protein. As for the protein with medium chlorine amino acid, when the concentration of the protein is 1mg/mL, the protein recovery is 100%; when the concentration is 300 μg/mL, the recovery is 80% to 90%; when the concentration is decreased to be 50 μg/mL, the recovery rate is only 60% to 70%.

(4) Identification of labels

1) Radiochemical purity refer to the percentage of the radiation intensity of a chemi-cal radioactive material to the total intensity. Identification method: Take a little la-belled protein or peptide liquid, add 1% to 2% of the carrier protein and an equal amount of 15% trichloroacetic acid, agitates and let it stand for a few minutes, then centrifuge for 15 min at 3,000 r/min, and, determine the radioactivity of supernatant liquid (containing free iodine) and precipitation (including the labelled antigen) respec-tively. Generally it is required that free iodine content is below 5% of the total activity iodine. If labelled antigen has been stored for a long time, there is still part radioactive iodine flake off from the marker which should be removed before use, otherwise, it will affect the accuracy of RIA.

2) Evaluation of immuno-chemistry activity: Usually iodine is utilized to mark the antigen, but due to the oxidant it may damage some activity, so ^{3}H, ^{14}C is used to mark antigen which does not change the chemical constitution of the antigen.

Method for examination of immunity activity: excessive antibody is added in the small amount labelled antigen, after the reaction at suitable condition, B and F is sepa-rated and determines the radioactivity separately, then calculate the binding percentage. This value should be above 80%. The bigger this value is, the less loss of immuno-chemistry activity it indicates.

3) Radioactivity intensity: Radioactivity intensity is expressed with specific activity which refers to the radioactivity intensity per unit weight antigen. The higher the spe-cific activity, the more sensitive is the method. Therefore the sensitivity of the determi-nation needs labelled antigen with suitable specific activity. The computation of specific active labelled antigen is based on the utilization percent of the radioactive iodine.

$$\text{Utilization rate} = \frac{\text{The total radioactive of labelled antigen}(\mu\text{ci/mL})}{\text{The total radioactivity}} \times 100\%$$

$$\text{Specific activity} = \frac{\text{The total radioactivity input }(\mu\text{ci/mL})\text{-Utilization rate}}{\text{The amount of labelled antigen }(\mu\text{ci/mL})}$$

$$\text{The radioactivity intensity of label} = \frac{\text{Specimen tube } CPm - \text{Background } CPm}{\text{Standard tube} CPm - \text{Background} CPm}$$
$$\times \text{radioactivity intensity of standard tube}$$

CPm is the pulse number per minute.

$$\text{Labelling efficiency} = \frac{\text{Tracer radioactivity intensity}(\mu ci/mL) \times \text{First peak volume (mL)}}{\text{Input isopote radioactivity intensity when labelling}(\mu ci/mL)}$$
$$\times 100\%$$

5.1.3.4 Separation of B and F

When the labelled antigen and unlabelled antigen combines with antibody, the antigen-antibody complexes are formed. It cannot automatically precipitate because of its low concentration. The end of RIA determination depends on the combination ratio of labelled antigen and competitors, so for RIA, it is crucial that antigen-antibody complex is isolated completely from free labelled antigen.

Choice of separation technology is based on the characteristics of antigen, the volume of biological fluids tested, the sensitivity needed of the determination, accuracy, and degree of technology proficiency[6].

For the more commonly used techniques of B, F separation, see Table 5-2.

Table 5-2　B, F separation method

Separation Method	(B) Antigen-antibody Complex	(F) Free antigen
Equilibrium dialysis	In the dialysis bag	Equal concentrations inside and outside the dialysis bag
Activated carbon adsorption	In solution after centrifugation	Adsorbed by activated charcoal in precipitation
Gel filtration	Being eluted earlier	Being eluted later
Millipore filter filtration	On the cellulose acetate membrane surface	Bing washed from the membrane
Electrophoresis	In the γ-globulin electrophoresis zone	Moves at its own electrophoresis mobility speed
Double antibody method	Bing precipitated by the second antibody	Being supernatant after centrifugation
Solid phase antibody method	Binds to solid phase	In the dissolved phase
Ammonium sulphate precipitation	Precipitate after centrifugation	In solution
Polyethylene glycol (molecular weight 6,000)	Precipitate after centrifugation	In solution

(1) Salting out method

33% saturated ammonium sulphate solution can precipitate antigen-antibody com-

plex. Saturated ammonium sulphate solution is added to the solution. and the last concentration reaches 36% of saturation. Then it is stirred and allowed to stand. Then the
labelled antigen-antibody complex is centrifuged, and free labelled antigen is remains in
solution. The disadvantage of this method is that free antigen may precipitate at the
same time as the labelled antigen complex is precipitated down.

(2) Double antibody method

Double antibody method is commonly used. After the antibody binds to the labelled
antigen. It still has the ability of recognizing anti-antibody, namely, secondary antibody.
When the primary antibody and labelled antigen encounter the secondary antibody, the
larger complex such as $* Ag-Ab_1-Ab_2$ (also $Ag-Ab_1-Ab_2$) forms and precipitates in
the solution. The purpose is to separate the free antigen. This method is moderate and
the separation is more complete (up to 80% -90%).

(3) Albumin (or dextran) active carbon absorption method

Active carbon is suspended in certain concentration dextran solution (or albumin),
and glucan molecules form a membrane with certain aperture mesh on carbon surface
which only allows smaller molecules adsorb onto the top and then adsorbed by activated
carbon. Macromolecular substances will not be adsorbed and can be excluded. The best
separation condition is at pH 6. 5-9 and 4 ℃ for 15-30 min.

This method is fast, easy and has good separation effect. But when labelled antigen
and antibody complex is not stable, they will be dissociated by the break-up of the
balance of antigen-antibody reaction when free antigen is adsorbed by activated carbon.
In this situation, this method cannot be applied.

5.1.3.5   The establishment of standard curve

The establishment of standard curve directly affects the sensitivity, precision and
the measure scope of RIA.

(1) The establishment of antibody titration curve

The antibody is diluted into series of concentration, and then is added to the same
amount of labelled antigen and unlabelled antigen. The complex and free labelled antigen are separated, and the B/F ratio is measured, and then the antibody titration curve
is established through plotting the antibody dilutions (horizontal axis) against the B/F
ratios (vertical axis). The dilution which can combine 50% labelled antigen will be regarded as the required amount of antibody.

(2) The establishment of standard curve

Based on antibody titration curves, the antibody amount to combine 50% of the labelled antigen can be calculated, then standard curve can be made (also known as antigen plus line).

The known different dilution of antigen and labelled antigen is added to the aforementioned amount of antibody. After reacting for a certain time, the B is separated from the F and the radioactivity of the B is measured. Take the pulse number or the bound rate of labelled antibody compound as a y-coordinate and the logarithm of unlabelled antigen concentration as x-coordinate. If the straight line cannot be obtained, it should be calculated through the logit computation to change the curve into a straight line. In the coordinate system, the region with the most curve slope value is the practical scope of RIA. The higher the sensitivity is, the higher the slope is and the smaller the measuring range is. If the slope is smaller and the practical scope is larger, the sensitivity is lower.

Under the same conditions, when measuring the unknown sample, as long as the radioactivity of labelled antigen-antibody complex is measured and the binding ratio is calculated, the amount of unknown antigen may be found from the standard curve.

5.2　Enzyme immune techniques

5.2.1　Summary

5.2.1.1　Principles

Enzyme immune technique is a developed field in modern times. It combines the immunological reaction of antigen-antibody with the highly effective catalytic activity of enzyme through connecting enzyme to antibody (or antigen) by chemical method and forming enzyme tracer. This enzyme tracer simultaneously has immunological and chemical reactivity. When it is bound to the antigen or the antibody to be tested, it can develop colour through decomposing the substrate of enzyme, thus demonstrating the immunological reaction. The amount of substrate decomposed is correlated to the colour development [7,8].

5.2.1.2　Classification and characteristics of enzyme immunoassay

Enzyme immunoassay can be classified as follows.

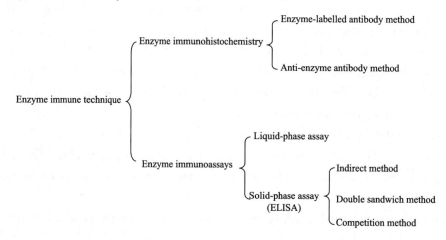

The characteristics of enzyme immunoassay technique are as follows [9]: ① High sensitivity: It is comparable with RIA, and the detection capacity of it can be up to ng level. ② Wide applications: It can detect antibodies or antigen, moreover, it can detect qualitatively or quantitatively, and implement location and antigen analysis. ③ Does not need special equipment: RIA requires special laboratory conditions, and IFA requires fluorescence microscopy. ④ Specimens can be stored for a long time. ⑤ Expiry period of enzyme tracer is long and generally up to more than one year under cold storage.

5.2.1.3　Commonly used enzyme labels and substrates

Selection of enzyme label is according to the requirements as follows: ① Having high activity and strong ability to decompose substrate. ② Specificity. ③ Enzyme activity can still remain after integrated to antigen or antibody. ④ Can transfer substrate into colour compounds. ⑤ High purity, and easy to be purified. ⑥ Good solubility (water-solubility), and stability in solution. ⑦ Determination method is simple. ⑧ Widely available and low price.

These four enzymes can meet these requirements: ① Horseradish peroxides(HRP), molecular weight 40,000. ② Alkaline phosphatase(AKP), molecular weight 82,000. ③ β-galactosidase, molecular weight 540,000. ④ Glucose oxidase, molecular weight 186,000. Among them, HRP is most commonly used, because it holds the advantages such as high activity, stability, low molecular weight, and is easy to be purified.

HRP is widely distributed in various plants, and the concentrations in horseradish is the highest. It is a glycoprotein consisting of zymoprotein and brown iron porphyrin prosthetic group, containing 18% of saccharide and 300 amino acids and 4 disulfide bonds. The isoelectronic point is 5.5 to 9.0. The optimal pH in catalysis varies along with different hydrogen donor and most of them are about pH 5.0. HRP is easily soluble in water and 58% or less of saturated ammonium sulphate solution.

The maximum absorption peaks of HRP enzyme protein and other impurities proteins are at 275 nm, and that of cofactor part is at 403 nm. The ratio of OD_{403}/OD_{275} expresses enzyme purity and is abbreviated as RZ value (Reinheit Zabl in German). RZ>3 RZ>2.5 and RZ<2.5 denote high quality, medium quality, and low quality that requires purification, respectively. Smaller RZ value indicates more foreign proteins. If the RZ is 0.6, it indicates 75% of foreign proteins.

There are two standards for evaluating the enzyme quality. One is purity, namely RZ value, and the other is enzyme activity. Thus, the purity can't indicate comprehensive quality of the Enzyme. Currently, an enzyme unit is the enzyme activity that can decompose 1 mmol/L substrate in a minute under certain conditions (pH, temperature and substrate concentration). The number of activity unit which 1mg enzyme contains is expressed as specific activity. Specific activity can be used to compare different enzymes[10].

The specific activity of peroxidase as label is generally greater than 250 units/mg. The substrate of HRP is H_2O_2, and hydrogen donor is required in the catalytic reaction. 3,3'-diamino-benzidine (DAB) is a hydrogen donor commonly used in immunohisto-chemistry. In reaction, its oxidized intermediates can rapidly aggregate to form insoluble brown phenazine derivatives which are suitable for optical microscopy observation. This polymer can reduce and chelate osmium tetroxide (OsO_4) to produce product with high electron density, which is suitable for electron microscopy observation.

Some hydrogen donors in ELISA are listed in the Table 5-3.

Table 5-3   The hydrogen donors for ELISA

Hydrogen donor	Reaction color	Wavelength /nm	Terminator	Termination color	Determination of wavelength/nm
Dianisidine (OD)	Orange-red	512	5 mol/L HCl	Yellow	400
O-phenylenediamine (OPD)	Yellow	444	2 mol/L H_2SO_4	Orange	492
O-benzyl aniline (OT)	Blue	636	2 mol/L H_2SO_4	Yellow	442
5-Aminosalicylic acid	Brown	475	1 mol/L NaOH	Brown	449

The OD and OPD are most commonly used. Due to carcinogenic effects, OPD should be paid attention in experiment.

Alkaline phosphatase can be extracted from calf intestinal mucosa, or *E. coli*. It is a phosphatase, and can catalyze phosphate hydrolysis to release phosphate acid and de-velop color reaction. Or through hydrolysis produce phosphoric acid and molybdate react to produced phosphomolybdic acid. Then generate blue product under the reducing agent which can be determined.

5.2.2   Preparation of enzyme-labelled antibody

5.2.2.1   Requirements for enzyme-labelled antibody

Enzyme-labelled antibody should meet the undermentioned requirements: ① high purity, and containing less impurity proteins, IgG is better than serum antibodies, and Fab is better than IgG. ②high specificity that only IgG has only one precipitation line in immunoelectrophoresis when running with anti-IgG or whole antiserum of IgG. ③ high titer, that is higher than 1 : 64 when identified with agar diffusion.

5.2.2.2   Enzyme labelling method

(1) Crosslinking method

Glutaric dialdehyde is a crosslinker commonly used and most of the reagents availables are 25% solutions. Through covalent binding of two symmetrical aldehyde groups of glutaric dialdehyde to free amino and phenyl of enzymes, enzyme labelling can be obtained. Glutaric dialdehyde should be as fresh as possible in labelling. If the value of $OD_{235\ nm}/OD_{280\ nm}$ of glutaraldehyde is less than 3, it is a good crosslinker.

　　Glutaric dialdehyde crosslinking method can be divided into one-step method and two-step method: ① For one-step method, substrates (enzyme and antibody)and linker agent(glutaraldehyde) are mixed together in one-step procedure to prepare labelled products. Then the excessive glutaric dialdehyde is dialyzed to prepare the enzyme complex with high molecular weight. Since antibody molecule is of high molecular weight (160,000) and enzyme molecule is small (40,000), resulting in the amino residue number of antibody are much more than that of zymoprotein, a large number of antibody-glutaraldehyde -antibody complex will be produced and enzyme-label obtained is of low activity For sample, an operation procedure in one-step method: ① Preparation of Anti IgG-AKP: 10 mg AKP are dissolved in 1mL PBS (0. 01 mol/L pH 7. 0) solution containing 5mg anti-IgG, and stirred lowly in the ice-bath, avoiding the air bubble as far as possible. After they are dissolved completely, 4 mL of 1% glutaric dialdehyde is added slowly in drops to 0. 2% final concentration of glutaric dialdehyde. Then they stand at room temperature for 2-3 h, and the excessive glutaric dialdehyde is removed at 4 ℃ by 0. 01 mol/L pH 7. 0 PBS dialysis or SephadexG-25 column. Then it is stored at 4 ℃ (also can be preserved in 50% glycerol at low temperature refrigerator); ② Preparation of Anti IgG-HRP: 12 mg HRP are dissolved in 1mL PBS (0. 1 mol/L pH 6. 8) containing 5 mg anti-IgG, 4 mL of 1% glutaric dialdehyde are added under slow agitation, and they stand at room temperature for 2-3 h, Then the excessive glutaric dialdehyde are removed. The label should be preserved at low temperature in refrigerator, avoiding freezing and thawing repeatedly, or stored after freeze-drying.

　　In two-step method: firstly enzyme will react with glutaraldehyde and unreacted glutaraldehyde is removed. Then the activated enzyme molecules can be bound to amino group of antibody. Finally, a few lysine are added to block the enzyme residue activated by glutaric dialdehyde. A sample for operation is given here: 10 mg enzyme are dissolved in 0. 2 mL of 0. 1 mol/L and pH 6. 8 PBS containing 1. 25% glutaric dialdehyde, and is allowed to for 18 h at the room temperature. After the unreacted glutaric dialdehyde are removed by sufficient dialysis or SephadexG-25 column, 1mL of physiological saline is added, and then 1mL of 5mg/mL antibody solution and 1 mL of 1 mol/L and pH 9. 6 carbonate buffer, are added. After incubated at 4 ℃ for 24 h, 0. 1 mL of 0. 2 mol/L lysine are added and is allowed to stand at room temperature for 2 h. After sufficiently dialyzed by 0. 15 mol/L and pH 7. 2 PBS and centrifuged, to discard precipitate, supernatant, namely enzyme conjugate, will be obtained. After further purification by ammonium sulphate precipitation, it can be applied. Generally, AKP is prepared not through two-step method but through one-step method.

Modified two-step method for HRP labelling: 10 mg HRP are dissolved in 0. 4 mL of 0. 25 mol/L, PBS (pH 6. 8) or 0. 05 mol/L carbonate buffer (pH 9. 6), 0. 1 mL of 25% glutaric dialdehyde is added at 37 ℃. After they are incubated for 2 h, 2 mL of cold absolute ethyl alcohol is added. The solution is centrifuged at 2,500 r/min for 10 - 15 min. After the supernate is discarded, the precipitation is suspended in 4-5 mL of 80% ethyl alcohol. Then they are centrifuged again, and the precipitation obtained is dissolved with 1 mL of 0. 05 mol/L the pH9. 6 carbonate buffer. After 0. 5-1 mL of IgG solution (about 15 mg) is added, the solution is kept in 4 ℃ overnight. Then some KH_2PO_4 is added to adjust the solution to near neutrality pH.

(2) Sodium periodate oxidation method

Sodium periodate can oxidize hydroxyl group of mannose into aldehyde group which can bind to free amino groups of antibody[11,12].

The protocol of this method is as follows. Resuspend 5 mg HRP in 1. 0 mL of 0. 3 mol/L carbonate buffer (pH 8. 1), and let it add 0. 3 mL of freshly prepared 0. 1 mol/L sodium periodate in 10 mmol/L sodium phosphate (pH 7. 0). Incubate at room temperature for 30 min, add 1 mL of 0. 16 mol/L glycol and let it stand at room temperature for 1 h. Dialyze the HRP solution against 1 mmol/L sodium acetate (pH 4. 0, 3×2 L) at 4 ℃. Prepare 0. 5 mL of 10 mg/mL antibody solution in 20 mmol/L carbonate. Remove the HRP from the dialysis tubing to the antibody solution. Incubate at room temperature for 2 h. The Schiff's bases that have formed must be reduced by adding 5mg sodium borohydride into HRP solution and incubate at 4 ℃ for 2 h.

Modified sodium periodate oxidation method: Dissolve 5 mg HRP in 0. 5mL double distilled water, and add fresh 0. 06 mol/L $NaIO_4$ (10 mL double distill water +128 mg $NaIO_4$) 0. 5 mL. After blended, the solution is kept at 4 ℃ for 30 min. Add 0. 5 mL of 0. 16 mol/L glycol peroxide solution (10 mL the H_2O+0. 09 mL glycol), and let it stand at room temperature for 30 min. Add 5 mg purified antibody, blend and dialyze against carbonate buffer (0. 05 mol/L, pH 9. 5) by slowly stirring for 6 h (or overnight), Add 0. 2 mL of $NaBH_4$ solution (5 mg/mL), blend and stand at 4 ℃ for 2 h. Slowly add an equal volume of saturated ammonium sulphate solution into the above mentioned solution, stir at 4 ℃ for 30 min. Centrifuge and discard supernatant. Dissolve the precipitate obtained by a few PBS buffer (0. 02mol/L, pH 7. 4), and dialyze by the same way aforementioned at 4 ℃ overnight. Centrifuge and remove the precipitate, then obtain enzyme-labelled antibody conjugate (HRP-IgG). Adjust the volume to 5 mL by adding PBS buffer (0. 02 mol/L, pH 7. 4). After determining the titer of the label, the same volume of glycerol is added. The mixture was then separated into small vials and stored at low temperature. The comparison between one-step method and two-step method is summarized in Table 5-4.

Table 5-4   The advantages or disadvantages of two crosslinking method in labelling with enzyme

Items	Glutaraldehyde method		Sodium periodate method
	One-step method	Two-step method	
Procedure	Simple	Fairly complicated	Fairly complicated
Utilization ration of enzyme	2%-4%	2%-4%	70%
Ability of penetrating tissue	Poor	Good	Common
Loss of antibody activity	Much	Less	Much
Loss of enzyme activity	Much	Less	Much
Staining effect	Poor	Good	Good

(3) Dimaleimide method

Dimaleimide can crosslink proteins by reacting with the hydrosulfunryl group of cysteine residue in protein. Firstly, protein (IgG) was reduced by α-cysteamine and activated by N,N'-O-phthalaldehyde maleimide. Then the excess reagents is removed, and finally β-galactosidase is conjugated with dimaleimide-IgG. The protocol of this method is as follows. The IgG antibody was dissolved in 0.1 mol/L sodium acetate buffer (pH 5.0) and dialyzed against the same buffer overnight. After centrifuged to remove insoluble substances, IgG (14 mg/2 mL) was incubated with 10 mL of α-cysteamine at 37 ℃ for 90 min. The 3.0 mg/mL reduced IgG was obtained through sephadex column purification and concentrating The saturated N,N'-O-dimaleimide solution was added at 1 : 1 ratio in ice-bath, mixed and incubated at 30 ℃ for 20 min. The unreacted dimaleimide was removed through sephadexG-25 column purification. The activated IgG was collected and concentrated to 1.0 (OD$_{280nm}$ value, light path 1 cm). To 1mL of the activated IgG solution, 20 μL (5 mg/mL) β-galactosidase was added and kept at 30 ℃ for 20 min. After neutralination with 0.10 mol/ L NaOH. the mixture was kept at 4 ℃ for 24-72 h. Then they were added into sepharose-6B column (1.5 cm × 40 cm) and eluted with pH 7.0 PBS. The enzyme conjugate collected was stored at 4 ℃ for use.

5.2.2.3   Purification of enzyme-labelled antibody

The enzyme-labelled antibody solution is a mixture of various crosslinked substances such as enzyme-antibody, enzyme-antibody-enzyme, antibody-antibody and enzyme-enzyme conjugate, and even free enzyme and antibody. Except the antibody-enzyme and enzyme-antibody-enzyme, the others should be removed.

(1) 50% saturated ammonium sulphate precipitation method

This method can only remove the free enzyme and enzyme polymers. But it cannot remove the other unwanted substances. But this method can minimize the activity loss of enzyme-labelled antibody.

(2) Purification through sepharose-6B or sephadex-G200 column

This method is more complicated and will cause high activity loss of enzyme-labelled antibody, but better purification result can be obtained.

5.2.2.4　Identification of enzyme-labelled antibody

Identification of the activity of enzyme labelled antibody is generally done by agar diffusion and immune electrophoresis. The enzyme labelled antibody will react with the corresponding antigen (1 mg/mL) to produce a precipitation line, and colour will be developed in substrate solution after washing. After rinsed with saline, precipitation lines do not fade, which indicates the activity of enzymes and antibody. Agar diffusion titer of good enzyme-labelled conjugate usually is higher than 1 : 16.

Quantitative analysis of enzyme conjugate includes determination of enzyme, IgG content, mole ratio and binding ratio of IgG and enzyme.

Quality standards of purified enzyme conjugate should include the following: ① Enzyme catalytic activity. ② Immunological activity. ③ Amount of free enzyme. ④ Amount of the original immune reactants. ⑤ The number of immune reaction molecules connected to an enzyme molecule. ⑥ Biochemical properties. ⑦ Effect on enzyme-labelled immunoassay experiments.

Quality of enzyme conjugate has big effect on the test sensitivity. For example, different HRP-IgG conjugates prepared by different methods will obtain different results in immunoassay.

5.2.2.5　The preservation of enzyme-labelled antibody

Enzyme labelled antibody should be separated and encapsulated in small vials, freeze-dried and preserved at low-temperature. It is better to add glycerol or bovine serum albumin (final concentration is 33%) and avoid repeated freezing and thawing. In general, the activity of label can be kept for 1-2 years.

5.2.3　Amplification of enzyme immunoassay

5.2.3.1　AP substrate amplification system

AP can catalyze $NADP^+$ dephosphorylating and generate NAD^+ (coenzyme 1), alcohol will be dehydrogenated under deoxy ethanol enzyme catalysis and then NAD^+ is reduced to NADH. NADH is dehydrogenated under diaphorase catalysis, and tetrazoliim is reduced to colour formazan at the same time. AP continuously catalyzes NAD^+ to generate NADH, and NAD^+ and NADH converse each other and continuously generates formazan. The entire reaction cycle is an enzyme cascade including AP, alcohol dehydrogenase, and diaphorase enzyme. Each AP molecule can generate 6×10^4 NAD^+ molecules per minute, and each NAD^+ can produce 60 formazan molecules per minute. The EIA amplification system, firstly reported by lcelf in 1984, can improve the determination sensitivity about 250 times.

But Brooks and others found that acetaldehyde produced in the cycle can inhibit the amplification effect in the entire enzyme cascade. When semicarbazide hydrochloride was added, the reaction cycle can be further extended, which thereby increases the measurement sensitivity. The reason is that semicarbazide hydrochloride can react with acetaldehyde and generate semicarbatone and water, thus eliminating the inhibition effect of acetaldehyde. This improved method was applied in determination staphylococcus aureus containing protein A in food. The sensitivity of the original method was improved from the self's $(4-6) \times 10^3$ colony forming units (cfu)/g to 20 colony forming units (cfu)/g, and, the determination of the sensitivity was improved about 200 to 300 times.

Self and others also adopted resazurin instead of tetrazolium salt in the original system. Resazurin could be transferred into fluorescent resorufin under the effect of diaphorase, and then fluorescence colorimeter was used to determine the result. The sensitivity of this approach was also much higher than the original method, and 350 AP molecules per well could be measured.

5.2.3.2　Dual-enzyme cascade amplification system

Dual-enzyme cascade amplification system is also known as enzyme inhibitory cascade system. Esterase inhibitor 4-(3-O-4,4,4-Trifluoro-Butyl) phenyl phosphate has no inhibitory activity to esterase because of its $-PO_4$ group. However, under the action of AP, the $-PO_4$ group can be eliminated, and the inactivated inhibitor will become activate inhibitor which can inactivate carboxylesterase. The residual esterase activity can be determined by the colour reaction, which can be used to reflect the original activity of AP. The sensitivity of this amplification system is increased about 125 times compared with ordinary method.

5.2.3.3　Enzyme activation cascade amplification system

Enzyme activation cascade is a kind of enzyme amplification system with AP as label. Flavin-adenine dinucleotide phosphate (FADP) is dephosphorylated into FAD under the catalysis of AP, and FAD make the amino acid oxidase without cofactor into a holoenzyme which can catalyze praline to generate H_2O_2. Thus in the presence of DCHBS, 4AAP and HRP, colour product will be obtained. The amplification system can determine 1 nmol/L AP.

In addition, catalyzed reporter deposition (CARD) is also a multi-enzyme cascade EIA amplification system. HRP can oxidize biotin or fluorescein-labelled tyramine to generate free radicals which can react with tyrosine and tryptophan of the protein on solid phase and deposit a large number of biotin or fluorescein in the solid region around HRP. The biotin can be determined using HRP and 6 -galactosidase or AP-labeled streptavidin. Fluorescein can be determined using enzyme-labelled anti-fluorescein. This method transfers one HRP molecule into a large number of labels, which greatly improves the sensitivity of EIA determination (about 30 times). Diamandis and others replaced the above-mentioned enzyme labelled with the Eu^{3+} chelate streptavidin-thyroglobulin reagents to determine AFP, which not only improved the sensitivity, but also eliminated the background signal about 6 to 8 times than the common method. CARD amplification system can be used directly in the HBsAg ELISA kit, and the sensitivity of the kit can be improved about 5 times without changing the composition and operating procedures.

5.2.3.4　Coagulation factor enzyme cascade amplification system

Lipopolysaccharide (LPS) can activate coagulation cascade of blood cell lysates of the horseshoe crab. Firstly, the factor C is activated by LPS, and the activated C will activate factor B, and the activated B can transfer former-coagulase into coagulase, and then the coagulase catalyzes substrate (Boc-Le, u-Gly-Arg-p) to produce colored p-nitroaniline (PNA) which can be determined at $A_{405\,nm}$. If antibody is labelled with LPS in immunoassay, the sensitivity can be improved to determine 10^{-11}-10^{-7} g/mL IgG and 10^{-11}-10^{-7} g/mL anti-IgG. Seki and others adopted this method in a double sandwich assay for HbsAg and obtained a sensitivity of about 10^{-12}-10^{-10} g/mL.

5.2.3.5　Immune complex transfer two sites enzyme immunoassay

Generally, in the enzyme immunoassay, the non-specific binding of labelling enzyme to the solid phase is an important limit factor of determination sensitivity. because this will increase the background and mask the specific signal of low concentration analytes. One way to solve this problem and improve the measurement sensitivity is to elute analytes-antibodies-enzyme complex from the solid phase to another solid phase and utilize a dual-labelled antibody [such as biotin and dinitrophenyl (DNP) labelled antibody and an enzyme-labeled antibody. In liquid phase, the immune complex (double-labelled antibody: antigen-antibody-enzyme) will be captured on the polystyrene beads coated with DNP-IgG. After washing, dinitrophenyl-L-lysine can replace the complex from the solid phase, and then the immune complex is captured by polystyrene beads coated with streptavidin. Finally, the enzymatic catalysis reaction is determined. Hashida and Shikawa used this method to detect ferritin and the sensitivity could reach 1 zeptomole (1×10^{-21} mol, about 600 molecules). The sensitivity was improved about 30 times compared with the conventional method. Many researchers used this method to measure anti-thyroglobulin, anti-IgC HTLV-1 IgG and anti-HIV-1 IgG, and the sensitivity was far more than those of conventional methods.

In addition, Domingo and Marco and others selected 4-chloro-1-naphthol as HRP substrate in spot immunobinding assay, and insoluble product produced could be observed under ultraviolet light for it has unique UV absorption and fluorescence characteristics. This method increased the sensitivity about 100 times compared with that of conventional method.

In conclusion, amplification approach of EIA determination method is varied under unceasing innovation One start point of researchers is to reduce non-specific colour, for example, immune complexes physical transfer two points enzyme immunoassay, which achieves the purpose by increasing experimental steps. The other one is to improve the quality of signal to improve the sensitivity, such as biotin-avidin, multi-enzyme cascade amplification system which increases response levels. But in a strict sense, only the latter is truly EIA amplification system. The former can only be called ultrasensitive EIA. As for changing detection approach by using characteristics of enzyme reaction products (such as chemiluminescence and fluorescence detection), they can improve the sensitivity of EIA just relying on instrument to widen the "vision".

5.2.4　Application and development trend of enzyme immunoassay

Enzyme Immunoassay (EIA) is an important labelling immunoassay technique which uses enzyme-labelled antibody (antigen) as the main reagent and combines enzymatic reaction efficiency and the specificity of antigen-antibody reaction. As one of three classical labelling technologies, enzyme immunoassay technology has been widely used in laboratory medicine and is continuously updated and improved with integration of other advanced techniques such as fluorescence and light-emitting technologies. Many automated devices are actually complex of EIA technology and other modern technologies, and their sensitivities have reached or exceed the sensitivity of radio immunoassay (RIA). Since the enzyme reagents haven't radioactive contamination and analysis forms is diversified, simple and flexible, extensive attention has been paid to EIA in clinical application [13-15].

5.2.4.1　Advancements of EIA technology

In recent years, EIA has made significant developments mainly in the following areas:

(1) Following the monoclonal antibodies, the third generation antibody, genetic engineering antibodies, have been applied widely. The improvement of antibodies preparation technology realizes the large-scale preparation of the reagent and expands the scope of detection, which realizes the detection of small molecule antigen and hapten.

(2) The sensitivity is improved through improvement of analysis approach. ① In addition to early competition and indirect method, the means such as double antigen

sandwich method to measure antibody, coupled enzyme label to measure antigen, bridged enzyme immunoassay and enzyme immunoassay amplification technology are further developed. Enzyme immunoassay amplification technology applies the biotin-avidin system, the substrate loop amplification system, enzyme-anti-enzyme and enzyme amplification system coupled amplification system, etc. ② With the progress of enzyme-labeled technology, the application of enzyme from the peroxidase has extended to alkaline phosphatase, β-galactosidase, urease, glucose-6-phosphate dehydrogenase, glucose oxidase and malate dehydrogenase. So far, there are over 20 enzymes used in EIA, but horseradish peroxidase(HRP) and alkaline phosphatase (ALP) are still the most widely used enzymes. ③ Combination of EIA and other labelling immunoassay, such as integrating fluorescence immunoassay (FIA) and EIA to form fluorescent enzyme immunoassay (FEIA), enzyme-amplified time-resolved fluoroimmunoassay(EATRFIA), and integrating EIA and chemiluminescent immunoassay (CLIA) to form enzyme-immunoassay analysis, enhanced chemiluminescence enzyme immunoassay (ECLEIA), and integrating EIA and polymerase chain reaction to form PCR-EIA analysis. ④ Improvement in solid-phase carrier technology Solid-phase carrier in traditional ELISA is polystyrene microtiter plate, polystyrene beads or strips. After that, nitrocellulose membrane, activated filter paper, silicon, nylon, and various solid-phase particles synthesized by polymer materials were developed. To improve the binding capacity and scope of solid surface, the surface of polystyrene was modified through various methods, such as chemical coupling method leading to functional aldehyde group, acyl and alkyl amine, to combine better with carboxyl group of protein and peptide. Using site-oriented covalent coupling method to introducing avidin, protein A or polylysine, proteins or peptides can be captured firmly. Preparation of ultra-smooth surface of polystyrene can overcome the heterogeneity caused by surface roughness, and an super smooth surface with average roughness of only $2A^+$ was achieved, which realized the large capacity of protein combination, low desorption rates and high uniformity between wells as well as 5% testing precision. In double antibody sandwich method, the polystyrene irradiated by ray can increase the absorption properties, especially on the immune globulin adsorption. In recent years the Corning company developed a succinimide ester activated microtiter plate which achieved the same level of amino activation. ⑤ Innovative methods: Simultaneous ELISA is to determine multiple antibodies. The antigen is coated in the nail matched the nail plate and can react with the required and porous specimens, reagents at the same time when being covered, which is a modified ELISA. Using the character that antibody and anti-antibody complex can form multi-layer properties, the conventional two-step incubation method is changed to a one-step incubation assay. This method significantly shortened the test time and now has been widely used.

5.2.4.2　The integration of EIA and modern technology

EIA is suitable for various levels of inspection department, not only due to advantages such as high sensitivity and specificity, stable reagents, simple and without radiation hazards of EIA, but also owing to the development of modern biology, the preparation of antigen-antibody and enzyme labelling technology, especially application of commercial kit standardization and the development of automation device. With the continuous development of laboratory medicine, EIA will develop because of integrating with modern technique.

(1) Dot-ELISA

Difference between Dot-ELISA principle and ELISA is that the former use nitrocellulose membrane as a solid phase carrier which has strong adsorption for proteins. After the reaction of enzyme and substrate, colour sediment will form on nitrocellulose membrane and stain the membrane. Its sensitivity can reach nanogram level. The method only needs small dosage of and doesn't need other equipments.

(2) Immunoblotting

Immunoblotting combines electrophoresis with the ELISA method. It divides into three steps including electrophoresis, transfer and EIA. Immunoblotting combines high resolution of electrophoresis and high sensitivity and specificity of EIA. It is an immunological method which can be used for the analysis of sample components.

(3) Luminescence enzyme immunoassay

Difference between luminescence enzyme immunoassay and general EIA is that the light-emitting agent is used as substrates of enzyme catalysis. In luminescence enzyme immunoassay, the product is light-emitting different with the colour product in EIA, and can be determined by specific instrument. HRP and AP are commonly used labelling enzyme in this technique.

(4) Enhanced luminescence enzyme immunoassay

Enhanced enzyme immunoassay luminescence enzyme immunoassay (ELEIA) is the new development of EIA. It features that enzyme can enhance luminescence signal, stabilize and extend the time of light signals. It not only maintains the high sensitivity of luminescence immunoassay, but also overcomes the shortcomings of short light-emitting signal time in the traditional enzyme immunoassay. Currently, this method has realized automated analysis.

(5) BAS-ELISA

BAS-ELISA combines the biotin-avidin (BAS) amplification system and the ELISA technology. Biotin (B) has two ring structures. One of them can combine with avidin, and the other can combine various substances including enzymes and antigens (antibody). Avidin (A) has 4 subunits which can bind to biotin stably and an avidin molecule can combine with 4 biotin molecules, which is a key point in amplification system. Avidin can also be labeled by enzyme.

Features of biotin and avidin are used in the BAS application and two types of reaction are designed. In one of them, free avidin is used as a "bridge" which is connected to biotinylated antigen-antibody and enzyme-labelled biotin, respectively. This type of methods includes ABC and BAB method. In another, reaction system, labelling avidin directly connects to biotinylated antigen-antibody, and BA method is this type. Through the BAS amplification, more enzymes can be gathered in the solid phase carrier, which further improves the sensitivity of enzyme immunoassay detection.

(6) Fluorescence enzyme immunoassay

Fluorescence enzyme immunoassay (FEIA) utilizes potential fluorescence substrate as display of enzyme catalysis amplification system. Due to the accumulation amplification and the high sensitivity of fluorescence, the sensitivity can be improved obviously.

(7) Enzymatic amplification of time-resolved fluorescence immunoassay (EAT-RFIA)

Its outstanding feature is the introduction of time-resolved fluorescence immunoassay technology, which will benefit eliminating the interference of unusual fluorescence and enhance the specificity of the measurement by amplification of the enzyme signal.

5.2.4.3　Automation of EIA

In the 1950s, biochemical automatic analyzer had been used in practical application. Owing to specificity of EIA technology, automated analysis system was published until the 1980s,. Based on different process of analysis, EIA can be divided into homogeneous and heterogeneous categories.

(1) Homogeneous EIA and its automation

Homogeneous EIA, also known as non-solid phase or liquid immunoassays, is used for the determination of small molecules, especially drugs. The type is competition method, in which labelling antigen and the antigen in specimens will compete the specific antibody and then the label will be determined. In certain circumstances, labels bound to the antibody will lose its characteristics, so it needn't separate the bound (B) and free (F) and the quantity of label in F can be directly determine.

The enzyme-multiplied immunoassay technique (EMIT) developed in 1972 by Syva Company and cloned enzyme donor immunoassay (CEDIA) developed in 1992 by Microgenics company are both homogeneous EIA. The detection of reagents in the two methods can be carried out with biochemical automatic analyzer.

(2) Heterogeneous EIA and its automation

Most of labelling immunoassay belong to heterogeneous immunoassay. The most commonly used separation method for B and F is washing solid-phase carrier, but this step is not there in clinical chemistry determination. In early 1970s, Engvall and Perlaman first developed heterogeneous enzyme immunoassay technique, also known as enzyme-linked immunosorbent assay (ELISA). In ELISA, usually polystyrene is used as solid phase carrier and has microtiter plate, pipes, beads, particles, magnetic particles and other types. Different forms of solid phase requires different cleaning devices. Therefore, the automation of heterogeneous immunoassays cannot rely on biochemical automatic analyzer; and various methods have their special automatic analysis system.

Some microplate automatic analyzers are introduced as follows.

ELISA reagents for plate-type of different manufacturers are different, but unified 96-well microplate, 8×12 type, is used. Therefore, plate-type ELISA automatic analyzer provided by suppliers are "open", which can be applied for any reagent. Thorough cleaning of each microplate well is a key to ELISA, and is also difficult in automatic analysis. Therefore, fully automatic plate-type EIA analyzers emerged in recent years. Such as CODA automated EIA analyzer and more intelligent EVOLIS EIA analyzer from Bio-Rad's company, and DSX, DIAS, Dias Ultra automated enzyme-labeled analysis system of Dynex's company can all simultaneously complete automatic measurement for multiple microplates. Swiss Tecan of Tecan company, a large automatic enzyme multifunction platform, has double mechanical arm working at the same time, and all the processes can proceed without human intervention. Robotic Microplate Processor (RMP) can complete the final results from the specimens assignment to the whole process, which is suitable for the laboratory with large-scale projects. Microlab F. A. M. E automatic ELISA processing system, produced by Swiss HAMILTON company, is fully automatic ELISA processing system and can perform all the necessary processing steps of the ELISA test at the same time. The instrument is an open system which can handle different projects tests of various ELISA from different manufacturers, and has advantages such as flexible operations, large capacity and easy maintenance.

Automated analyzer with other forms of solid phase.

In these automated analyzers, except the type with microporous plate as solid-phase carrier, reagents should match instrument.

Tube carrier has character that the small tube can be used as reaction and also colour container at the same time. Such as ES300 analyzer and ES600 automatic EIA analyzer developed by Boehringer Mannheim belong to this type. Since the company is merged into Roche company, the production of various types of ES instruments has been stopped.

Abbott Company firstly used beads, 0. 6 cm in diameter, as a carrier in a semi-automatic analyzer. The beads were placed into a corresponding size plastic plate and was rinsed when rolling. COBAS CORE II of Roche company, a fully automatic analysis system, utilized small-sized beads as solid-phase carrier, horseradish peroxidase (HRP) as labelling antibody and TMB as colour-displaying substrate, and can detect more than 40 projects.

MEIA (microparticle enzyme immunoassay) automatic analysis system of Abbott Company utilized latex particles as solid carrier and was more advanced type than beads. The latex can be adsorbed onto glass fiber membrane and be washed. This kind of automatic analysis system is called IMx, and now combined with TDx to form the versatile Axsym system.

Magnetic particle is an ideal carrier because its surface area is large and it can be separated by magnet attraction. SEROZYME enzyme immunoassay analyzer of Swiss Serono company belongs to this type. This carrier has been used by many other immunoassays.

5.2.4.4　Application of combination of EIA technology and other analysis technology in the automation devices

(1) Chemiluminescence enzyme immunoassay

This is a kind of enzyme immunoassay which uses chemical light-emitting agents as enzyme substrates. Duo to two-level amplifier through enzyme and chemiluminescence, it has very high sensitivity. Amersham company early developed AMERLITE, a chemiluminescence enzyme immunoassay system, which adopted peroxidase as label and luminal as luminescent substrate, and added luminescence enhancing reagent to improve the sensitivity and luminescence stability. After improvement by Johnson & Johnson company, it was named VITROS Eci, an automatic optional enhanced chemical luminescence immunoassay analysis system. This instrument combined enzyme immunity technology, biotin-avidin technology and enhanced chemiluminescence technology. It adopted horseradish peroxidase to label antigen or antibody, plastic tube with well as solid phase carrier and luminol as chemiluminescent agent. The key technology was the application of luminescence enhancers which increased luminescence intensity thousand times and also extended the life-time and stability of luminescence. Access automatic microparticle enzyme amplification chemiluminescence EIA system of American Becman Coulter company was collaboratively deigned and produced by American Beckman Company and France Pasteur institute. The system adopted alkaline phosphatase to label antigen or antibody, magnetic particles as solid phase carrier and AMPPD (Diomums) as chemiluminescent reagent which was stable, long duration and, easier to be controlled than flashing light. Similar analysis system was IMMULITE2000 type of Diagnostic Production corporation, which adopted automatic optional for enzyme-labelled antigen or antibody, AMPPD (Dioxeten) as a light-emitting compound and plastic beads as solid phase carrier. The measuring principle was basically the same with the Access system of Beckman company.

(2) Enzyme-enhance fluorescence amplified immunoassay technique

VIDAS automatic fluorescence EIA system of French Merieux company applied double-antibody sandwich method with alkaline phosphatase-labelled antigen or antibody, a plastic straw (SPR) as the solid phase carrier, 4-methyl ketone phosphate (4-MIjP) as luminescence reagent. Aura Flex of U. S. Nexct company used beads and labelling enzyme to direct catalyze luminescence substrate (AKP as label, 4-MUP as substrate).

(3) Enzyme-enhanced time-resolved fluorescence amplified immunoassay (EATRFIA)

Microplate-type fluorescent microplate reader SPECTRAmax GEMINI XS of United States MD corporation (Molecular Devices corporation) firstly applied double monochromatic continuous wavelength system in porous plate and integrated fluorescence, luminescence and time-resolved fluorescence triple models. It can complete endpoint reading, dynamic reading, spectral scanning and cell counting experiments.

(4) Microparticle enzyme immunoassay (MEIA)

This method has advantages such as stable reagent and instrument, high sensitivity and wide linearity range. IMX automated immunofluorescent analyzer of U. S. Abbott Company belongs to this type. In this system, a sandwich complex, antibody coated on particles, target and alkaline phosphatase-labelled secondary antibody, is formed which is finally transferred into a glass fiber column washed with buffer. Then substrate will be added for assay. This company also supplies the whole set of reagents for this instrument. AXSYM automatic rapid immunoassay system of Abbott company has integrated MEIA and fluorescence polarization immunoassay. This system can measure hormones, tumor markers, hepatitis, virus markers, concentration of vitamins and various drugs, etc.

5.2.4.5 Limitation of EIA

In fact, specificity of EIA depends on antigenic determinant recognized by monoclonal antibody. Moreover, it is influenced by purity, specificity, stability and affinity of coated antigen-antibody, and preparation technology of enzyme label. Of course, other labelling immunoassays also have such problem. It should be noted that there is largest number of suppliers for EIA Kit, so the test kit should be strictly compared and selected.

Currently, the insulin or C peptide, commonly measured by EIA, are found to be the mixture of them and their precursors in recent years. The mixture is difficult to be separated because of their cross-reaction with antibodies, which is a good example of limitation of antibody. What have been detected usually are only insulin or C peptide with immune response, which sequires us to reevaluate the understanding of insulin or C peptide measured in the past and undergo the determination of "true" or "pure" insulin and C peptide. Similar samples might be found in future.

Since EIA is based on solid phase reaction, we should pay attention not only to the surface effect caused by inconsistent amount of coating antigen (antibody) in the different part of solid phase, but also to the edge effect during incubation caused by different reaction conditions between edge wells and center wells. The hook effect caused by the proportion mismatch between antigen and antibody also should also be prevented. It must be noted that the simple one-step method is more inclined to produce hook effect than the two-step method.

Some problems including that solid phase materials have non-specific adsorption, hemolysis, that refrigerator storage can release peroxidase, and that long refrigerator storage time can lead to aggregation of serum IgG all are prone to cause high background or even seriously interfere with the determination. It should be noted that these defects are not very obvious in highly automatic instruments and large workload laboratory. But in the situation that equipment is semi-automatic and specimens are too less to be stored in refrigerator, these defects are obvious. In short, with the development of science and innovation of technology, EIA technology will become more perfect, with higher degree of automation, more accurate and precise, which will make greater contribution for human health.

5.3　Fluoroimmunoassay

5.3.1　Brief review

Immunofluorescence technique, also known as fluorescent antibody technique, is the earliest labelling immunoassay technology. It is based on immunology, biochemistry and microscopy. For long some scholars have been trying to combine antibody molecules with some tracer material to position the tissue or cell antigens through antigen-antibody reaction. For the first time, coons firstly successfully accomplished fluorescence labelling in 1941. This technique which uses fluorescent material to label antibodies for antigen localization is called fluorescent antibody technique.

The method of using fluorescent antibody to trace or check the corresponding antigen is called fluorescent antibody method. The method of using fluorescent antigen to trace or check the corresponding antigen is called fluorescent antigen method. These two methods are generally called immune fluorescence technology. The fluorescent pigments can combine not only with antibody globulin to detect or position various antigen, but also with other proteins to detect or positioning antibody. Since in practice fluorescence antigen technology is rarely used, usually fluorescent antibody technique is called immune fluorescence technique. Presenting and examining cell or tissue antigen or antigenic substances with immunofluorescence is known as immune fluorescent cells (or tissue) chemical technology.

The main features of the technology are of high specificity and sensitivity and also that it is fast. The main disadvantages include: nonspecific staining problem has not been fully solved, the result of determination can't be judged impersonally and technical procedures are also fairly complex[16].

Fluorescence immune method can also be further divided into several kinds based on reaction system and quantitative methods. Compared with the radioimmunoassay, fluorescence immunoassays will not risk radioactive contamination, and most of them are more simple and easy to be worked with. Due to the problem of high background in fluorescent determination, fluorescence immune technology can hardly realize measure quantitatively. In recent years, several special fluorescence immunoassay methods have been developed and applied in clinical diagnosis.

5.3.1.1 Principles

Immune fluorescence technology, based on highly specific antigen-antibody reaction, adopts fluorescence labelling antibody to which fluorescent pigment is bound and will not affect the activity of antibody. After fluorescence labelling antibody combines with corresponding antigen, specific fluorescence reaction will be presented under a fluorescence microscopy[17].

Indirect fluorescence immunoassay methods are usually applied in practice. In general the fluorescence labelling secondary antibody is anti-IgG globin. Due to the specimen specificity of IgG globin, for example anti-chicken IgG globin only reacts with chicken IgG globin, the fluorescence -labelled secondary antibody is suitable for detecting various antigen from chicken.

5.3.1.2 Representative fluoroimmunoassay

(1) Immunofluorometric assays

Since 1940, fluorescent labeled antibodies have been used to detect specific receptors, microbes, or antibodies with a fliorescence microscopic staining techqiue . In the immuno fluorescence staining technique tissue sections or microbe fixations generally are the solid matrix , and the products are visual detected with fluorescence microscopes.

(2) Fluorescence polarization immunoassay(FPIA)

when a fluorescent antigen conjugate is excited with polarized light , the polarization of the resulting emission depends inversely on the decay constant of the probe and on the rotation motion of the conjugate. With small molecules, random rotation decreases the polarization signal; when bound to specific antibodies, their rotation slows , and the polarization signal increases.

5.3.1.3 Fluorescent probes

In optimizing detection sensitivity , the probe muse have a hgih fluorescence intensity, the fluorescence signal must be distinguishable from the background ,and the binding of the probe to antibody or antigen should not adversely affect their properties. The fluorescenceintesnsity depends on both the absorption of the excitation and the quantum yield . The molar adsorptivity of the probe must be high (more than 10^{-4}), and the quantum yield in the solvent used (buffered aqueous solution) must as high as possible.

(1) Organic probes

1) Fluorescein isothiocyanate (FITC) is yellow or orange yellow crystalline powder, and easy to be soluble in water or alcohol solvent. Its molecular weight is 389. 4, the maximum absorption wavelength and the maximum emission wavelength of it are at 490-495 nm, and 520-530 nm, respectively. It can emit bright yellow-green fluorescence, and its structure is shown in Figure 5-1.

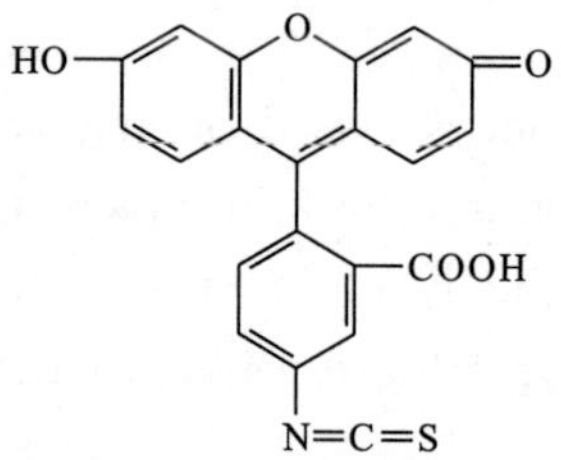

Figure 5-1   Structure of fluorescein isothiocyanate

It has two isomeric structures in which the type I isomers is better in efficiency, stability and protein binding capacity. And it can be stored for years in cold, dark, dry place and is the most widely used fluorescein. Its main advantages include that the human eye is more sensitive to yellow-green and that usually green fluorescent of sliced specimens is less than red.

2) Rhodamine (RIB200) is an orange powder, insoluble in water and soluble in alcohol and acetone. It is stable and can be preserved over a long period. Its structure is shown in Figure 5-2.

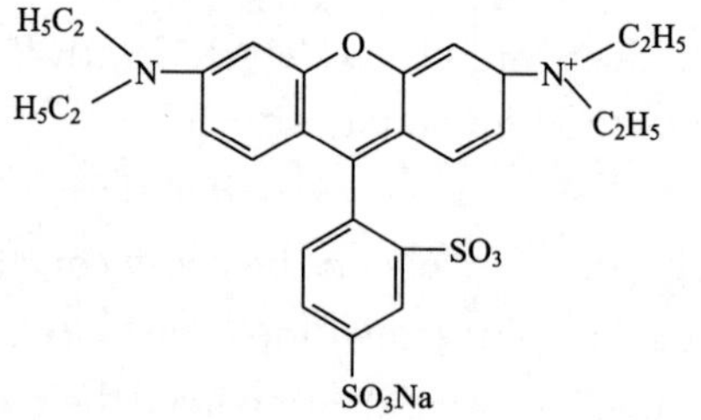

Figure 5-2   The structure of rhodamine

Its maximum absorption wavelength is at 570 nm, and the maximum emission wavelength is at 595-600 nm. It shows orange red fluorescence.

3) Tetramethyl rhodamine isothiocyanate (TRITC Figure 5-3).

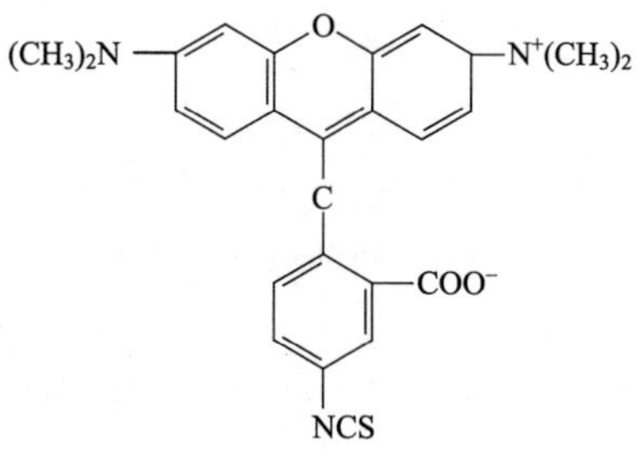

Figure 5-3   Structure of tetramethyl rhodamine isothiocyanate

Its maximum absorption wavelength is at 550 nm and the maximum emission wavelength is at 620 nm. It shows salmon pink fluorescence which can be distinctly contrastd with the emerald fluorescence of FITC, and can cooperate with FITC for double labelling or contrast staining. The isothiocyanate can conjugate with proteins, but its fluorescence efficiency is low.

(2) Other fluorescent substances

1) Compounds which can generate fluorescent after catalytic reaction of enzyme.

Some compounds themselves have not fluorescence effect, but they will emit strong fluorescence after the action of enzyme. For example, 4-methyl ketone-β-D galactoside is broken down under the action of β-galactosidase into 4-methyl ketone. The product can emit fluorescenc, and the excitation wavelength and emission wavelength are 360 nm and 450 nm, respectively. Others include 4-methyl ketone phosphate and hydroxyphenyl acetic acid which are the substrates of alkaline phosphatase and horseradish peroxidase, respectively.

2) Metal chelates.

Fluorescent rare earth ions and their chelates are especially interesting because of their unique fluorescent properties , high detection sensitivity (to as little as 10^{-14} mol/L). Some lanthanides , especially quropium[Eu(Ⅲ)]and terbium[Tb(Ⅲ)], form highly fluorescent chelates with some organic ligands. The ligand's absorption of light is followed by efficient energy transfer from its excited singlet state through triple state to the resonance levels of the rare earth ion. The metal ion emits energy as narrow band line-type emission. The fluorescence of the chelate is characterized by broad excitation in the absorption region of the ligand (250 to 360nm), a large Stokes shift(>250nm), emission lines typical of the metal and an exceptionally long fluorescence lifetime (100 to 100 μs).

5.3.2　Labelling method

5.3.2.1　Direct method

This is the most simple and basic method for fluorescent antibody technique. Fluorescent antibody is dropped on the specimen chip. After reaction and washing, it is observed under a fluorescence microscope. If the corresponding antigen is present in the specimen, it will specifically bind to the fluorescent antibody and the fluorescence antigen-antibody complex is obviously found under the microscope. The advantages of this method are simple and specific. But its shortcomings are that the technique needs various corresponding specific florescence antibodies for checking any antigen, and the sensitivity is lower than the indirect method.

5.3.2.2　Indirect method

It is based on the anti-globulin test principle to label anti-globulin antibody with fluorescein (referred to as the labelled anti-antibody). Detection protocol is divided into two steps. Firstly, the antibody to be tested (first antibody) is added to sample chip containing known antigen. After reacting for a certain time, unbound antibodies are washed away. Secondly, labelled anti-antibody is dropped. If the antigen-antibody binding reaction has occurred in the first step, the added labelled anti-antibody will combine with antibody (first antibody) molecules which have been fixed on the antigen molecules, thus, the antigen-antibody-labelled anti-antibody complex will form and show specific fluorescence. The advantages of this method are that it is more sensitive than the direct method and can be used to detect variety of antigen-antibody from the same species of animals without preparation of fluorescein labelled anti-globulin antibodies. Staphylococcus A protein can also be utilized to label fluorescein to replace labelled anti-immunoglobulin antibodies in indirect method. Moreover, it is not restricted by the source of the first antibody species, but the sensitivity is lower than labelled anti-antibody method. Its shortcoming is that sometimes non-specificity fluorescence is produced easily in this method. This method is commonly used in a variety of auto-antibodies detection. If the first antibody is known, the indirect method can also be used to identify unknown antigens.

5.4　Chemiluminescence immunoassay

5.4.1　Principle

Chemiluminescence immunoassay(CLIA) combines highly sensitive chemiluminescence determination technology and highly specific immune reaction, and can be utilized for detection and analysis of a variety of antigens, haptens, antibodies, hormones, enzymes, fatty acids, vitamins and drugs, etc. It is a new developed immunoassay technology after RIA, ELISA, FIA and TRFIA.

In the middle of 1970s, Arakawe firstly reported CLIA which by none has become a mature and advanced testing technology for ultra trace active material and has been widely applied. The past decade has witnessed an almost explosive growth of CLIA which is currently one of the most popular and advanced immunoassay methods. It advantages mainly include high sensitivity, specificity, cost-efficient and stable reagent, long period of validity of reagents (6 to 18 months), stable and fast method, wide detection range, simple operation and high automaticity, etc. Highly sensitive chemiluminescent detection technology is recognized by many researchers and is gradually replacing the traditional bio-detection technology [18].

CLIA and RIA both are regarded as most accurate and mature detection methods for tumor marker and a variety of hormones. They all have been applied widely in clinical diagnosis for tumor and hormone detection. The reagents and equipment of both methods are certified by the U. S. FDA and China SFDA. Although RIA is highly sensitive, it is difficult to be popularized in primary health care institutions due to the problems of radioactive protection and isotopes pollution and that reagents are expensive and shelf life is short. CLIA inherits all the advantages of RIA, and, it overcomes the shortcomings of RIA and ELISA. Thus, it is one of best new clinical immunoassay method.

CLIA consists of two parts, the immune reaction system and chemiluminescence analysis system. In chemiluminescence analysis system, chemiluminescent compound is catalyzed by catalyst and oxidized by oxidant, and then an excited state intermediates will form. When this excited state goes back to a stable ground state, it will emit photons at the same time, and the quantum yield of luminescence can be measured using luminescence-emitting signal instruments. In immune reaction system, luminescence-emitting compound, which can generate intermediate in excited state with the action of reactant or enzyme, is directly labelled on antigen (CLIA) or antibodies (immune chemiluminescence analysis). The reaction principle is as follows.

Start illumination reagent $+$ L$\rightarrow$Electronic excited state intermediate$\rightarrow$Ground state$+hv$

$$Ag + Ag\text{-}L + Ab \rightarrow Ag{-}Ab + Ab\text{-}Ag\text{-}L \xrightarrow{\text{Start luminescence reagent}} +hv$$

(or SP-Ab $+$ Ag $\rightarrow$ SP-Ab-Ag) SP-Ab-Ag $+$ Ab-L $\rightarrow$ SP-Ab-Ag-Ab $\xrightarrow{\text{Start luminescence reagent}} hv$

$$Ab\text{-enzyme} + Ag \rightarrow Ag\text{-}Ab\text{-enzyme} \xrightarrow{\text{Start luminescence reagent}} hv$$

Among them, L is the luminescence tracer or luminescence substrate, Ag is antigen and Ab is antibody and Sp is solid phase.

5. 4. 2　Classification of CLIA method

Based on different labelling methods, CLIA can be divided into two different methods[19].

(1) Chemiluminescence labelling immunoassay

In chemiluminescence labelling immunoassay, also known as chemiluminescence immunoassay (CLIA), utilizes chemiluminescent reagent to label antigen or antibody directly. The commonly used chemiluminescent label is acridine ester (AE). As a valid luminescence label, it is exitated directly and emits luminescence through the action of starting luminescence reagent ($NaOH_2 H_2O_2$). The luminescence is strong and can be accomplished in one second (Figure 5-4), thus it is scintillating luminescence. Acridine ester is used as label in immunological analysis and the chemical reaction is simple, fast and needs no catalyst. The detection of antigenic molecules is based on competitive method (Figure 5-5), and for the macromolecules antigen, the detection is based on sandwich method (Figure 5-6). The detections have, low non-specific binding, a low background. The binding to macromolecules will not reduce the amount of light produced, which benefits high sensitivity of the method.

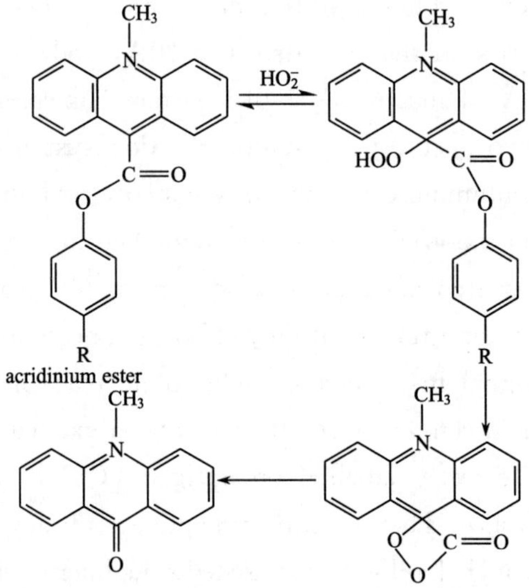

Figure 5-4　Chemiluminescence reaction of acridinium ester

(2) Chemiluminescence enzyme immunoassay

From the aspect of labelled immunoassay, chemiluminescence enzyme immunoassay (CLEIA) belongs to EIA. The only difference with EIA is that the substrate is luminescence agent, so the operating procedure of CLEIA is almost the same with EIA. In CLEIA, enzyme is used to label bioactive substances (such as enzyme-labeled antigen or

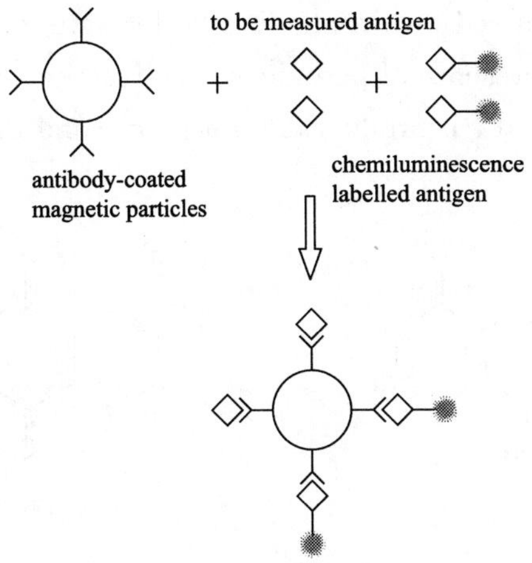

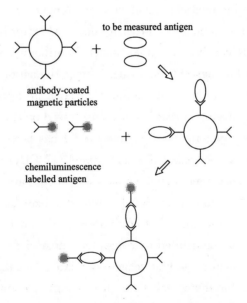

Figure 5-5　CLIA- Competition law

Figure 5-6　CLIA- Sandwich method

antibody), the enzyme on immune reaction complex can act on the luminescence-emitting substrate which will emit luminescence under the action of the signal reagent and the determination is implemented by luminescence signal detector. The commonly used enzyme for labelling include horseradish peroxidase (HRP) and alkaline phosphatase (ALP) which have corresponding luminescence substrates.

　　1) Chemiluminesence detection techniques using horseradish peroxidase (HRP).

　　The luminol (3-amino-phthaloyl hydrazide, luminol) and its derivatives, such as isoluminal (4-amino-phthaloyl hydrazide) are commonly used substrates which are a kind of important luminescence reagents. Their structures are shown in Figure 5-7.

Oxidation of luminal is carried out in alkaline buffer solution, and it produces excited state intermediate in the presence of peroxidase and reactive oxygen. When the intermediate returns to ground state, it produces luminescence and the maximum wavelength is at 425 nm.

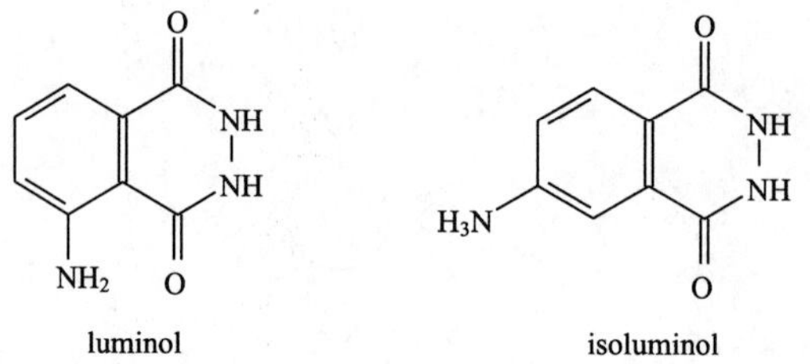

luminol　　　　　　　　　　isoluminol

Figure 5-7　The chemical structure of luminol and isoluminol

At early stages of luminescence analysis, antigen (or antibody) was directly labelled with luminol, but luminescence intensity decreased after being labelled, which affected the sensitivity. Recently, peroxidase-labeled antibody was adopted in CLIA and luminol was utilized as luminescent substrate. Under the action of peroxidase and trigger reagent($NaOH_2H_2O_2$), the chemiluminescent label produces light which intensity depends on the level of enzyme in reaction system. Based on this mechnism, a marketed semi—automatic analysis system (Kodak Amerlite™) has been developed.

　　2) Enhanced luminescence enzyme immunoassay (ELEIA).

　　Enhanced luminescence agents such as para-iodophenol are added into the luminescent system to enhance the luminous signal, and, stabilize the signal for a long time, which facilitates repetitive measurement and improves the analysis sensitivity and accuracy. In automatic analyzer controlled by computer, most of the operations can be automatically implemented, such as adding reagents, mixing, incubation, washing, adding luminescence reagents, luminometer, data processing, plotting the standard curve, completing the analysis of serum samples and printing out the results. Amerlite™ ELEIA system utilizes enhancers such as fluorescein and thiazole, and, the luminescence can last up to 20 mins. Kits with thyroid function test can determinate thyrotropin, 3,5,3′-triiodothyronine, thyroxine, thyroxine combining globulin and free thyroxine. Kits for sex hormones related such as luteinizing hormone, follicle-stimulating hormone, human chorionic gonadotropin, alpha-fetoprotein, estradiol and testosterone are also available. Others such as carcinoembryonic antigen, ferritin and digoxin, etc. , also have been developed.

　　3) ALP-labelled CLEIA.

　　This CLEIA utilizes some special substrates. For example, ring 1, 2-dioxane ethane derivatives, which is a very promising class of luminescent substrates for chemical luminescence enzyme immuoassay. The molecular structure contains groups, adamantane, with stabilizing effect. Its luminophore are aromatic group and the groups which enzyme can act on. Luminescence is caused by the effects of enzymes and starting luminescence reagent. Under the effect of alkaline phosphatase (ALP), phosphate group will be hydrolyzed and a phosphate will break off from it, and then a medium stable intermediate, AMPD (half-life is 2-30 min), is produced. The intermediate is cleaved to an adamantine ketone molecule and a molecular inter-oxygen methyl benzoate anions in the excited state through intra-molecular electron transfer. When it returns to the ground state, it will produce 470 nm light which can be sustained for tens of minutes (Figure 5-8). AMPPD is a direct chemiluminescent substrate of phosphatase and used to detect alkaline phosphatase or enzymes, antibodies, nucleic acid probes and other ligand conjugates. The concentration of alkaline phosphatase can be detected is 10^{-15} mol/L.

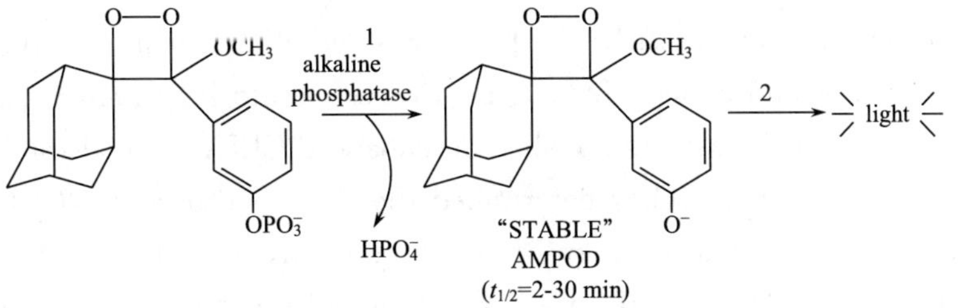

Figure 5-8　Reaction principle of AMPPD luminescence

5. 4. 3　Applications and prospects

CLIA doesn't need external light source and has higher signal to noise ratio than IFA. The limit of its detection can reach 100 molecules and the sensitivity is higher than RIA or EIA 1 up to 2 orders of magnitude. The detection range can be up to six orders of magnitude, and high degree of automation improves the analysis precision. CLIA has become an advanced trace or ultra-trace detection technology.

5.4.3.1　Applications

(1) The application in veterinary science

The application of CLEIA in veterinary science is still in the initial stage. However, it has been authorized by the European Union as one of rapid detection method for bovine spongiform encephalopathy. It was also listed in plan to spread the application of CLEIA in the detection of scapie for small ruminant like sheep.

Buschmann and others applied 4 rapid detection methods (CLEIA, indirect ELISA, sandwich ELISA and Western blot) to detect bovine spongiform encephalopathy in 48 sheep from Germany and 209 sheep and 19 goats from France. The results from CLEIA was in good agreement with the confirmatory test, and as for the other 3 methods, 53 sheep didn't match the confirmatory test. Bolea compared the two rapid methods, CLIA and western blot, with immunohistochemical technique known as gold standard to test sheep scrapie antigen, namely prion protein. The three detection methods all can detect prion antigen protein in the sheep brain, cervical spinal cord and thalamus, but only the immunohistochemical method can detect prion protein in the brain obex. Spanish National Reference Center considered that CLIA can be used as a supplementary screening tool for TSE.

Greening and others used CLEIA and RT-PCR to test RNA viruses. They added the DIG–dUTP (digoxigenin-labeled deoxyuridine) in RT-PCR. The RT-PCR products and biotin-labelled intestinal virus-specific nucleic acid probe was added to streptomyces avidin microplate, and then peroxidase labelled anti-digoxin antibody was added, and finally the virus was detected by ELISA and CLEIA. The results showed that CLEIA had a very high sensitivity and specificity for poliovirus and other intestinal virus. They considered that CLEIA could be used for enteric viruses monitoring in environmental water sample. Vidziunaite and others established an enhanced CLIA method for measurement of brucellosis and tularemia. They determined the best conditions for CLIA inspection system which were: the concentration of coated antibody was 20 $\mu g/mL$, paired antibody concentration was 200 $\mu g/mL$, the linear range of the detection for brucellosis and tularemia were 10-2,500 ng/mL and 1-500 ng/mL, respectively.

Zamora and others applied CLIA to detect salmonella in slaughter pock samples from different laboratories of Germany and compared with indirect ELISA method. More than 1,350 samples were tested. The sensitivity and specificity of the CLIA method were 97.3% and 95.1%, respectively and the sensitivity and specificity of the indirect ELISA were 96.2% and 94.6%, respectively. However, the detection of the lipopolysaccharide antigen of $E.\ coli$ by indirect ELISA had cross reactions with Yersinia pestis, but this cross reaction did not exist in CLIA. Zammora considered that CLIA had a wider detection spectrum, which benefited specific immune reaction. In short, CLIA had a high correlation with indirect ELISA. Thus, Zammora considered CLIA can be used as reference method to examine antibody of salmonella in slaughter pork. Zammora also reported the applications of indirect ELISA and CLIA to detect salmonella in chicken houses in order to establish a monitoring salmonella enteritidis infected serological test system in chicken farms. Studies suggested that indirect ELISA and CLIA could be used to monitor the salmonella enteritidis infection of chicken without being immunized by salmonella enteritidis.

Kovacs and others established a CLEIA test kit for $E.\ coli$ O_{157} : H_7. The culture broth for $E.\ coli$ incubated for 24 h was 10-fold diluted, and the detection sensitivity of the CLEIA was 10^3-10^4 cfu/mL and also very specific. After raw beef culture treated with novobiocin was used to incubate *E. coli* O_{157} : H_7 bacteria at 42 ℃ for 4 h, CLEIA detection sensitivity could reach 10-10^2 cfu / mL and also had a high specificity. Therefore, Kovacs thought that the test kit was suitable for screening *E. coli* O_{157} : H_7 by selective cultivation. Singh detected toxoplasma IgG (Toxoplasma IgG, T-IgG) from serum samples in pigs and cats by ELISA and liquid CLEIA. Liquid CLEIA could sensitively detect the T-IgG of clinical samples and also feed samples from pigs and cats. But ELISA can't sensitively detect cat T-IgG in clinical samples.

Almeida and others developed a CLEIA method for Chagas disease detection. The confirmed sensitivity of the method can was 100% when detecting 100 cases of Chagas disease. Purified flagella body sugar complex and unpurified flagellar antigen were used to establish CLEIA, respectively. 1,288 sera which had not been vaccinated and had no other infection were tested, and the sensitivity of aforementioned two CLEIA method were 100% and 99.7%, respectively. Almeida considered the CLEIA method could be used for the screening of blood bank and the monitoring of human disease and epidemic after the treatment.

(2) The application in medicine

Relative to the veterinary field, CLIA has been widely used in basic and clinical medicine fields as first choice technology to replace RIA. Based on luminescence enhanced EIA system of Amerlite™, Amersham company developed kits for thyroid function testing such as the thyrotropin, 3,5,3'-triiodothyronine thyroxine, thyroxine, thyroxine binding globulin and free thyroxine, and sex hormone-related kits such as luteinizing hormone, follicle stimulating hormone, human chorionic gonadotropin, alpha fetoprotein, estradiol and testosterone, and other aspects such as carcinoembryonic antigen, ferritin, digoxin, etc. Although Amersham company developed more than 10 relative kits, they were not widely used in clinical diagnosis since the operation was complicated and test items were limited to protein macromolecular compounds.

ACS : 180 automated CLIA systems developed by U. S. Ciba Corning could very accurately measure the luminescence. Through continuous improvement, the ACS : 180 CLIA system achieved full automation and led to new products "ADVIA Series". The developed examination items include 47 items such as thyroid system, the gonadal system, blood system, tumor markers, cardiovascular system and blood concentration etc. Meanwhile, some new projects are still under development.

After the modification, the adamantane derivatives may send out high-intensity glow under the effect of alkaline phosphatase and the light signals can last for 1-2 h. Many manufacturers developed kits with alkaline phosphatase as label, and the matching system included IMMULITE automatic CLIA system of DPC and ACCESS automated microparticle CLIA system of Beckman. IMMULITE fully automated CLIA systems can test heart disease, thyroid function, gonadal hormones, infectious diseases, drug, serology, blood diseases, addictive drugs, diabetes, allergy testing and tumor markers. ACCESS automated microparticle CLIA system is mainly used to detect thyroid function, blood system, endocrine hormones, drugs, tumor factors, cardiovascular system and diabetes etc.

The ECL automatic analysis system of Roche company is the only commercialization analysis system at present. The main features include minimal background signal, high specificity and that the minimum detectable value is below 1pmol as well as the operation is very simple and fast. It is a perfect CLIA technique with integrated advantages. The provided kit can test tumor markers, thyroid function, endocrine, infectious diseases, cardiac markers and vitamins etc.

(3) The application in food analysis

Lin and others established a CLEIA system used for detecting chloramphenicol in food. The sensitivity, specificity, precision and accuracy of the system has reached and exceeded the level of the EIA, and the sensitivity reached 0. 05 ng/mL, the variation coefficient of inter-batch and intra-batch were less than 8. 0% and 20%, respectively. The linear regression analysis was from 87% to 100%. The dilution correlation coefficient r was 0. 999 and the detection range was 0. 1-10 ng/mL. Zhang and theirs established a CLEIA to test chloramphenicol in chicken, the detection limit was 6 pg/mL, and the limit of detection for chloramphenicol in chicken increased by 10 times of the ELISA method. When the chloramphenicol in the chicken reached 0. 05-5 μg/kg, the CLEIA determination value via regression was 97%-118%, and the coefficient of variation was 6%-22%. The results of CLEIA test and gas chromatographic microporous board electronic capture test were very high correlated. Zhang thought that this method was very suitable for the screening of chloramphenicol in chicken samples.

Liang and others reported the successful application of ECL for the detection of β-lactam antibiotics and β-lactamase in milk. Pang and others also used ECL for the determination of tetracycline antibiotic residues, when maintaining terpyridine ruthenium $[Ru(bpy)_3^{2+}]$ and tripropylamine (TPA) at the voltage of 1. 05 V in pH 8. 0 carbonate buffer fluid, the detection limits for tetracycline and terramycin were 4. 0 ng/mL and 3. 8 ng/mL, respectively, which showed that the sensitivity of this method was higher than most of the methods reported.

Xu and others employed indirect competition CLEIA to test clenbuterol in pig urine samples and compared with ELISA. Although the ELISA was simple and fast, the limit of detection was only about 0. 1 μg/L, it was difficult to reach the requirement of test, and it usually needed further confirmation by gas chromatography-mass spectrometry (GC-MS) or liquid chromatography-mass spectrometry (LC-MS). And the sensitivity of CLEIA for clenbuterol detection can achieve 0. 01 μg/L, and the linear range is 0. 04-25. 8 μg/L. Therefore, the CLEIA for clenbuterol had a higher sensitivity than the ELISA.

Magliulo and others established a rapid and ultrasensitive CLEIA to detect aflatoxin M_1 in milk. The standard matrix of aflatoxin M_1 was a buffer prepared with milk and did not need any further processing when undergoing determination. The limit of quantification was 1ppt, the precision of inter-batch and intra-batch was very good, the coefficient of variation was less than 9% and the recoveries was in the range of 96% to 122%. This method was of very high specificity, and had no obvious cross-reaction with other aflatoxins. The CLEIA method and a HPLC method were applied to detect 24 milk samples, and the results obtained had good correlation ($y=0.98x + 1.71$, $R^2=0.98$, $n=24$). Magliulo thought that CLEIA was suitable for aflatoxin M_1 screening of massive milk sample examination.

Wu and others developed a new rapid and simple detection method for botulinum toxin in contaminated food and applied this method in the food botulinum toxin inspection. They used western blot combining with chemiluminescence technology and density analysis to analyze the content and type of botulinum toxin in food. The method established could quickly detect the concentration and type of botulinum toxin in contaminated food. The detection sensitivity could reach 1pg/mL and, detect 4 pg/g of type B botulinum toxin in food within 6h. It was found that the three kinds of botulinum toxin antibodies had no cross-reaction with each other in western blot. They completely believed that this method could be used to rapidly detect botulinum toxin in the food with high sensitivity and specificity.

Rivera and others created and ECL analysis method for the detection of clostridium botulinum toxin A, B, E and F in human serum and urine and some foods (whole milk, apple juice, raw beef and raw egg). The sensitivity for different types of clostridium botulinum toxin was different. For type A and type E they were 50 pg/mL, for type B it was 100 pg/mL and for F type it was 400 pg/mL. This method detection had no cross-reaction with other method and greatly shortened the measuring time, and the limit of detection was at the similar level with the gold standard method of mice biological monitoring. Rivera et al. believed that this analysis had a very extensive application in clostridium botulinum toxins screening of clinical medicine, and food.

5.4.3.2 Prospects

At present, veterinary determination methods are still limited to the traditional serological methods which is time-consuming, and, difficult to meet the requirements of rapid diagnosis. CLIA methods have been reported one after the other. Chemiluminescence enzyme immunoassay and electrochemical immunoassay technology featuring high sensitivity will enable the detection of veterinary to achieve a qualitative leap.

The methods of food analysis mainly include capillary gas chromatography and high-performance liquid chromatography and their coupling techniques. The current standards of food determination need expensive and large analysis instruments, and some time-consuming pre-treatments such as organic solvent extraction and purification, and can't implement rapid detection at anytime and anywhere. CLIA in clinical medicine has been matured and even has a trend to replace RIA and ELISA to become a mainstream product in diagnostics market. However, the CLIA detection systems with sensitive and reliable detection are still expensive. It still needs continuous study on improving the sensitivity and specificity of this immunoassay and developing new portable analysis system. CLIA method is of high sensitivity and specificity and widely used for immunoassays against antigen, antibody and hapten and is also of wide linear range. Thus, it meets the need for rapid detection of biomedical and food safety for which it provides an ultra-trace non-radioactive immunological detection methods. With the development of magnetic nanoparticle technology, new immune reaction enhancers and luminescence reagents, and the improvement of flow injection chemiluminescence immunoassay, capillary electrophoresis chemiluminescence immunoassay, high performance liquid chromatography chemiluminescence immunoassay, CLIA will have wide application prospects in various fields such as veterinary science, medicine and food analysis, etc.

References

[1] Causse JE, Ricoredel I, Bacheier MN and others. Radiommunological determination of total thyroxin by antibodies immobilized on polystren tubes coated with styrene-male-ic anhydride copoplymer. Clin Chem, 1990, 36: 525

[2] Osmans S, Turpeinen U, IskonenO and others. Optimization of a time-resolved immunofluorometric assay for tumor-associated trypsin inhibitor (TATI) using the streptavidin-biotin system. Journal of immunological methods, 1993, 161(1): 97-106

[3] Tian GL. Immunology and inspection technology. Shijiazhuang: Hebei Education Press, 1997: 103

[4] Dai BM. Monoclonal antibody and zoonosis. Beijing: Science Press, 1989: 172

[5] Guo JY. Enzyme immunoassay technology of immunological test. Journal of Laboratory Medicine, 2005, 28 (2): 221-224

[6] Sanchczc-carbay M, Espzsa A, Chinchll V and others. New electrochemilumiluminescent immunoassay for the determination of CYRA 21-1: Analytical eraluation and clinical diagnostic performance in urine samples of patients with beadder cancer. Clin Chem, 1999, 45:1944

[7] Iwata, Hayashi T, Nakao Y and others. Direct dertermination of plasma ET-1by chemiluminescence enzyme immunoassay. Clin Chem, 1996, 42:1155

[8] Meng FP. Basis of clinical immunology. Zhengzhou: Zhengzhou University Press, 2004: 247

[9] The national health profession technology qualifications test Expert Committee. Health profession technology qualifications test instruction Clinical medicine examination and technology (intermediate). Beijing: People Health Press, 2006: 519

[10] Li YQ. Modern enzymatic analysis. Beijing: Beijing Medical University and Peking Union Medical University Joint Publishing House, 1994: 157

[11] Pang ZJ. Free radical medical research methods. Beijing: People Health Press, 2000: 97

[12] Chishima S. A radioimmunoassy for TA-0910, a new metabolically stable thyrotrophin- releasing hormone analogue. Journal of pharmaceutical and biomedical analysis, 1994, 12(6):795-804

[13] Zhang NM. Immunosorbent assay detect hepatitis B anti-HBC affected by the time difference. Chinese New

Medicine Forum, 2008, (4): 67-67

[14]   Li XS, Zhang GJ. Estabishment of chemiluminescence enzyme immunoassay determinated triiodothyronine. Chinese Journal of endemic disease prevention, 2008, 23 (3): 179-181

[15]   Zhang M, Wang CL. Determination of FK506 in blood by microparticle enzyme immunoassay. Science Technology and Engineering, 2008, 8 (17): 4976-4978

[16]   Palfreyman JW, Thomas DG, Ratcliffe JG. Radioimmunoassay of human myelin basic protein in tissue extract, cerebrospinal fluid and serum and its clinical application to patients with head injury. Clin Chim Acta. 1978, 82(3): 259-270.

[17]   Li DH, Liu BL. Application of immunofluorescence technique in biomaterial in vitro cytological research. PLA Medical Journal, 1997, 22 (2): 97-98

[18]   Wang X, Lin JM. New progress on chemiluminescence immunoassay. Analysis Laboratory, 2007, (26) 6: 111-122

[19]   Gao R, Zhao YZ, Zhao JZ. Progress on chemiluminescence immunoassay application. Journal of Biological Products of China, 2008, 21 (4): 338-342

Chapter 6   Time-Resolved Fluorescence Immunoassay

6.1   Principles

Time-resolved fluorescence immunoassay (TRFIA), developed for trace analysis in recent years, is the most sensitive analysis technique and the sensitivity is up to 10^{-19} mol/L and three order of magnitude higher than radioimmunoassay (FIA) [1].

TRFIA is a special fluorescence analysis method and in fact developed based on the fluorescence analysis (FIA). FIA utilizes the great differences between the luminescence wavelength and excitation wavelength, and overcomes the impact of heterogeneous light in common UV-Vis analysis. Meanwhile, FIA is different from ordinary spectroscopic analysis. In this method, excitation light is not in the same line with optical receiver and cannot directly reach the receiver, thus greatly improving the sensitivity of the optical analysis. However, during the ultra microanalysis, the impact of stray light from excitation light is more serious. Thus, how to eliminate the impact of stray light from excitation light is the bottleneck to improve the sensitivity of FIA.

The best way to solving the impact of stray light is when there is not excitation light in the measurement. But the lifetime of ordinary fluorescent markers is very short, when the excitation light disappeared, fluorescence also disappeared. However, the fluorescence lifetime of rare earth metals (Eu、Tb、Sm、Dy) is longer up to 1-2 ms and can meet the measurement requirements. Time-resolved fluorescence immunoassay was generated based on long-lifetime fluorescent markers by determining the fluorescence intensity after excitation light is turned off.

The mainly used rare earth metals are europium and terbium. The fluorescent lifetime of europium is 1ms. but europium is unstable in water, which can be overcome by addition of enhancer. The fluorescent lifetime of terbium is 1.6 ms. Although terbium is stable in water, it has some drawbacks such as short fluorescence wave length, serious scattering, that complex of it is easy to be decomposed under high energy. Therefore, it is not suitable for biological analysis.

The reinforcing agents are important components as Eu is commonly used as fluorescent labelling. The principle of the reinforcing agents is to make Eu stable in water with chelate agent and surfactant which are hydrophilic and lipophibic.

With the development of laboratory medicine, more and more trace determination for microgram or nanogram amount are needed. Meanwhile, RIA's pollution problems will be paid more and more attention. Therefore, TRFIA will be applied in more fields.

6.1.1   Homogeneous fluorescence analysis

Theoretically, FIA is the most sensitive detection method. It has been regarded as a new homogeneous analysis method for long before since many intra-molecular and inter-molecular changes could also influence the intensity of emitting fluorescence. Polarization, quench, time-related, and lifetime changes of fluorescence and fluorescence resonance energy transfer (FRET) have been widely applied in the researchs of inter-molecular interaction. However, these applications were severely restricted by a number of technical restrictions including low amplitude of signal, signal interference from environmental components, rapidly varying fluorescence background and light scattering induced by molecules and polymers in analysis media. Nevertheless, FRET is a very interesting among these techniques.

Foster supposed that energy transfer rate is dependent on the distance between donor in excited state and adjacent acceptor molecules. For a known donor-acceptor pair, the value of R_0 (the distance when transmission efficiency is 50%) is 1-7 nm. Using these features, FRET can be used as an optical ruler to measure the relationship between biological macromolecules, which have been confirmed by Stryer's pioneer work and the followed up researches(Figure 6-1).

Only a few techniques, one of which FRET is, are feasible in investigating the distance and interaction between molecules in solution. FRET can be used widely in biology, such as to analysis of enzymatic activity, protein-DNA interactions, interaction of receptor and ligand induced by exogenous lectin of cell surface. DNA hybridization has attracted more attention and become a research focus for it can provide models for the principle of FRET, structure, oligonucleotide hybridization and gene translocation, etc. FRET occurs after antigen-antibody interaction happens, based on which homogeneous fluorescence analysis were carried out widely. Recently, FRET was applied to monitor receptor oligomerization which is the key step of signal transduction pathway. Tsien and others developed a new fluorescent donor/acceptor indicator for measuring surface potential of membrane and observing the process of gene transcription[2].

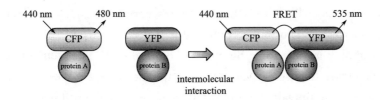

Figure 6-1   Schematic diagram of the FRET principles(See Colored Figures)

CFP's excitation wavelength is 436 nm and its emission wavelength of cyan fluoresce is 476 nm. The excitation and emission wavelengths of YFP are 516 nm and 529 nm, respectively. When the distance between two fluorescins is $\leqslant$10 nm, CFP can transfer the energy to YFP without radiation when stimulated by excitation light at 436 nm, which results in YFP emitting yellow fluorescence.

However, these very attractive characteristics can't hide the problem in FRET technique. For example, the auto fluorescence in system, which requires that the fluorescence intensity measured must be corrected is a problem. Chromophore unbound decreases the sensitivity of the analysis severely and complicates the analysis of results. If donor / acceptor emission spectrum are filtered off, the detection sensitivity for long-life fluoresce emitted by receptor will be decreased. For fixing the problem, CIS bio international company developed HTRF technique which provided the probability for studying the interaction thoroughly between receptor and donor. This technique utilizes the marker of Eu chelate, whose structure is cave-like, and XL665 as receptor-donor pair. The characters of HTRF technique includes special spectrum of emission light and time-resolved monitoring mode which can be selected. When the fluorescent acceptor and donor contact closely, part of the energy of the donor will transfer to the receptor after the donor is stimulated. These features make HTRF technique overcome various constraints of FRET technique.

6.1.2　Time-resolved fluorescence and rare earth element

Many complex and proteins of biological fluid and serum can emit fluorescence of themselves, and, so the sensitivity will seriously decline when using traditional chromophore for fluorescence detection. As most of the background fluorescent signals are of short duration, combining marker of long lifetime and time-resolved fluorescence can minimize the instantaneous fluorescence interference.

Specific photophysical and spectral properties of rare earth complexes make them a focus of fluorescence applications in biology. The electron cloud structure of trivalent rare earth ions protects some electron in active energy levels and forms its own unique emission spectrum. According to quantum mechanical law, the electron transfer is restricted to some extent which leads to the fluorescence of these elements decaying very slowly (from microsecond to millisecond). For the same reasons, light collection efficiency of its complex is very low. So chelates of rare earth elements are designed as a light absorbing device to adsorb the ordered part. The energy collected makes rare earth elements transfer from univalent state to trivalent state and continuously from trivalent state to the emission levels, which lead to the generation of long lifetime fluorescence[3].

As fluorescent probes for biological detection, the rare earth complexes should have some properties such as stability, optional complex sites, high fluorescence quantum yield as well as conjugation ability with biological molecule. Moreover, low susceptibility to fluorescence quenching is prerequisites for study in biofluid.

In fact, not all the known rare earth chelates possess these characters. And even these molecules can be used in the homogeneous analysis, there are many constraints including stability, competitive compounds with chelating ability and sensitivity, which are issues to be addressed.

6. 1. 3   Adjustment of fluorescence resonance energy transfer(FRET)

The homogeneous technology developed by CIS Bio International company is based on the fluorescence resonance energy transfer between Eu chelate donor and acceptor (the second fluorescent marker). Acceptor marker is phycobiliprotein of 105 kDa molecular weight modified chemically with a chromophore (XL665) group in the middle of it. Acceptor marker can match Eu^{3+} chelate and also shows the unique optical physical properties : ① can absorb chelate's emission efficiently; ② interference of chelates to XL665 is negligible in the range of XL665 emission spectrum; ③ high quantum yield (about 70% of fluorescence will not be quenched in the biological fluid). During the fluorescence resonance energy transfer process, fluorescence lifetime of the acceptor is equivalent to the donor since TBP-Eu (donor) of long decay period induces XL665 acceptor to emit fluorescence for a long time. Thus, the FRET signal can be easily distinguished from the short-lived auto-scattering fluorescence background in the end[4].

6.2 Applications

6. 2. 1   Application in immunology

(1) Determination of cytokines

Cytokines not only involves in immune response but also plays an important role in the occurrence and development of tumor and inflammation. Therefore, accurate determination of cytokine levels is very important for the understanding their pathophysiological function. In 1992, TRFIA was used to determine tumor necrosis factor (TNF) and interleukin 6 (IL-6). Gradually TRFIA replaced RIA and EIA in terms of cytokine determination.

(2) Determination of complement

Components of complement play a role in various autoimmune diseases. Frequent detection of complement C3 in clinical expevience was used to evaluate the function of immune system. TRFIA kit for complement C3 detection appeared in 1993 and can determine complement C3 in cerebrospinal fluid.

6. 2. 2   Applications in microbiology

After the theory of TRFIA formation in 1979, researchers focused on development and use of the relative reagents. The major development is microorganism diagnostic reagents in early period. TRFIA for rubella virus antibody was developed in 1982. A TRFIA method for rapid diagnosis of influenza was created in 1986. A TRFIA for simultaneous detection of toxoplasma IgM and IgA antibodies was announced by Sorensen on Journal of Clinical Chemistry in 2002.

6.2.3   Applications in molecular biology

Europium-labelled streptavidin-biotin probe can make the whole nucleic acid hybridization quick and simple. At present, the probes labelled with rare earth elements have been applied successfully to detect HIV, the mRNA of prostate specific antigen (PSA), the adenovirus' DNA, the DNA of streptococcus pneumoniae and the allelic gene of HLA-27 cell.

6.2.4   Applications in hormone detection

Storch detected insulin in serum, Stenman detected the hCG by TRFIA and Wu determined thyroid hormones in serum. Besides, hormones of animals also can be detected by TRFIA. Ogine detected relaxin of porcine. Parra reported the detection of cortisol and FT_4 in dog serum by TRFIA. Kohek analyzed the differences of concentration between hormones (LH, FSH, PRL, T_4, GH, Ins, FT_4, TSH and TT_3, etc.), and found that there is a difference between EDTA and sodium citrate anticoagulated serums when analyzing thyroid stimulating hormone (TSH), insulin (Ins), TT_3, estradiol (E_2), Testo (testosterone) and Prog (progesterone). Harma improved the TRFIA method for determination of PSA. Nowadays, TRFIA have replaced RIA and EIA gradually in terms of determination thyroid hormones, gonadal hormones, hypothalamic hormones, insulin and gastrin etc.

6.2.5   Other applications

Qin reported a TRFIA method for measuring urinary albumin and it can be done within only 12 min. Kurogochi and others reported the mechanism and application of fluorescence-labeled β-galactosidase. It is reported that quantitative detection of perforin protein in the cell lysosome was completed by TRFIA[5].

6.3   Multi-labelled time-resolved fluorescence immunoassay

6.3.1   Principle

Since fluorescence decay time and wavelength of lanthanide ions are significantly different, fluorescence of lanthanide ions can be measured simultaneously, which makes multi-labelled fluorescence immunoassay possible. Multi-labeled TRFIA is a unified reaction system, in which bind reaction of two or more analytes and their corresponding specific ligands will take place, and after introducing two or more lanthanide ion markers corresponding to analytes, the amount of the analysts can be obtained through time-resolved fluorescence measurement [6, 7].

6.3.1.1 General immunoassay formets

Solid-phase double-epitope sandwich pattern similar to IRMA, which is adopted by most of multi-labeled fluorescence immunoassay, holds the advantages that eventual simple separation of free and combined components and high specificity because it is based on double-epitope recognition. However, small molecules with single epitope can't adopt this pattern and should utilize solid-phase competitive immunoassay. As shown in Figure 6-2, second antibody is usually adsorbed onto solid-phase in this case.

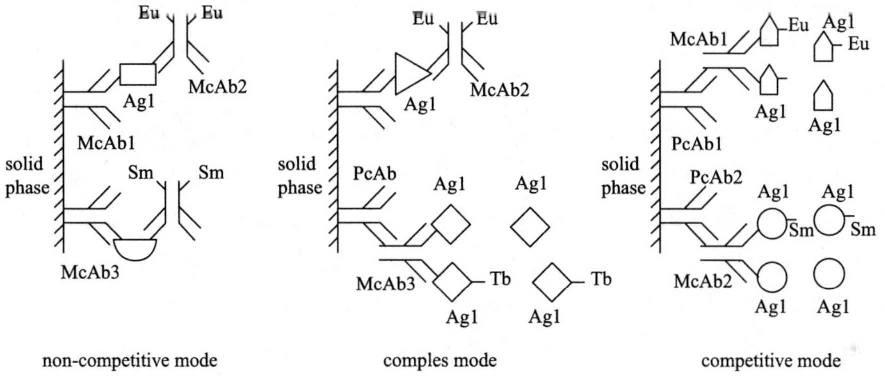

Figure 6-2　Modes of multi-labeled TRFIA method

Non-competitive mode is carried out as followed. Firstly, several monoclonal antibodies with different specificity are coated on the wells of plastic microplate by physical adsorption or chemical combination, then standards or samples to be tested are added and bound to the specific antibodies on the solid phase. After washing, the other specific monoclonal or polyclonal antibodies labelled by lanthanide ion are added which combine with analytes bound to the solid phase, which lead to a sandwich structure. Then enhancement solution is added to dissolve the lanthanide ions and the fluorescence of each well is measured. As for competitive mode, lanthanide ions are labelled not on the specific antibody, but on the antigen. After antibody is coated to the solid phase, standards or samples as well as lanthanide ions labels are also added in the well to react with the antibody simultaneously. Therefore, the reaction mechanism is basically similar to competitive solid-phase RIA.

6.3.1.2　Characters of common lanthanide ions and their compatibility

Current lanthanide ions in use, including Eu^{3+}, Sm^{3+}, Tb^{3+} and Dy^{3+} all have a characteristic narrow emission spectrum and are significantly different from each other in terms of decay time and wavelength. Eu^{3+} is the most frequently used lanthanide ions in TRFIA. Its fluorescent chelate is of high photon yield and its measurement is not interfered by the background. Sm^{3+} can replace of Eu^{3+}. Sm^{3+} and Eu^{3+} is a common pair of lanthanide ions used in dual-labelled immunoassay. Fluorescent chelate of Sm and β-diketone (β-NTA) can be measured on the effect of enhancement solution suitable for Eu^{3+} fluorescence measurement. Emission wavelength of Sm^{3+} is 643 nm which can differentiate from background interference. But Sm^{3+} has low photon yield and its decay time is about 50 μs which is shorter than Eu^{3+} and Tb^{3+}. Tb^{3+} and Eu^{3+} is another pair of lanthanide ions used in dual-labelled immunoassay. Chelate of Tb^{3+}, whose emission wave length is 543-545 nm, also has a high yield and long fluorescence decay time, but is sensitive to the interference derived from plastic and needs short excitation wavelength.

Dy^{3+}, whose emission wave length is 570 nm, is of shortest decay time in these lanthanide ions. Its photon yield was higher than Sm^{3+}, but the background interference on the measurement is greater. As shown in Table 6-1 and Table 6-2, Eu^{3+} can be used to analyze as little as 1-10 amol antigen (10^{-18}-10^{-17} mol), and Dy^{3+} can only implement detection of 1- 10 fmol antigen molecules (10^{-35}-10^{-34} mol). Furthermore, the interference between the other lanthanide ions is almost negligible, except that Sm^{3+} can interfere Dy^{3+} to fairly high extent. The interference from Sm^{3+} to Dy^{3+} in measurement can be corrected by mathematical model.

Table 6-1　Characteristics of simultaneous measurement of four lanthanide ions

Excitation wavelength 315 nm	strongest emission peak/nm	Decay time/μs	Fluorescent number/s		Lowest detection value/(pmol/L)
			nmol ion/L	background	
Eu^{3+}	612	820	3,640,000	760	0. 035
Tb^{3+}	544	323	1,880,000	4,700	0. 34
Sm^{3+}	647	88	11,800	410	7. 9
Dy^{3+}	574	27	19,500	6,200	46

Table 6-2　Interference between four lanthanide ions in simultaneous measurement

Interfering ion	Measuring ion			
	Eu^{3+}	Tb^{3+}	Sm^{3+}	Dy^{3+}
Eu^{3+}	—	0. 0017	0. 0032	0. 0033
Tb^{3+}	0. 0366	—	0. 0049	0. 0169
Sm^{3+}	0. 0176	0. 0000	—	0. 0593
Dy^{3+}	0. 0000	0. 0000	0. 0031	—

6.3.1.3 Choice of enhancement solution in multi-labeled TRFIA

In TRFIA, fluorescence intensity emitted from lanthanide ions in the complex is weak under the effect of the excitation light when the binding reaction is completed. Thus, enhancement solution is required for improving fluorescence efficiency. Direct enhancement solution can be used in single marker immunofluorescence analysis and its composition include β-NTA (β-b ketone), TOPO (three-octyl phosphine oxide), acetic acid and Triton X-100. At pH 2-3, lanthanide ions chelate with amidocyanogen polycarboxylic acid dissociates and then forms a new β-NTA chelate whose structure is as follows: $Ln^{3+}(\beta\text{-NTA})_3(TOPO)_2$.

β-NTA can effectively absorb and transfer excitation energy to the Ln^{3+}. But water molecules surrounding Ln^{3+} can also absorb energy transferred from β-NTA and decrease the fluorescence intensity of Ln^{3+}. Tirton X-100 can help liposoluble β-NTA form vesicles and separate Ln^{3+} from water after it is added to the enhancement solution. The TOPO can be joined to form collaborative coordination around Ln^{3+} and reduce the number of water molecules in the coordination layer, which improves the fluorescence intensity of Ln^{3+} by several times and the sensitivity of measurement greatly.

In the dual-labelled immunofluorescence analysis, attention should be paid in a situation when direct enhancement solution is used. When Tb^{3+} and Eu^{3+} as dual-label Ln^{3+}, β-NTA should be replaced with PTA because Tb^{3+} can't bind to β-NTA in enhancement solution which is suitable for Eu^{3+}, Sm^{3+} measurements. Such a mixed enhancement solution obtained is suitable for Eu^{3+}, Tb^{3+} as well as Sm^{3+}, Tb^{3+} measurement.

The enhancement solution in single or dual-labelled fluorescence analysis is mainly based on β-NTA or PTA as enhanced ligand, which cannot be used in simultaneous measurement based on more than two fluorescent labels. Although the PTA-based direct enhancement solution can be applied to measure simultaneously all the four lanthanide elements in theory, the values of lowest detection obtained for the Sm^{3+} and Dy^{3+} have been greatly restricted. Recently, Xu and others developed several common fluorescence enhancement solutions (CFES), which are suitable for simultaneous measurements of multiple labels selected from lanthanide ions such as Eu^{3+}, Sm^{3+}, Tb^{3+} and Dy^{3+} without decreasing the sensitivities. The following is the PTA-Y (yttrium) based CFES, which is composed of two parts: ① Dissolved part containing PTA, yttrium, Triton X-100 and ethanol and adjusted to pH 3.5 by acetic acid. ② Enhanced part containing phenathroline and Tris. The enhancement solution should be stored at 4 ℃ and darkness. In the CFES, intermolecular energy transfer was used to increase the excitation efficiency and the fluorescence intensity and the decay time of the four Ln^{3+}. Especially for Sm^{3+} and Dy^{3+}, they reach practical sensitivity. In immunofluorescence analysis, an important effect on detection sensitivity is the fluorescent background which depends on the wavelength and the decay time of the element. CFES can extend the decay time at all fluorescence ions. In particular, the decay time of Dy^{3+} is extended from 0. 1 μs to 27 μs, which greatly reduces the background of measurement and realizes high sensitivity.

6.3.2　Application

6.3.2.1　Dual-labelled time-resolved fluorescence immunoassay

To simultaneously analyze two components in a system. Leinonen and others utilized a dual-labelled TRFIA of Eu^{3+} and Sm^{3+} to measure PSA and PSA-ACT. Firstly they coated a PSA monoclonal antibody to solid phase, and then labelled another anti-PSA monoclonal antibody with Eu^{3+} to measure the total PSA levels, and in addition labelled an anti-ACT polyclonal antibody with Sm^{3+} to measure the amount of PSA-ACT. In a result, the lowest detection value of total PSA and PSA-ACT is 0. 03 $\mu g/L$ and 0. 16 μg / L, respectively. Originally PSA has an important value only in monitoring prostatic cancer after prostatectomy and is of low value in early diagnosis, because PSA is also high when the prostate is enlarged. Now it is found that monitoring the ratio of PSA-ACT/PSA can improve the value of early diagnosis. When PSA-ACT/PSA is higher than 0. 66, is indicative of cancer, which is particularly significant for the diagnosis limitations of some early prostate cancers[8].

To analyze two substances with common structure. Hemmila and others analyzed LH and FSH with Eu^{3+} and Tb^{3+} dual-labelled TRFIA method. Firstly, an anti-LH and a FSH α polypeptide-specific monoclonal antibody was coated on the solid phase, and then monoclonal anti-FSH β-chain labelled with Tb^{3+} and the anti-LH β-chain monoclonal antibody labelled with Eu^{3+}, were added and kept at room temperature for one hour, and at last direct enhance solution based on PTA was added to the reaction system and measured by time-resolved fluorescence instrument. The sensitivity for LH and FSH of the method was 0. 1 Iu/L and 1 Iu/L. This method has excellent correlation with a single LH and FSH TRFIA method.

To determinate two substances with completely different structures, Pettersson and others simultaneously measured the amount of AFP and free β-HCG in maternal blood with a Eu^{3+} and Sm^{3+} dual-labelled fluorescence analysis in order to screen fetal Down's syndrome and neural tube defect during prenatal period, and achieved very good results. An anti-AFP monoclonal antibody and an anti-β-HCG monoclonal antibody were coated on solid phase at the same time at first, and allowed the serum to be tested react with the specific antibodies on the solid phase for one hour, and then another anti-AFP monoclonal antibody labelled with Eu^{3+} and another monoclonal antibody anti-β-HCG labelled with Sm^{3+} were added and kept at room temperature for 30 min. At last, β-NTA-based direct enhancement solution was added and the fluorescence intensities were measured at 613nm and 643nm respectively. The method led to fairly high sensitivities which were 0. 02 kIu/ L for AFP and 0. 2 Iu/L for free β-HCG.

6.3.2.2 Four-labelled fluorescence immunoassay

The method is established for simultaneous measurement of four different substances needed to be analyzed regularly. Xu and others developed simultaneous fluorescence immunoassay method for TSH, immune-active trypsin (IRT), creatine kinase-MM (CK-MM) and 17-α hydroxyprogesterone (17-α -OHP) by using Eu^{3+}, Sm^{3+}, Dy^{3+} and Tb^{3+} four lanthanide ions as labels. The whole reaction system consisted of three non-competitive solid-phase immune reactions and a competitive solid-phase immune reaction. After the reactions, the combined parts were dissolved by a common fluorescence enhancement solution and the fluorescence counts were measured at the various characteristic wavelengths. The method achieved a very satisfied sensitivity. This method can obtain four indicators of the neonatal congenital hypothyroidism (TSH), cystic fibrosis (IRT), adrenal hyperplasia (17-α-OHP) and Duchenne-Becker muscular dystrophy (CK-MM), by a small amount of blood samples from the newborn. It could realize relative early screening of the aforementioned diseases and fully demonstrated the advantages of multi-labelled analysis including its being time-saving, effort-saving and reagents-saving.

6.4   Enzyme-amplified time-resolved fluoroimmunoassay

6. 4. 1   Principles

The core idea of enzyme-amplified time-resolved fluorescence assay (EATRFA) is to measure fluorescence in liquid phase without enhancment solution through combining the advantages of biotin-streptavidin of high affinity and amplification, enzyme amplification, lanthanide chelates and time-resolved measurement technique [9].

The basic principle is as follows. A McAb on solid phase reacts with an antigen and a biotinylated McAb at the same time to form a immune response complex. And then an ALP-SA complex, which is a streptoavidin (SA) labelled with alkaline phosphatase (ALP), is added and a complex (solid McAb-antigen-McAb-biotin-SA-ALP) is produced. Then an ALP substrate, 5-fluorosalicyl phosphonate ester, is added which is transferred into 5-fluorosalicylic acid. At last, Tb^{3+}-ethylenediaminetetraacetic acid (EDTA) chelate is added at high pH and a ternary complex FSA-Tb^{3+}-EDTA with long fluorescence lifetime and fluorescence yield will be generated. Determination of the fluorescence intensity at 546 nm wavelength can obtain the concentration of the antigen. In this method, environmental changes, quenching bodies and lanthanide pollution have very small impact on the fluorescence emission characteristics. Moreover, the fluorescence can be measured in liquid phase without separation.

A second enzyme-amplified time-resolved fluoroimmunoassay system was also proposed for nucleic acid hybridization analysis in recent years. The principle is as follows. Firstly, an antibody is coated on micro-titration well and then reacts with target DNA labelled with specific antigen, which connects the target DNA with the antibody on the solid phase[10]. Then biotinylated probe hybridizes with target DNA and combines streptavidin-labelled horseradish peroxidase (HRP) to the hybrid through the reaction of streptavidin and biotin-avidin. When a substrate solution with H_2O_2 and biotinylated tyramine (BT) of HRP is added, the oxidation of tyramine is catalyzed in the presence of H_2O_2 by HRP and free radical groups are generated, whilst the oxidation of tyramine acyl of the protein (such as albumin) pre-fixed to the micro-titration well surface is also catalyzed. Due to the formation of dimer, B-T integrate covalently onto the solid phase. Then ALP-SA is added and ALP is bound to the solid phase through the reaction of SA and B. Consequently, it realizes the hybrid and the quantitative relationship of the hybrid. After washing, the ALP substrate, 5-FSAP, is added. The further steps are the same as EATRF analysis system. The main purpose of secondary analysis of enzyme-amplified time-resolved fluorescence system is to increase magnification and thereby greatly improve analytical sensitivity. Report about this system declared that it can increase 30 times of specific signal, 10 times of signal/noise ratio and the sensitivity up to 3×10^{-16} mol.

6.4.2 Application

Ultrasensitive enzyme-amplified time-resolved immunofluorescence analysis combines enzyme-amplified immunoassay with TRFIA based on lanthanide ion chelate and have vasied applications.

Chen Panzao and others established an enzyme-amplified time-resolved immunofluorescence analysis (TRIFMA) for β-HCG. The method utilized alkaline phosphatase to enlarge the reaction and introduced the affinity reaction of biotin and streptoavidin (SA), immune reaction as well as Tb^{3+} chelate TRF analysis. The β-HCG enzyme-amplified TRIFMA is a sensitive, rapid and automated method.

Firstly, β-HCG-McAb was coated to activated well of microplate, then the -McAb reacted with the standard or β-HCG to be tested in samples as well as biotinylated anti-β-HCG McAb, and subsequently an affinity reaction was perfor med with streptoavidina-lkaline phosphatase (SA-ALP). The substrate, 5 fluoride phosphate (5-FSAP), was hydrolyzed by the ALP bound to the well and 5-fluoro salicylic acid (5-FSA) was generated. 5-FSAcan absorb the energy of exciting light, pass to a certain amount of Tb (Ⅲ)-

EDTA in the solution and bring a very strong and specific fluorescence which is recorded by a time-resolved fluorescence instrument. The amount of β-HCG in samples is proportional to the relative fluorescence intensity (signal to noise ratio), from which a standard curve can be produced and the amount of β-HCG can be obtained.

Zhao Qiren adopted 5-fluoro acid phosphonate ester as the enzyme substrate in an enzyme-amplified lanthanide luminescence method to determine alkaline phosphatase and optimized various methodological factors. The results showed that the stability of the signal / noise ratio could be improved by methyl silicone (I). The method sensitivity for alkaline phosphatase was 4U/L, the precision was 10% at the concentration range from 2. 00 U/L to 300 U/L, the relative standard deviation of measurement was less than 10% and the recovery was 93% to 95% when measuring alkaline phosphatase in serum[11].

Zhou Weiling[1] developed an enzyme-amplified lanthanide luminescence time-resolved fluorescence analysis based on nucleic acid hybridization for determination of pBR322 NDA. Standard curve obtained had wide dynamic range within three magnitude levels. The sensitivity was 6 pg and the accuracy is ranged from 90% to 115%. When pBR322 DNA content was about 0. 05-50 ng, the detection precision was 10%. Various factors affecting the stability and sensitivity, such as the exposure time under UV light, the concentration of biotinylated probe, alkaline phosphatase labelled streptavidin, 5-fluoro acid phosphonate ester and Tb^{+3}-EDAT as well as washing methods, were investigated and optimized. The phenomenon that fluorescence declined with extended measurement time was observed in the analysis system, a theoretical explanation was given and this problem was fixed[12]. The analysis method, holding the advantages of high sensitivity, simple system, rapid detection and stable label, etc. , demonstrated the potential in clinical diagnosis and scientific research and has a bright future.

References

[1]　Zhou WL, Zhao QR. The development and application of time-resolved fluorescence assay. International Journal of Radiation Medical and Nuclear Medical, 2006, 30(2): 103-106

[2]　Almeida IC, Covas DT, Soussumi LM and others. A highly sensitive and specific chemiluminescent enzyme-linked immunosorbent assay for diagnosis of active Trypanosoma cruzi infection. Transfusion, 1997, 37(8): 850-857

[3]　Jiang ZC. Analytical Chemistry of Rare Earth Elements. Beijing: Science Press, 2000: 247

[4]　Bronstein I, McGrath P. Chemiluminescence lights up. Nature, 1989, 338(6216): 599-600

[5]　Wu XC, He L, Zhou KY. The Research and Clinical Application of Time-resolved Fluoroimmunoassay. Medical Recapitulate, 2006, 12(7): 434-436

[6]　Qin QP. Multi-Labeled Time-resolved fluorescence immunoassay and application. Labelled Immunoassays and Clinical Medicine, 1994, 1(4): 233-236

[7]　He GP. Double labelled time-resolved fluorescence immunoassay. Foreign Medical Sciences: Clinical Biochemistry and Laboratory Medicine, 1994, 15(2): 69-70,79

［8］　Buschmann A, Biacabe AG, Ziegler U and others. Typical scrapie cases in Germany and France are identified by discrepant reaction patterns in BSE rapid tests. J Virol Methods, 2004, 117(1): 27-36

［9］　Song NL, Zhao QR. Advances and applications of signal-amplified Time-resolved fluorescence technology. Journal of Chinese Medicine, 2006, 6(9): 997-1000

［10］　Bolea R, Monleon E, Schiller I and others. Comparison of immunohistochemistry and two rapid tests for detection of abnormal prion protein in different brain regions of sheep with typical scrapie. J Vet Diagn Invest, 2005, 17(5): 467-469

［11］　Zhao QR, Li MJ, Liu J. Enzyme-amplified lanthanide luminescence for determinating alkaline phosphatase. Biochemistry and Biophysics, 1998, 25(1): 71-74

［12］　Zhao QR, Li MJ, Liu J. Enzyme-amplified time-resolved fluorescence analysis based on nucleic acid hybridization. Acta Academiae Medicinae Sinicae, 2002, 24(2): 84-88

Chapter 7　Molecular Immunology and Immunogenetics

7.1　DNA isolation and purification

7.1.1　Introduction

In molecular biology, a vector is a DNA molecule used as a vehicle to transfer foreign genetic material into another cell. Plasmid is one type of vectors widely used in recombinant DNA technology[1].

Plasmids are extra-chromosomal circular DNA molecules that can stably function and replicate independently of chromosomal DNA. They are double stranded and, in many cases, circular, and their size varies from 1 to over 1,000 kbp. Plasmids usually occur naturally in the domain Archaea, Bacteria, and Eukarya. Duplication and transcription of plasmids depend on enzymes and protein encoded by host cell. Plasmids cannot survive without host cell. Plasmids may constitute a substantial amount of the total genetic content of an organism. They usually contain genetic information determining resistance to antimicrobial agents or for degradation of additional substrates, etc[2]. The common natural plasmids include Col plasmid, conferring the ability of producing bacteriocin on their host, F plasmid, containing the F or fertility system required for conjugation, and R plasmid, carrying genes encoding antibiotic resistance.

Based upon the number of copies per cell, plasmids are classified into two types: stringent control and relaxed control. Stringent plasmids, which is under the control of bacterial genome for replication and segregation, exist in small numbers, i. e. <100 copies/cell. Generally, conjunctive plasmids are mostly stringent plasmids. Relaxed plasmids, which are not under the control of bacterial genome for replication and segregation, exist in large numbers, i. e. >100 copies/cell. Generally, relaxed plasmids are of low molecular weight and most of them are of non conjugative type.

Plasmids containing the same origin of replication are generally considered incompatible. That is, they cannot stably co-exist in a cell together. In this concept of incompatibility, it is believed that competition for replication factors leads to competition between plasmids, with those having growth advantages, such as faster replication (due to smaller size, for example), or less toxicity, rapidly outgrowing other plasmids in the cell. This is particularly important at low copy numbers.

Different phenotype of plasmids usually contain genes encoding certain enzymes, including resistance to antibiotics, producing antibiotics, degradation of complex organic compounds, producing enterotoxin, certain restriction enzymes and modification enzymes, etc.

Plasmid vector is constructed based on the natural plasmid. Compared with the natural plasmid, plasmid vector often contains one or more selective marker genes (such as antibiotic resistance genes) and a synthetic sequence containing multiple restriction enzyme recognition sites and most of the non-essential sequence is removed to reduce molecular weight and facilitate genetic engineering. Most plasmid vectors have auxiliary sequence with multi-purpose including identification of recombinant clones by histochemical method, generation of single-stranded DNA for sequencing, transcription of exogenous DNA *in vitro*, identification of the direction of inserted sequences, large scale expression of foreign genes. An ideal cloning vector should have the following features: ① small molecular mass, multi-copy, and controlled relaxation type; ② with a variety of endonuelease sites for common single restriction enzyme; ③ can be inserted with large foreign DNA; ④ with easy-to-detect phenotype. The size of plasmid vector is general 1 to 10 kb, such as PBR322, PUC series, PGEM series and pBluescript (pBS) and so on.

Isolation of plasmid DNA from bacteria includes three basic steps: harvests, lyses of bacterial cell and separation and purification of plasmid DNA. Cell walls and membranes could be degraded by lysozyme and surfactants, such as sodium dodecyl sulphate (SDS) and Triton X-100, respectively. Chromosomal DNA, which is much greater than plasmid, attaches to the cell debris and is segmented due to mechanical forces and DNase during these treatments. Chromosome would denature, after heat, acid and alkali treatment, and is difficult to refolding when the external conditions return to normal. The two chains of covalently closed circular DNA (cDNA) will not be separated from each other, and is entangled with denatured proteins and cell debris. Double-strand of plasmid DNA, however, would renature and dissolve in liquid.

cDNA shows supercoiled form in bacterial cells and generate other forms of plasmid DNA during extraction. When severed at one or more site, one chain would rotate to eliminate the tension, and form lax ring molecules, called open-loop DNA. Linear DNA would be formed if two chains are cut open in one place. Supercoiled form swims faster than the open-loop and linear ones during electrophoresis.

7.1.2　Several common methods for enrichment and purification of DNA

7.1.2.1　Bacteria culture and collection

DH5α, a plasmid containing bacteria, could be cultured on LB agar (containing 50 μg/mL Amp), for 12-24 h. Pick a single colony with a sterile toothpick and inoculate into 5 mL LB medium (containing 50 μg/mL Amp), and then incubate with shaking under 37 ℃ for about 12 h to late logarithmic growth phase [3].

7.1.2.2 Rapid extraction of a small amount of plasmid DNA

Method for extraction of small amount of DNA is useful for partia preparation of plasmid DNA from a large number of transformants. Common feature of these methods is simple, fast and high-throughput. The purity of obtained DNA meets the need for restriction enzyme digestion and electrophoresis analysis.

(1) Isolation of plasmids from *E. coli* by boiling lysis

1) Transfer 1.5 mL of cultures to an eppendorf tube. Centrifuge for 1 min (12,000× g, 4 ℃) and remove supernatant.

2) Add 700 μL TE (10 mmol/L Tris-Cl, 1 mmol/L EDTA, pH 8.0). Add 25 μL lysozyme stock solutions. Vortex and place tubes on ice for 5-10 min.

3) Boil for 1 min.

4) Centrifuge for 10 min.

5) Pull pellet out by a toothpick.

6) Add 700 μL isopropanol (RT). Mix tubes by inversion.

7) Centrifug for 10 min. A small white/clear pellet will form.

8) Remove supernatant. Wash with 70% ethanol. Dry pellet and resuspend in 50 μL water containing 20 μg/mL RNase A.

(2) DNA Plasmid Miniprep using Alkaline Lysis

1) Centrifuge 1.5 mL cells in 1.5 mL Eppendorf tube at top speed for 1 min. Aspirate supernatant. Pour off the supernatant and blot the inverted tube on a paper towel to remove excess media.

2) Resuspend cell pellet in 100 μL of GTE buffer (50 mmol/L Glucose, 25 mmol/L Tris-Cl, 10 mmol/L EDTA, pH 8). Vortex gently if necessary.

3) Add 200 μL of NaOH/SDS lysis solution (0.2 mol/L NaOH, 1% SDS). Invert tube 6-8 times.

4) Immediately add 150 μL of 5 mol/L potassium acetate solution (pH 4.8). This solution neutralizes NaOH in the previous lysis step while precipitating the genomic DNA and SDS in an insoluble white, rubbery precipitate. Centrifuge at top speed for 1 min.

5) Transfer supernatant to a new tube, being careful not to pick up any white flakes. Precipitate the nucleic acids with 0.5 mL of isopropanol on ice for 10 min and centrifuge at top speed for 1 min.

6) Aspirate off all the isopropanol supernatant. Dissolve the pellet in 0.4 mL of TE buffer (10 mmol/L Tris-Cl, 1 mmol/L EDTA, pH 7.5). Add 10 μL of RNAse A solution (20 mg/mL stock stored at −20 ℃), vortex and incubate at 37 ℃ for 20 to 30 min to digest remaining RNA.

7) Extract proteins from the plasmid DNA using PCIA (phenol/chloroform/

isoamyl alcohol) by adding about 0. 3 mL. Vortex vigorously for 30 sec. Centrifuge at full speed for 5 min (RT). Note organic PCIA Layer will be at the bottom of the tube.

　　8) Remove upper aqueous layer containing the plasmid DNA carefully avoiding the white precipitated protein layer above the PCIA layer, transferring to a clean 1. 5 mL epindorf tube.

　　9) Add 100 mL of 7. 5 mol/L ammonium acetate solution and 1 mL of absolute ethanol to precipitate the plasmid DNA on ice for 10 min. Centrifuge at full speed for 5 min (RT).

　　10) Aspirate off ethanol solution and resuspend or dissolve DNA pellet in 50 μL of TE. Dissolve 5 μL in 995 μL of water, and spec (blank spectrophotometer to water). The absorbance at 260 nm multiplied by ten is the concentration of the DNA in units of mg/mL for a 1 cm pathlength cuvette (i. e. 50 mg/mL/OD$_{260nm}$)[4].

　　(3) The Wizard Minipreps DNA Purification System

　　The entire miniprep procedure can be completed within 30 min or less, depending on the samples number. Purified plasmids can be used for molecular biology techniques, such as automated fluorescent DNA sequencing, without further manipulation.

Production of a cleared lysate

　　1) Harvest 1-5 mL (high-copy-number plasmid) or 10 mL (low-copy-number plasmid) of bacterial culture by centrifugation for 5 min at 10,000 $\times$ g in a tabletop centrifuge. Pour off the supernatant and blot the inverted tube on a paper towel to remove excess media.

　　2) Add 250 μL solution and completely resuspend the cell pellet by vortexing or pipetting. It is essential to thoroughly resuspend the cells. If they are not already in a microcentrifuge tube, transfer the resuspended cells to a sterile 1. 5 mL microcentrifuge tube(s).

　　Note: To prevent shearing of chromosomal DNA, do not vortex after step 2) Mix only by inverting the tubes.

　　3) Add 250 μL of cell lysis solution and mix by inverting the tube 4 times (do not vortex). Incubate until the cell suspension clears (approximately 1-5 min).

　　Note: It is important to observe partial clearing of the lysate before proceeding to addition of the Alkaline Protease Solution (step 4)); however, do not incubate for longer than 5 min.

　　4) Add 10 μL of Alkaline Protease Solution and mix by inverting the tube 4 times. Incubate for 5 min at room temperature. Alkaline protease inactivates endonucleases and other proteins released during the lysis of the bacterial cells that can adversely affect the quality of the isolated DNA. Do not exceed 5 min of incubation with Alkaline Protease Solution at step 4), as nicking of the plasmid DNA may occur.

5) Add 350 μL of Neutralization Solution and immediately mix by inverting the tube 4 times (do not vortex).

6) Centrifuge the bacterial lysate at maximum speed (around 14,000 $\times$ g) in a microcentrifuge for 10 min at room temperature.

Centrifugation protocol

1) Transfer the cleared lysate (approximately 850 μL) to the prepared Spin column by decanting. Avoid disturbing or transferring any of the white precipitate with the supernatant.

Note: If the white precipitate is accidentally transferred to the Spin Column, pour the Spin Column contents back into a sterile 1.5 mL microcentrifuge tube and centrifuge for another 5-10 min at maximum speed. Transfer the resulting supernatant into the same Spin Column that was used initially for this sample. The Spin Column can be reused but only for this sample.

2) Centrifuge the supernatant at maximum speed in a microcentrifuge for 1 min at room temperature. Remove the solumn from the tube and discard the flowthrough from the collection tube. Reinsert the spin column into the collection tube.

3) Add 750 μL of column wash solution previously diluted with 95% ethanol, to the spin column.

4) Centrifuge at maximum speed in a microcentrifuge for 1 min at room temperature. Remove the spin column from the tube and discard the flowthrough. Reinsert the spin column into the collection tube. Repeat the wash procedure using 250 μL of column wash solution.

5) Centrifuge at maximum speed in a microcentrifuge for 2 min at room temperature.

6) Transfer the spin column to a new, sterile 1.5 mL microcentrifuge tube, being careful not to transfer any of the column wash solution with the spin column. If the spin column has column wash solution associated with it, centrifuge again for 1 min at maximum speed. Transfer the spin column to a new, sterile 1.5 mL microcentrifuge tube.

7) Elute the plasmid DNA by adding 100 μL of Nuclease-Free Water to the spin column. Centrifuge at maximum speed for 1 min at room temperature in a microcentrifuge.

8) After eluting the DNA, remove the assembly from the 1.5 mL microcentrifuge tube and discard the spin column.

9) DNA is stable in water without addition of a buffer if stored at -20 ℃ or below. DNA is stable at 4 ℃ in TE buffer. To store the DNA in TE buffer, add 11 μL of 10 $\times$ TE buffer to the 100 μL of eluted DNA. Do not add TE buffer if the DNA is to be used for automated fluorescent sequencing. Cap the microcentrifuge tube and store the

purified plasmid DNA at -20 ℃ or below.

7.1.2.3 Large scale extraction and purification of plasmid DNA

Alkaline lysis large scale plasmid preparation

1) Incubate 250 mL of the bacteria at 37 ℃ with shaking (12-16 h at 225 rpm) until the culture reaches log phase ($OD_{600} = 0.4$).

2) Centrifuge at 5,000 rpm for 10 min. Pour off supernatant and resuspend pellet in 20 mL of 50 mmol/L Tris, 5 mmol/L EDTA solution.

3) Centrifuge at 5,000 rpm for 10 min and freeze pellet at -70 ℃ for 15-30 min or at -20 ℃ for 2-3 days.

4) Resuspend the pellets in 5 mL cold filter sterilized solution I (50 mmol/L sucrose, 10 mmol/L EDTA, 25 mmol/L Tris, pH 8.0). Allow tubes to sit at room temperature for 5 min. Add 10 mL of fresh solution II (0.2 mol/L NaOH, 1% SDS). Invert tubes to an upside down position and vortex gently. Place tubes on ice for 15 min.

5) Add 7.5 mL of cold autoclaved solution III (11.5 mL of glacial acetic acid, 28.5 mL of ddH$_2$O, 60.0 mL of 5 mol/L potassium acetate per 100 mL) per tube. Invert tubes vigorously until the solution becomes cloudy. Place tubes on ice for 10 min. Centrifuge at 12,000 rpm for 40 min.

6) Transfer supernatants to two 50 mL centrifuge tubes, discard pellets. Add 0.6 volumes of isopropanol (in comparison to the amount of supernatant) and 1 mL of 5 mol/L NaCl. Allow DNA to precipitate at room temperature for 15-30 min.

7) Centrifuge at 10-12,000 rpm for 15 min. Remove supernatant and dry pellet. Add 5 mL TE (pH 8.0) to each tube, resuspend pellet, and combine into one tube.

8) Add 10 μL of 10 mg/mL RNase A under the fume hood with disposable tips. Use 1 μL of 10 mg/mL stock for every mL of solution in the tubes. Incubate at 37 ℃ for 1 h.

9) Add an equal volume of phenol, vortex to form an emulsion, and leave on ice for 5 min. Centrifuge at 8,000 rpm for 15 min to separate the phases.

10) Remove supernatant and extract it with an equal volume of phenol:chloroform (1 : 1). Centrifuge at 6,000 rpm for 15 min. Repeat extraction and separation with chloroform alone.

11) Remove supernatant and add it to an equal volume of isopropanol and 0.1 volume of NaCl. Place at -20 ℃ for 1-2 days or at -70 ℃ for 2-3 h to precipitate DNA. Centrifuge at 12,000 rpm for 45 min. Discard supernatant and wash pellet with 70% ethanol. Remove as much ethanol as possible.

12) Cover the tubes with parafilm. Poke holes in the film so that the pellets can be vacuum dried for about 10 min. Resuspend the pellets in 1 mL of TE (pH 8.0). Aliquot 5 μL and save to run on a 0.8% gel to check for plasmid DNA. Bring volume up to

7. 6 mL with TE (RT).

13) Add 8. 2 g of cesium chloride per tube. Make certain that it goes into solution. Heat tubes at 55 ℃ if necessary. Cover tubes with aluminum foil and add 0. 42 mL of 10 mg/mL EtBr. Put solution into an ultracentrifuge tube with pasture pipettes.

14) Add immersion oil to the top of the solution in the ultracentrifuge tube. Use a syringe (best) or a pastens pipette and avoid getting oil into the neck of the tubes. Balance the tubes (with caps if using that type) to within 0. 01 g. Note tube positions in rotor as labels may wear off while spinning. Ultracentrifuge at 45,000 rpm for 46-48 h at running temperature of 15 ℃, maximum temp of 30 ℃. Check tubes with a hand held UV light for band of plasmid DNA (top band is genomic DNA, towards bottom of the tube is RNA and junk).

15) Puncture tubes at the bottom center with a heated needle, or from the side with a heated needle or syringe to isolate the band. Remove EtBr with a butanol/ water saturated solution. Dilute DNA first with 2 volumes of TE (pH 7. 6). Extract with butanol until solution becomes clear.

16) Place DNA into dialysis tubing and dialyse in 2 L of dialysis buffer. Cover buffer with aluminum foil and allow DNA in tubing to Centrifuge at 4 ℃ for two hrs, then change buffer solution and allow DNA to spin overnight. Collect DNA from tubing and determine concentration with a spectrophotometer. Run gels to determine purity.

Wizard® Plus Megapreps DNA Purification System

Large-scale plasmid preparations, such as cesium chloride purification, can be both laborious and time-consuming, often requiring an overnight centrifugation. The Wizard® Plus Megapreps DNA Purification System is simple and rapid and requires only a centrifuge, a vacuum source and a vacuum manifold.

1) Pellet cells at 1,500 × g for 20 min at room temperature.

2) Suspend pellet in 30 mL cell resuspension solution.

3) Add 30 mL Cell lysis solution. Invert to mix.

4) Add 30 mL neutralization solution. Invert to mix.

5) Centrifuge lysate at 14,000 × g for 15 min.

6) Filter supernatant into a graduated cylinder. Measure volume, then transfer liquid to centrifuge bottle.

7) Add 0. 5 volume isopropanol. Invert to mix.

8) Centrifuge at 14,000 × g for 15 min.

9) Resuspend pellet in 4 mL TE buffer.

10) Resuspend resin. Add 20 mL resin to DNA from step 9). Swirl to mix.

11) Attach Megacolumn to vacuum manifold. Transfer resin/DNA mixture to Megacolumn. Apply vacuum and release when all liquid has passed through Megaco-

lumn.

12) Add 25 mL column wash solution containing ethanol. Apply vacuum, pulling liquid through column. Release vacuum and repeat.

13) Add 5 mL of 80% ethanol. Apply vacuum; continue for 1 min after liquid has passed through column.

14) Transfer Megacolumn to 50 mL centrifuge tube. Centrifuge at $1,300 \times$ g for 5 min, using a swinging bucket rotor.

15) Place Megacolumn on manifold. Apply vacuum and continue for 5 min.

16) Place Megacolumn in reservoir tube provided. Add 3.0 mL preheated water. Wait 1 min, then centrifuge at $1,300 \times$ g for 5 min to elute DNA. For plasmids $\geqslant 10$ kb, use water preheated to 70 ℃; for plasmids $\geqslant 20$ kb, use water preheated to 80 ℃.

17) Filter eluted DNA. Centrifuge filtrate at $14,000 \times$ g for 1 min.

18) Transfer supernatant containing DNA to a new centrifuge tube and store at -20℃ or below.

DNA Purification on Sephrose 2B chromatography column

1) Balance the Sepharose 2B using 0.1% SDS in TE (pH 8.0).

2) Add no more than 1 mL of DNA solution on column.

3) Connect reservoir, containing 0.1% SDS in TE (pH 8.0), on top of column after all DNA solution entered.

4) Collect every 1 mL of effluent.

5) Determine of OD_{260} to confirm DNA containing tube.

6) Combine all positive tubes, extract using same volume of phenol/chloroform (1 ∶ 1), and centrifuge at $1,2000 \times$ g for 2 min at 4 ℃. Transfer water phase into a new tube.

7) Add 2 volume cold absolute ethyl alcohol, Place at -20 ℃ for 1-2 days or at -70℃ for 2-3 h to precipitate DNA.

8) centrifuge at 12,000 rpm for 45 min. Discard supernatant and wash pellet with 70% ethanol. Remove as much ethanol as possible.

9)Cover the tubes with parafilm. Poke holes in the film so that the pellets can be vacuum dried for about 10 min. Resuspend the pellets in 1 mL of TE (pH 8.0). Cap the microcentrifuge tube and store the purified plasmid DNA at -20℃ or below.

7.2　Nucleic acid probe technology

7.2.1　Overview

Principle of nucleic acid probe technology is the base pairing, namely, nucleic acid hybridization. The two complementary single-stranded nucleic acid duplex is formed by annealing. Nucleic acid probe is DNA sequence for specific detection of gene. Each pathogen has a unique DNA fragment that could bind with specific probe for diagnosis of the disease [5].

7.2.1.1   Type of nucleic acid probe

(1) Classification by source and nature

There are 4 types of nucleic acid probes including genomic DNA probes, cDNA probes, RNA probes and synthetic oligonucleotide probes. Genomic DNA probes and cDNA probes are widely used as diagnostic reagents. The former is the most widely used one. The specific DNA sequence could be prepared by enzyme digestion or polymerase chain reaction (PCR) and cloned into plasmid or phage vector. Large number of high-purity DNA probe could be obtained with proliferation of plasmid in bacteria and phage then. cDNA is the product of RNA reverse transcription. cDNA probe was used for the detection of RNA viruses. A large number of cDNA probes can be obtained by cloning into the plasmid or phage. Message RNA (mRNA) can also be used as nucleic acid hybridization probe. It is seldom used, however, due to inconvenient natuse of the source and easy degradation of the RNA by nuclease in environment. Synthetic oligonucleotide fragments are widely used as nucleic acid probe. The probes could be arbitrarily synthesized according to the corresponding sequences need, even for only dozens bp probe sequence. It was particularly applicable for the detection of point mutations and small pieces of base deletion or insertion.

(2) Classification by markers

These types of nucleic acid probes include radioactive labelled and non-radioactive labelled probes. Radioisotope, such as ^{32}P, ^3H, ^{35}S, is the first and most widely used marker currently And ^{32}P is the most commonly used one. Radiolabeled probe is very sensitive and can detect at pg levels. However, it is easy to cause radioactive contamination and isotopes are expensive, unstable and have short half-life. Therefore, the radioactive-labelled probe can not be commercialized. Many laboratories are currently committed to develop non-radioactive labelled probes. Haptens, such as biotin and digoxin, are presently used non-radioactive markers. Biotin is a small molecule water-soluble vitamin, is known for a unique affinity for avidin, can form a stable complex with avidin connected with colour material (such as enzymes, fluorescent etc.) and used for detection. Digoxigenin is a steroid hapten molecules can bind with its antibody for detection based on similar principle. Digoxigenin-labelled nucleic acid probe is increasingly widely applied because its sensitivity is equal to radioisotopes and it is more specific than biotin-labeled probe [6].

7.2.1.2   Markers of probes

Ideal markers should be highly sensitive, specific, cheap, and safe, should not affect the main physical and chemical properties of probed molecules, and should have long shelf life. The detection method should be simple.

(1) Radioactive marker-radioisotope

Radioisotope is a highly sensitive tracer for hybridization. Detection based on radio-nuclide-labeled probe is highly sensitive, simple and stable. It could directly photosensi-tize x-ray sensitive latex particles on the film. It can be used to detect 1-10 μg of single copy genomic DNA of higher organisms. The commonly used radioisotopes are ^{32}P, ^{3}H, and ^{35}S, etc. ^{32}P has high radioactivity, can release high-energy β particles, and is the most commonly used DNA markers. It is widely used in various membrane hybridization detected by autoradiography. Labelled probe should be best used within 1 week because the high energy β particles can damage the probe structure. ^{3}H has low specific activity with low free energy of β particles. The low background and high quality result could be obtained by extending the exposure time and resolution in autoradiography. ^{3}H labelled probe is only used in situ hybridization, due to the short half-life. The labelled probe can be stored for a long time. Radioactivity of ^{35}S is slightly lower than that of ^{32}P. Its ray scattering is weak. It is especially used in situ hybridization. Its half-life is longer than ^{32}P, and is labelled probe could be stored for 6 weeks under -20 ℃.

(2) Non-radioactive markers

Since radioactive labelled DNA probes have many limitations, research in non-radioa-ctive labelled DNA probes developed rapidly, and replaced the radioactive labells in many field, promoting the wide application of molecular hybridization.

1) Biotin, a small molecule water-soluble vitamin, can link with 5th carbon and do not affect its ability and specificity to form hydrogen bonds in base pairing. Biotinylated nucleotides can be integrated in DNA probe by enzymatic polymerization, and could be displayed and detected with biotin conjugated heterozygotes after hybridization reaction.

2) Photoreactive biotin is a synthesized biological derivatives of biotin. Generally, biotin is connected with an active light photosensitive group, such as aryl azides, through a spacer (Figure 7-1). The labelling of it is simple, fast and reliable, and does not need nucleic acid synthesis enzyme, primer or substrate. When the photosensitive biotin is mixed with nucleic acid solution, irradiated by strong visible light for 10-20 min under certain conditions, aryl azides will be transformed into activated aromatic nitro and directly bind to specific nucleotide sites by covalent bond. Biotin-labelled DNA probe can be extracted with sec-butyl alcohol and by ethanol precipitation.

3) Digoxin, a steroid hapten, can bind 5th carbon of pyrimidine ring on uracil nucle-otide through a cross arm containing 11 carbon atoms. This marked uracil nucleotides can bind DNA like unlabeled one. Hybridization signals could be detected by immunoas-say based on a colour reaction of corresponding hapten antibody.

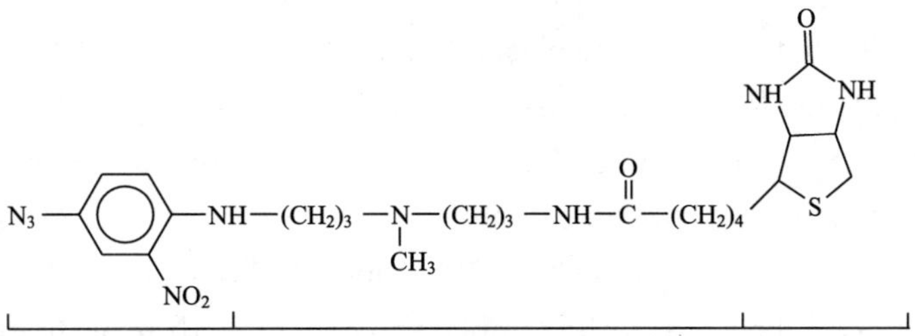

Figure 7-1   Structure of photoreactive biotin

4) Different fluorescents can emit different colours of fluorescence under excitation. Fore example, fluorescein isothiocyanate (FITC), resorufin and hydroxy coumarin show yellow-green, red and blue fluorescence, respectively. Using this feature, different fluorescent-labelled probes could be used to detect multiple locations on the same target nucleotide sequence.

7.2.1.3   Common method of probe labelling

(1) DNA nick transition

Nick translation is a fast, simple, cheap method to produce uniformly labelled DNA with high specific activity. The technology can be used to prepare sequence-specific probe. Although any form of double-stranded DNA can be labelled by nick translation when using the recombinant plasmid probe, the insert fragments are commonly digested by restriction endonuclease and purified by gel electrophoresis. In addition, hybridization with this probe produce a low background signal.

(2) DNA random pairing

This method uses oligonucleotide primers and Klenow fragment *E. coli* polymerase I to label DNA. Except for producing uniform labelling, this method is better than the nick translation method in many ways, such as the incorporation rate could reach more than 50%. Because the added DNA fragment is not degraded during the reaction, the random primer DNA added to the reaction may be less than 200 ng, even 10 ng can be effectively labelled. The size of DNA fragments does not affect the results. Markers could be incorporated along the full length of DNA. Labelled probes can be used directly without removing of unincorporated nucleotid. Both single-stranded and double-stranded DNA can be labelled as a template. Random primers can be used for smaller DNA fragments (100-500 bp), and the nick translation method would be best used for large DNA fragments (> 1,000 bp). However, productivity of random priming is less than nick translation and cyclic DNA must be linearized by DNA restriction enzyme I endonuclease or alkalinet to generate gap.

（3）Transcription of RNA *in vitro*

Commercial RNA transcription plasmid could be used to label RNA probe. These plasmids should include RNA start sites of SP6, T3, or T7 RNA polymerase, and these sites should be adjacent to multiple cloning sites (multiple cloning sites, MCS).

（4）Oligodeoxyribonucleotide of RNA

Cloning vectors, such as pSP64, pGEM-3 and pGEM-4, have SP6 RNA promoter neighbouring MCS can be used as the template to generate RNA probe from DNA oligonucleotide based.

（5）Oligonucleotide "tailing"

Deoxyuridine triphosphate can be added to free 3'-OH of DNA molecule by terminal deoxynucleotidyl transferase forming extended "tail" on 3'end of probe. In this way many genes can be added to produce probes of high specific activity to identify cloning sequences of gene library, to detect point mutations in genomic DNA, and to conduct in situ hybridization.

（6）T4 polynucleotide kinase active on 5'end of oligonucleotide

Synthetic oligodeoxyribonucleotides are usually not phosphorylated, and thus the 5' end contains a hydroxyl group. γ-phosphate of γ-^{32}P-ATP could be transferred to free 5'-OH side by T4 polynucleotide kinase. Because the probe molecule contains only one isotopic molecule, the probe activity is dependent on the length. Short oligonucleotides can be labelled with high specific activity, while decreased with increase of the length of the probe. This method is most commonly used for DNA sequencing.

（7）Polymerase chain reaction

PCR is highly specific, can produce large number of synthetic DNA fragments within 1-2 h with incorporation rate as high as 70% to 80%, which is suitable for large-scale testing and non-radioactive marker. The disadvantage is the need to synthesize a pair of specific PCR primers. Prepared from the probe using a small fragment of DNA primers can also obtain a better labelling effects [7].

7.3　Nucleic acid molecular hybridization

7.3.1　Principles and types

Sequence homology between nucleic acid could be studied by nucleic acid hybridization technology. As long as there is a certain degree of complementarities between sequences, single-stranded nucleic acids from different sources can link with hydrogen bond by base pairing rules. Sequence of a nucleic acid can be detected by other nucleic acid labelled by radioisotope, such as ^{35}S and ^{32}P, or biotin, etc. Hybridization can be generated between DNA and DNA, RNA and RNA, or between DNA and RNA. This technology has been widely used in biology, medical research, clinical diagnosis of infectious and genetic diseases, due to high specificity and sensitivity [8].

In chemical and biological terms, probe is a molecule with suitable detectable marker that can only react with the specific target molecule. Hybridization between antigen and antibody, lectin and carbohydrate, avidin and biotin, receptor and ligand and between the complementary nucleic acid belongs to reactions between probe and target molecule. Protein probes (e. g. antibodies) hybridize with specific target molecules in a few locus through the mixed force such as hydrophobic, ionic and hydrogen bonds. Complementary strand of nucleic acid and probe react with each other basing on hydrogen bonds in the dozens, hundreds or even thousands of sites along the length of molecules. When modified by an atom, functional group or long side chain, probe can still possibly hybridize with other nucleic acid by base pairing, depending on the location and nature of modifiers. About 30 modifiers only label 1 kb of probe bases, and only 4% to 12% of the single base is replaced by modified analogues.

7.3.1.1　Solid-phase hybridization

For solid-phase hybridization, denatured DNA is fixed on the solid substrate (nitrocellulose or nylon membrane), and then is hybridized with the probe. It is also known as blot hybridization.

7.3.1.2　Dot hybridization

Denatured DNA or RNA is fixed on membrane, and then hybridized with excess of labelled DNA or RNA probe. The method is characterized by simple operation, without prior restriction enzyme digestion or gel electrophoresis separation of DNA and detection of multiple samples ever available simultaneously. Positive copies could be calculated by the results of hybridization. However, this method cannot measure molecular weight, and the specificity is poor, a certain percentage of false positives occurs during hybridization.

7.3.1.3　Blotting hybridization

Southern blot: restriction enzyme digested DNA is separated by gel electrophoresis, denatured and transferred in situ to nitrocellulose membranes or other solid phase support material, fixed in hot air drying oven. Single-stranded DNA fragments are then hybridized with labelled probe with corresponding structures, and molecules of a specific size could be detected using autoradiography or enzymatic colour reaction. This technology can be used for cloned gene restriction map analysis, qualitative and quantitative genome analysis, gene mutation analysis and restriction fragment length polymorphism (RELP), and so on.

Northern blot: this technology is used to detect RNA and was developed from Southern blot. RNA is denatured by formaldehyde or polyglyoxal, separated by electrophoresis, and transfer to the solid phase support material to carry out hybridization reactions to identify quantity and size of specific mRNA in the genome. This method is the commonly used gene expression studying method that can calculate the extent of oncogene expression.

7.3.1.4　Differential hybridization

Genomic library DNA from recombinant phage is transferred to nitrocellulose membrane, hybridized with two different cDNA probe mixture (such as cDNA of mRNA from non-metastatic and metastatic cancer), respectively. Differentially expressed genes can be separated from the hybridization information of corresponding location on the membrane. This technology is suitable for comparison of gene expression in less complex genomes of eukaryotes (e. g. yeast) with low false positive rate. It is of little value for eucaryota with nuclear complex genome (such as human), due to the heavy workload and a lower percentage of expressed genes (only about 5%).

7.3.1.5　cDNA microarray hybridization

The cDNA clone or PCR products are arrayed and fixed on solid support material (such as: the activation of the glass slide or nylon membrane) with a high degree of arrangement to form microarray, and hybridized mixture of different DNA probe. Information on microarray is scanned by fluorescence, chemiluminescence and confocal microscopy. The technology has been commercialized and is efficient, high speed and low cost for large-scale analysis. However, it cannot discriminate homologous or repeated sequence.

7.3.1.6　Oligonucleotide microarray hybridization

Oligonucleotide is synthesized in situ on and covalently bound to the surface solid supports, and hybridized with RNA or cDNA probe mixture with the average length of 20-50 nt to improve the specificity and sensitivity of hybridization. Information could be detected by confocal microscopy across three orders of magnitude. This technology could be used for detection of low abundance mRNA, to distinguish gene family expression pattern between different species, or identify the same transcript in different tissues and cells. But it is of a larger workload, high cost, and slow weaknesses.

7.3.1.7　Solution hybridization

The denatured single-stranded nucleic acid known single-stranded nucleic acid (probe) labelled with radioisotope are incubated in solution to form hybridization complexes. Un-hybridized nucleic acid is removed with hydroxyapatite or enzymatic treatment. Nucleic acid quantity could be calculated by detection of probe in hybridization complexes.

7.3.1.8   Subtractive hybridization

Subtractive hybridization is a powerful technique to study tissues, cells and stage specific gene expression. mRNA (or reverse transcription into cDNA) is extracted from two sources of cells of different functions, to react with excessive driving mRNA or cDNA by liquid-phase hybridization under certain conditions. Hydroxyapatite chromatography columns are used to remove homologous hybrids. Same genes between two cells are removed after several rounds of hybridization selection, retaining the specific targeting genes or gene fragments. Full-length cDNA sequences could be obtain after screening cDNA library of the latter.

7.3.1.9   Nucleic acid hybridization in situ

For nucleic acid hybridization in situ analysis, specific DNA probe hybridize with DNA or mRNA on tissue sections to detect DNA, chromosomes distribution and tissue specific gene expression. This technology was also used in non-specific detection of bacteria or virus in cells and tissues.

This method could precisely and specifically localize target DNA independent of the abundance in even one cell of a tissue section, without impacting other components and damaging organization and cell morphology, and extraction of DNA. The technology is widely used in medical biology research, such as gene localization, gene deletion, gene translocation, identifying specific genes integrated location. In recent years, the method has been improve and developed from qualitative to quantitative study.

7.3. 2   Application

Southern blot could be used to study occurrence specific gene, the size and distribution of restriction sites of DNA, with very high specificity, although its operation is more complicated. Northern blot technology is used to study transcription of specific genes in genome. Dot blot can quickly and easily detect the occurrence of specific gene even using a single drop of liquid samples without obtaining the information of gene size and repeat level. Bacterial colony hybridization in situ technology is used to screen positive clones from a number of clones.

Davis prepared complementary DNA probe of IBDV genome by reverse transcription using random primer. Although, the probe did not distinguish between vaccine, wild and mutants virus, dot blot showed certain specificity under very strict conditions.

Vakharia used primers of Australian IBDV sequence and pGEM system, prepared clone with known start site. He transferred RNA of the *IBDV Delaware E*, *IBDV Grayson* and *IBDV STC* on agarose gel to the membrane and detected using autoradiography of radioactively labelled probe with 1 h. But the probe did not distinguish between the different virus strains. Better hybridization results were obtained when digoxin, non-radioactive ligand, was used.

Zhao and others cloned synovial membrane *Mycoplasma* WVU1853 (MS) into plasmid vector pUC18, and transformed into *E. coli* to construct the gene library. In dot blot, four transformed clones could hybridize with radioactive phosphorus labelled MS chromosomal DNA, but not with that of MGS6. In Southern blot, all the cesium chloride purified plasmid clones contained two MS DNA fragments of 1. 0-2. 3 kbp. In dot blot 4 probes prepared from recombinant plasmid hybridized 9 strains of WVU1853 and wild strains of MS, but not with MG and other 15 kinds of avian *mycoplasma*.

Khan and others have prepared DNA probes to distinguish the pathogenic and vaccine MG strain. For the rapid diagnosis of MG infection, they directly applied strain and species specific DNA probes in wild strain sample in dot blot. One mL of culture medium was bound on membrane, and then checked with the probe. 6 kbp of specific probe has the same sequence of MG vaccine and can be used for detection of MGF immunized chickens. Species-specific probe, containing a 9 kbp insert fragment, can detect all of the MG, no cross-hybridization of mycoplasma in the other birds.

Jenkins studied the gene expression, genome composition and strain distribution of genome in *coccid*ia by Southern blot, and developed species-specific probes to study specific developmental stages and other species-specific probes for diagnosis.

Henderson and others used cDNA probes to detect chickens infected with *IBDV* by dot blot and compared with agar diffusion and immunofluorescence test. Virus was detected 2 days after inoculation and in the duration of 3 days and 4 days using the two latter methods. Dot blot detected virus 1 day after inoculation and in the whole experimental period in (24 days) meaning that the dot blot was more sensitive than the other two methods.

Snipes and others applied of REA and Southern blot differed pathogenic from vaccine strains (M9) of *Pasteurella multocida* isolated from turkeyand, although the two strains have same serotypes, capsules type, biochemical characteristics, drug resistance, whole cell protein PAGE and plasmid, providing a useful tool for epidemiological research of fowl cholera. In addition to these applications, the molecules hybridization can also be used for cell in situ hybridization to study the location of some genes in the cell or even a single chromosome when keeping their morphology. It has been widely used in diagnosis biology, developmental biology, cell biology, genetics and pathology studies. Factors that affect the test include the nature of genes and the size of the probe.

Embedded sections or frozen samples should be treated with crosslinking fixatives, such as formaldehyde, which can reduce the permeability of cells. Only smaller probe can penetrate into the cell reducing the sensitivity of this method. Therefore, the choice of fixed agents and probes is the key. Probe can be labelled with fluorescent or radioactive nucleotides. In situ hybridization is a difficult technique to master because many of the experimental conditions are critical.

Tanizaki and others used biotin labelled probes to detect distribution of fowlpox virus (FPV) in formalin fixed skin epithelial cells of chicken by in situ hybridization. The results showed that the FPV DNA was limited in cytoplasm.

Tripathy developed a gel in situ hybridization. Hybridization can be accessed on the agarose gel, without transfering to the solid carrier. This method is simple, timesaving and economical[9].

7.4   Construction and screening of cDNA library

7.4.1   cDNA library and construction

Since the first case of cDNA cloning in the 1970, construction of cDNA library has become the basic method in functional genomics study. Expressing of genome gene is difficult due to occurrence of introns. However, it is possible and easy to express cDNA in clone in large scale. Desired target gene can be screened from cDNA library and directly used for the purpose of gene expression. Rare biological resources could be protected by constructing their cDNA library. cDNA library is also used to provide probes for constructing linkage maps of molecular markers. What more, full-length genes could be obtained by cDNA library to study their function and expressing pattern in different cells. It has broader application in studying the individual development, cell differentiation, cell cycle regulation, regulation of cell ageing and death and other biological phenomena [10].

Since the first successfully construction of cDNA library in 1976 by Hofstetter, this technology developed gradually. The initial cDNA library, which was constructed in plasmid vector, was inefficient and difficult to amplify and preserve due to technical limitations. Using of YOUNG and DAVIS recombinant vector made a key development for the expression and preservation of library. In recent years, many new approaches were developed for cDNA library construction. Their common purpose is to make the cDNA library to meet the researching needs more quick and efficient.

cDNA library is a collection of all cDNA fragments clones connected in vector for reverse transcription of mRNA of an organism in specific development period. Classical cDNA library construction is based on reverse transcription of mRNA using Oligo (dT) or random primers. Appropriate connectors were added to synthesized cDNA, which would be connected to the appropriate carrier to obtain library. There are several basic steps for construction of cDNA library. RNA could be extracted by guanidine isothiocyanate, guanidine hydrochloride-organic solvent, hot phenol, etc. The selection of extraction methods was dependent on difference in samples. High-quality mRNA is essential for construction of a high-quality cDNA library. So be careful when dealing with mRNA samples. RNases exist in all organisms and are resistant to the physical environment such as boiling. Therefore, establishment of RNase free environment is critical for preparation of high quality RNA. First chain of cDNA would be synthesized by reverse transcriptase using high quality mRNA and oligo (dT) as primer. 2nd chain of cDNA could be synthesized by RNase H and E. coli DNA polymerase I, with repairing reaction supported by T4 bacteriophage polynucleotidase and E. coli DNA ligase. Then, double-stranded DNA would be cloned into the vector, after addition of synthetic joints, to analyze the cDNA insertion. cDNA library would be amplified and identified. λ phage was usually chosen because the λ DNA have sticky ends of 12 nucleotides that can be used to construct coemid which can accommodate large fragments of exogenous DNA.

7.4.1.1 Type of cDNA library

So far, based on the characteristics and differences in library, methods for construction of cDNA library include normal cDNA libraries, subtractive cDNA library, normalization cDNA library, full-length cDNA library, normalization and subtraction of full-length cDNA library, such as on[11].

(1) Normal cDNA library

Normal cDNA library firstly appeared in early 1980s. Oligo (dT) (or random primers) is served as reverse primer. cDNA is synthesized with the appropriate connectors and vectors. However, inserted cDNA is too small to meet the needs of large-scale gene study.

(2) Subtractive cDNA library

This method was first reported in early 1980s. It was often used to clone differently expressed genes of an organizations in two physiological or pathological states. This library enriches cDNA clones expressing only in one organization or state and significantly reduces the workload of library screening.

(3) Normalization cDNA library

Refers to a particular tissue or cells, each gene in the library has the same amount of cDNA in normalization cDNA library. This method is mainly used to resolve problem of low frequency of low abundance expressed sequence. The construction method is divided into two categories: first, using genomic DNA for selective hybridization, abundance of the resulting cDNA is similar with genomic DNA. But about 20% of non-single-copy gene in the construction show higher abundance; the second is based on refolding kinetics of cDNA annealing, abundance of all the cDNA is normalized because refolding or annealing of low-copy cDNA is slower than high copy cDNA.

(4) Full-length cDNA library

Normalization and subtractive cDNA library only produce gene segments although eliminating too many high abundance sequences. While studying gene structure and function, both complete coding and non-coding region are needed. Therefore, high-quality full-length cDNA library construction is focused by researchers. It is designed by utilizing the eukaryotic mRNA 5′end of the cap structure.

(5) Normalization and subtraction of full-length cDNA library

This method was first proposed by Carninci and others (2000). They integrated the advantages of full length library, subtractive library and normalization library, and further improved the quality. Firstly, full-length cDNA is obtained using cap-trapper method, followed by normalization/subtractive hybridization. The full-length single-stranded cDNA is served as tester and hybridized with excess biotinylated drivering mRNA to remove high expressed cDNA. Finally, the second chain is synthesized to construct the library.

7.4.1.2　Method for construction of cDNA library

cDNA library construction includes six stages [12].

Stage one: Synthesis of first strand cDNA by reverse transcriptase.

1) Following reagents should be mixed in sterile microcentrifuge tube and placed on ice bath for the synthesis of first strand cDNA:

Poly (A) + RNA (1 $\mu g/\mu L$)　　　　　　　　　　　　　　　　10 μL
Oligonucleotide primers 1 $\mu g/\mu L$)　　　　　　　　　　　　1 μL
1 mol/L Tris-HCl (pH 8.0, 37 ℃)　　　　　　　　　　　　2.5 μL
1 mol/L KCl　　　　　　　　　　　　　　　　　　　　　3.5 μL
250 mmol/L MgCl$_2$　　　　　　　　　　　　　　　　　　2 μL
dNTP solution (containing four kinds of dNTP which all are 5 mmol/L) 10 μL
0.1 mol/L DTT　　　　　　　　　　　　　　　　　　　　2 μL
RNase inhibitor (optional)　　　　　　　　　　　　　　　25 units
Add H$_2$O to　　　　　　　　　　　　　　　　　　　　　48 μL

2) Remove 2.5 μL reaction solution to another 0.5 mL microcentrifuge tube after blended at 0 ℃ and add 0.1 μL [$\alpha^{-32}P$] dCTP(400 Ci/mmol, 10 mCi/mL)to the reaction tube.

3) Large-scale and small-scale reaction tubes are incubated at 37℃ for 1 h.

4) Add 1 μL of 0.25 mol/L EDTA into the small-scale reaction tube containing the isotope and transfer the tube onto ice bath. Large-scale reaction tube is incubated at 70℃ for 10 min, and transferred to ice bath.

5) Determine total radioactive activity in small-scale reactions tube and the precipitate obtained by trichloroacetic acid (TCA). It is worth analyzing the product of small-scale by alkaline agarose gel electrophoresis with appropriate DNA molecular weight marker.

6) Calculate yield of the first-strand cDNA by the following formula:

[Incorporated activity (cpm) / total activity] $\times$ 66 (μg) = first strand cDNA synthesis (μg).

7) Carry out the next step of cDNA synthesis as quickly as possible.

Stage two: *cDNA second strand synthesis*

1) Add the following reagents directly into the large-scale first-strand reaction mixture:

10 mmol/L MgCl$_2$	70 μL
2 mol/L Tris-HCl (pH 7. 4)	5 μL
10 mCi/mL[α-^{32}P] dCTP(400 Ci/mmol)	10 μL
1 mol/L (NH$_4$)$_2$SO$_4$	1. 5 μL
RNase H (1,000 U/mL)	1 μL
E. *coli* DNA polymerase I (10,000 U/mL)	4. 5 μL

Mix the reagents carefully and centrifuge for a short while in a microcentrifuge to remove all air bubbles. Incubate at 16 ℃ for 2-4 h.

2) Add the following reagents to the reaction mixture:

β-NAD (50 mmol/L)	1 μL
E. *coli* DNA ligase (1,000-4,000 U/mL)	1 μL

Incubate at room temperature for 15 min.

3) Add 1 μL mixture containing 4 dNTP and 2 μL T4 phage DNA polymerase and incubate at room temperature for 15 min.

4) Remove 3 μL reactant, and determine the quality of second strand DNA according to the method described in the udermentioned steps 7) and 8).

5) Add 5 μL 0. 5 mol/L EDTA (pH 8. 0) to the remaining reaction mixture and extract the mixture once by phenol/ chloroform and chloroform, respectively. Recover DNA by ethanol precipitation in the presence of 0. 3 mol/L sodium acetate (pH 5. 2) and dissolve the DNA in 90 μL TE (pH 7. 6).

6) Add the following reagents to the DNA solution:

10 $\times$ T4 polynucleotide kinase buffer	10 μL
T4 polynucleotide kinase (3,000 U/mL)	1 μL

Incubate at room temperature 15 min.

7) Determine the radioactivity of 3 μL reaction mixture removed from the step 4) above. According to the methods described in Appendix 8 of "Laboratory Manual of Molecular Cloning" (3rd edition), and determine the radioactivity of 1 μL of TCA per-

ceptible products in second strand synthesis.

8) Calculate cDNA product of second strand synthesis reaction according the following formula, taking into account the incorporation of dNTP into first strand cDNA. The second strand cDNA synthesis often accounts 70% to 80% of first chain.

[Activity incorporation in the second chain reaction (cpm)/total activity (cpm)] × (66 μg-x μg) = synthesis of second cDNA strand/μg.

Where the x means the amount of first strand cDNA.

9) Extract the phosphorylated cDNA (from step 6)) with the same amount of phenol/chloroform soluiton.

10) Equilibrate Sephadex G-50 using solution containing 10 mmol/L NaCl in TE (pH 7. 6) and separate the un-incorporated dNTP from cDNA by the chromatography.

11) Add 0. 1 volume of 3 mol/L sodium acetate (pH 5. 2) and 2 volumes of ethanol to precipitate the cDNA eluted from the column chromatography. Incubate the samples on ice bath for at least 15 min, and then centrifuge at maximum speed under 4 ℃ for 15 min for recovering the precipitated DNA. Ensure whether the entire radioactivity is precipitated by checking the DNA with a portable mini-monitor.

12) Wash the precipitate with 70% ethanol, and repeat the centrifugation again.

13) Carefully aspirate all liquid and dry the ediment by air.

14) If undergoing cDNA methylation by EcoR I methylase, cDNA should be dissolved in 80 μL TE (pH 7. 6). If cDNA need to directly link Sal I and Not I adapter or oligonucleotides jiont, it can be suspended in 29 μL TE (pH 7. 6). Dissolve the re-precipitated cDNA and carry out the next step as soon as possible.

Stage three: cDNA methylation

1) Add the following reagents into cDNA sample:

2 mol/L Tris-HCl (pH 8. 0)	5 μL
5 mol/L NaCl	2 μL
0. 5 mol/L EDTA (pH 8. 0)	2 μL
20 mmol/L S-adenosylmethionine	1 μL
Add H$_2$O to	96 μL

2) Transfer two pieces of sample (each 2 μL) into 0. 5 mL microcentrifuge tubes, number them as 1 and 2, respectively, and place them on ice bath.

3) Add 2 μL EcoR I methylase (80,000 U/mL) to the remaining reaction mixture, kept at 0 ℃ until the end of step 4).

4) Aspirate two more pieces of sample (each 2 μL) from total reaction solution into 0. 5 mL microcentrifuge tube, number as 3 and 4, respectively.

5) Add 100 ng plasmid DNA or 500 ng of λ phage DNA into each of four small samples (from step 2) and step 4)). These unmethylated DNA are used in the pre-experiment to determine the methylation efficiency.

6) Incubate all four small samples and the remained large volume sample at 37 ℃ for 1 h.

7) Heat at 68 ℃ for 15 min, and extract large volume reaction solution once with phenol: chloroform, and extract with chloroform again.

8) Add 0. 1 volume of 3 mol/L sodium acetate (pH 5. 2) and 2 volumes of ethanol into the large volume of reaction solution, mix and store at −20 ℃ until the results of small volume reaction solution can be obtained.

9) Analyze the reaction in 4 small volume samples following the methods bellow.

① add the following regent into control tube:

0. 1 mol / L $MgCl_2$	2 μL
10 × *Eco*R I buffer	2 μL
Add H_2O to	20 μL

② Add 20 units of *Eco*R I into sample 2 and 4.

③ Incubate at 37 ℃ for 1 h and analyze by electrophoresis in 1% agarose gel.

10) Centrifuge at maximum speed for 15 min (4 ℃) to recover the precipitated cDNA. Add 200 μL 70% ethanol to wash the precipitate, and repeat centrifugation again.

11) Check whether all radioactive materials are precipitated using portable mini-detector. Aspirate ethanol out carefully. Dry the precipitate in air, and dissolve DNA in 29 μL TE (pH 8. 0).

12) Carry out the further procudures as quickly as possible.

Stage four: the connection of joint or adaptor

1) Incubate the cDNA at 68 ℃ for 5 min.

2) Cool the sample to 37 ℃ and add the following regents:

Phage DNA polymerase repair buffer	10 μL
dNTP solution containing 5 mmol/L each of four Deoxyribonucleic acids	5 μL
Add H_2O to	50 μL

3) Add 1 to 2 U of T4 phage DNA polymerase (500 U/mL), and incubate at 37 ℃ for 15 min.

4) Add 1 μL of 0. 5 mol/L EDTA (pH 8. 0) to terminate the reaction.

5) Extract the sample by phenol: chloroform, and remove un-incorporated dNTP by Sephadex G-50 column chromatography.

6) Add 0. 1 times volume of 3 mol/L sodium acetate (pH 5. 2) and twice volume of ethanol into the eluate and store the eluate at 4 ℃ for at least 15 min.

7) Centrifuge at maximum speed for 15 min (4 ℃), recover the the cDNA precipitate Air-dry it, and tehn dissolve in 13 μL of 10 mmol/L Tris-HCl (pH 8. 0).

8) Add the following reagents:

10 × T4 phage DNA polymerase repair buffer	2 μL
800-1,000 ng of phosphorylated joint or adaptor	2 μL
T4 phage DNA ligase (105 Weiss units / mL)	1 μL
10 mmol /L ATP	2 μL

Mix and incubate at 16 ℃ for 8-12h.

9) Aspirate 0. 5 μL of reactant out and store at 4 ℃, and incubate the remaining reaction mixture at 68 ℃ for 15 min to inactivate ligase.

Stage five: purification of cDNA by Sepharose CL-4B chromatography

Preparation of Sepharose CL-4B column

1) Push half of the cotton into the tip of 1 mL sterile pipette with hypodermic needles, cut off the part of cotton outside the pipette with a sterile scissors and discard it, and then blow the remaining cotton to the narrow end of the pipette by filtrated compressed air.

2) Connect a sterile PVC pipe to the narrow end of the pipette. Immerse the wide end into TE (pH 7. 6) solution containing 0. 1 mol/L NaCl. Connect PVC pipe to a cone bottle of vacuum device. Aspirate gently until the pipette is filled with the buffer. Close the pipe with the clamp.

3) Connect a vinyl foam tube to the wide end of the pipette. Stand for a few minutes, open clamp, and then a column form when the buffer drops from the pipette. If necessary, add more Sepharose CL-4B, until the pipette is almost filled with the gel.

4) Wash the column with several times volume of TE (pH 7. 6) containing 0. 1 mol/L NaCl and then close the soft tube at the bottom of the column.

5) Aspirate the liquid on the top of column with a Pasteur pipette, add cDNA to the column (50 μL or less), and loose the clamp to lead the cDNA into the gel. Wash the tube with 50 μL TE (pH 7. 6) and add the TE to column. Fill the foam tubes with TE (pH 7. 6) containing 0. 1 mol/L NaCl.

6) Monitor the cDNA through the column with a small portable detector. When the radioactive cDNA flows to 2/3 length of column, begin to collect with microcentrifuge tube, 2 drops per tube, until all the radioactivity are eluted from the column.

7) Measure radioactivity of each tube with the Cerenkov counter.

8) Remove a small portion from each tube, analyze by 1% agarose gel electrophoresis with end-labeled DNA fragment of the known size (0. 2-5 kb) serving as mark, and store the remaining at −20 ℃, until the autohistoradiograph is received.

9) Move the gel to a Whatman filter paper (Grade 3, 6 m) after electrophoresis, cover with a Saran packaging film, and dry on gel dryer. 20 min to 30 min before drying, heat the gel at 50 ℃ and then stop heating and dry in vacuum for 1-2 h.

10) Cover intensifying screens on the dried gel at −70 ℃ to X-ray exposure.

11) Add 0.1 volume of 3 mol/L sodium acetate (pH 5.2) and two volume ethanol to collection tube containing cDNA of bigger than 500 bp. Precipitate the cDNA at 4 ℃ for 15 min, centrifuge at 4 ℃ (12,000 ×g) for 15 min, to recover the precipitated cDNA.

12) Dissolve DNA in 20 μL of 10 mmol/L Tris-HCl (pH 7.6).

13) Measure radioactivity of each small portion. Calculate the total radioactivity of selected elute. Calculate the quantity that will connect to λ phage DNA.

cDNA needed= [total activity (cpm) / activity incorporated to the second chain (cpm)] × 2xμg cDNA second strand synthesis

Stage Six : connection of cDNA to phage λ

1) Reaction mixture:

Link mixture	A/μL	B/μL	C/μL	D/μL
λ phage DNA(0.5 μg/μL)	1.0	1.0	1.0	1.0
10×T4 DNA ligase buffer	1.0	1.0	1.0	1.0
cDNA	0 ng	5 ng	10 ng	50 ng
Phage T4 DNA ligase(105 Weiss U/mL)	0.1	0.1	0.1	0.1
10 mmol/L ATP	1.0	1.0	1.0	1.0
Add H₂O to	10	10	10	10

Incubate the link mixture at 16 ℃ for 4-16 h.

2) According to methods provided by the packaging extract manufacturer, add 5 μL of each reactant into phage particles.

3) Add 0.5 mL SM medium to each reactant mixture after packaging reaction.

4) Prepare overnight culture of *E. coli* and dilute packing mixture, by 100 times, spread 10 μL and 100 μL on plates and incubate at 37 ℃ or 42 ℃ for 8-12 h.

5) Calculate recombinant plaques and non-recombinant plaques. A mixture shouldn't have recombinant plaques, and the B, C and D should produce recombinant plaques, the number of which increases along with B, C and D.

6) Calculate cDNA cloning efficiency according to the number of plaques.

7) Pick 12 plaques of phage λ, and culture lysate in small-scale to prepare DNA,for the appropriate restriction endonuclease digestion.

8) Analyze the size of cDNA insert by 1‰ agarose gel electrophoresis with DNA marker of 500-5,000bp.

7.4.2 cDNA library screening

7.4.2.1 Spread plate of λgt11 cDNA library

(1) Preparation of host bacteria

Inoculate a single colony of *E. coli* into 2 × 5 mL LB medium (Y1088 for plaque hybridization, Y1090 for immunization screening), and shake overnight at 37 ℃. Centrifuge at 3,000 ×g for 5 min, and resuspend both cell pellet in 2 mL λ-dil (10 mmol/L tris-Cl, pH 7.5; 10 mmol/L MgSO$_4$; autoclave). Cell suspension can be kept for a week at 4 ℃.

(2) Spread prefabricated phage suspension

Dilute cDNA library or other phage suspension with titer known (for example, the number of phage particles per mL known) with λ-dil to achieve the required number of phage per plate. Spread 100 μL (5 × 10^3 phage) on 90 mm plate and begin screening

(3) Spread plate

1) Preheat LB plates at 42 ℃.

2) Melt top agar of LB agar (at least 2.5 mL per plate) in the microwave oven, and then set in water bath at 49 ℃.

3) Prepare a 12 mL test tube with a screw cap for each plate, and set on the right shelf at room temperature.

4) Add 100 μL of λ-dil diluted suspension of host cells to each tube by pipette.

5) Add 100 μL of diluted phage suspension to each tube and vortex.

6) Incubate at 37 ℃ for 25 min, and set at room temperature.

7) Heat a 10 mL sterile glass pipette in Bunsen burner flame, and take 2.5 mL top agar melted at 49 ℃ into a tube containing the infection mixture.

8) Scroll the tube immediately, and then evenly pour onto a warmed LB plate.

9) Stay at room temperature until the top agar is solidified (at least 5 min).

10) Repeat steps 7) to 9), until all bacteria and phage were spread. Incubate at 42 ℃ (for λgt11) until the plaques grow.

7.4.2.2   Hybrid screening with cDNA sequences as probe

1) Spread library on plates as described above, and use *E. coli* Y1088 to conduct plaque hybridization.

2) Incubate at 42 ℃ overnight.

3) Remove the plate from the incubator, and cool it at 4 ℃ at least 1 h.

4) Membrane stripping: Firstly, number the plate on dry nitrocellulose membrane with a soft pencil. Carefully spread it from the center to the edge of the plate. When the membrane becomes wet completely (become gray), then wait for further 30 s. Meanwhile, mark 3 asymmetric sites on the membrane using a needle plate. Carefully strip the membrane from the plate, and put the membrane, with contact surface upward, on a Whatman 3 MM filter paper. Before proceeding to the next step, repeat step 4) for all the flat plates.

5) Float the membrane, with the contact surface upward, on denaturing solution (0.5 mol/L NaOH; 1.5 mol/L NaCl) for 5 min.

6) Neutralize for 5 min.

7) Wash with 2 × SSC for 5 min.

8) Put the membrane on a new 3 MM Whatman filter paper, and wrap them with a lead foil.

9) Bake in a vacuum oven at 80 ℃ for 2 h.

10) Wet the film with 6 × SSC briefly, and wash it with pre-wash solution (50 mmol/L Tris-Cl pH 8.0; 1 mol/L NaCl; 1 mmol/L EDTA; 0.1% SDS) at 68 ℃ for at least 1h in a plate on shaking bath.

11) Transfer 25 films to a storage tank (diameter of at least 90 mm) with 50 mL hybridization solution (5×SSPE; 1×Denhardt's solution; 100 μg/mL cT-DNA; 0.1% SDS) and pre-hybridize at 68 ℃ for 2 h.

12) For each mL hybridization solution, denature 2 times 10^5 to 1×10^6 cpm of double-stranded probe on boiling water bath for 10 min, and rapidly put on ice bath for cooling.

13) Immediately add denatured probe to pre-hybridization mixture.

14) Hybridize at 68 ℃ overnight in a sealed storage tank with continuous shaking to prevent membrane sticking together.

15) Wash membrane with 2 × SSC and 0.01% SDS for 3 times at room temperature for about 1 h.

16) Put wet membrane on a piece of cardboard or X-ray film, and mark the needle label with the radioactive ink.

17) Wrap the membrane with a plastic bag (for example, Saran), add intensifying screen and expose on X-ray film overnight at −70 ℃.

7.4.2.3　Immunological screening

(1) Preparation of bacterial lysate

1) Inoculate single colony of a temperature-sensitive *E. coli* strain (BTA 282 (λgt11amp3) or BNN 97) in 2 × 7.5 mL medium and grow overnight at 32 ℃.

2) Dilute the culture with LB medium by 100 times and grow at 32 ℃ to 0.5 OD_{600} value.

3) Incubate at 45 ℃ for 15 min to induce phage expression.

4) Incubate further at 39 ℃ for about 2 h.

5) Determine the phage concentration for the collection: add 3 drops of chloroform into 1 mL culture, mix and incubate at 37 ℃. Take another culture without chloroform as control. After incubation for 5 min, if test samples become clear compared with control culture, the culture can be collected at this time.

6) Harvest bacteria by centrifugation at 4,000×g for 10 min at 4 ℃.

7) Re-suspend bacteria in coupling buffers and centrifuge again.

8) Re-suspend cell pellet in 5 mL of cold coupling buffer.

9) Ultrasonic process the cell suspension, until the viscosity increases to no longer release the nucleic acid.

10) Couple the bacteria lysate to CNBr pre-activated Sepharose 4B at 4 ℃ overnight.

11) Centrifuge at 3,000×g for 5 mins to recover conjugates.

12) Suspend the pellet in coupling buffer and centrifuge again.

13) Suspend the pellet in cold (4 ℃) saturated buffer and incubate on rotating wheel at 4 ℃ for 1 h.

14) Remove unabsorbed protein by consecutive washing three times with washing buffer (pH 4.0), buffer (pH 8.0) and buffer (pH 4.0).

15) Store conjugates in PBS containing 0.01% thimerosal tribute at 4 ℃ (can keep for several weeks).

(2) Pre-adsorption of antibodies for screening

1) Mix 0.5-1 mg/mL of undiluted primary antibody and secondary antibody with 1.7 volume of Sepharose-coupled bacterial lysate mixture, and incubate overnight at 4 ℃ on rotating wheel. Each screening cycle need approximate 300 μL processed primary antibody (1 : 100 dilution) and 15 μL processed seconday antibody (1 : 2,000 dilution).

2) Add suspension to a pipette sucker filled with small amount of glass wool. Elute unblocked antibodies with cold blocking solution (at least 5 column volumes).

3) Add cold blockning solution to eluate to 1/20 of original concentration of. Add thimerosal (final concentration 0.01%).

(3) Spread flat plate

1) Spread the plate as described above with the following modification:

Take another 12 mL test tube with screw cap containing 2.5 mL agar on top in 49 ℃ water bath as induction control.

For one infection mixture, induction control is 100 μL suspension of wild type phage containing β-galactosidase with known percentage of concentration.

In order to spread induction control plate, adding 20 μL X-gal to the top of 2.5 mL agar, before adding the appropriate infection the mixture.

2) Incubate at 42 ℃ for 3-4 h (with *E. coli* Y1090) until plaques appear.

3) Move the plate to laminar flow hood, cover with nitrocellulose membrane soaked by IPTG on the top to induce phage lacZ gene expression number. Dry nitrocellulose membrane in each plate with a soft pencil, and then soak them into IPTG solution within a Petri dish. Drain excess liquid of plate edges, and then carefully place the membrane on plate with growing bacteria from the center to edge of plate. Membrane should be always operated with flat forceps.

4) Incubate upside down at 37 ℃ overnight.

5) Plaque on induction control plate will become blue after overnight culture, due to the complete β-galactosidase expression, indicating successful induction by IPTG.

6) Mark 3 asymmetry positions in the membrane with a needle before stripping it from the plate.

7) Carefully strip membrane from the plate, and put on a 3 MM Whitman filter paper with the contact side upward, until all of the membrane was stripped off.

(4) Immunological staining

1) In order to wash away the residue and cell debris on the top of agar, transfer maximum of five nitrocellulose membranes to dishes (10 cm×10 cm) on the rocking bed, and wash with 25 mL TBS for 10 min and 2 times at room temperature.

2) Incubate membranes with 25 mL blocking buffer (at least 15 mL for each 90 mm diameter membrane) at room temperature for 30 min, to block non-specific binding sites on the membrane. Continuously shake to prevent membranes sticking together.

3) Put each membrane in a Petri plate, and incubate with 3 mL blocking buffer containing primary antibody for 2 h at room temperature on a shaking table. Dilution of the classic work solution of primary polyclonal antibody is 1 ∶ 100. This work solution can be reused to 5 times, and be stored at 4 ℃.

4) Wash the membrane with 25 mL TBT at room temperature for 3 times (10 min/time), with continuous shaking to remove unbound primary antibody.

5) Operation is similar to step 3). Dilution of secondary antibody in 3 mL blocking buffer is 1 ∶ 2,000. The secondary antibody working solution can be used not more than once.

6) Wash the membrane with 25 mL TBT room temperature for 3 times, and 10 min/time, with continuous shaking to remove unbound second antibody.

7) Mix 45 mL of cold (4 ℃) enzyme reaction buffer with 5 mL of cold (−20 ℃) chloronaphthalene and 50 μL H_2O_2 in a plastic tray. Put a maximum of five membranes in this developing solution and slowly stir. Positive plaques will become purple within minutes.

8) Put membranes in exposure box and record the results if using Polaroid photography (665-type film, aperture 4.5, and shutter speed 1/15, orange filter).

9) Mark the hyaline film with a needle to help picking out positive bacteria.

(5) Purification of plaques

1) Pierce a sterile blunt Pasteur pipette into agar at the position corresponding to the positive plaques, and suspend agar obtained in 0.5 mL λ-dil in an eppendorf tube.

2) Vortex the plaques for at least 1 h to release phage, and store under room temperature for at least 1 h before spreading on agar again.

3) Suspend at dilution ranging from 10^{-3} to 10^{-5}.

4) Repeat steps 1) to 3), until picking out a positive plaque from the original phage infected agar.

(6) Phage proliferation

1) In order to obtain high titer of uniform phage suspension, add 150 μL to 250 μL to *E. coli* Y1088 suspension and spread plante plat for proliferation.

2) Culture at 42 ℃ overnight, add 7 mL λ-dil into each lysate of phage clones and continuously shake at room temperature for 5 h to extract phage particles.

3) Carefully aspirate phage supernatant with a sterile 10 mL pipette and avoid cell debris. Add a few drops of chloroform, and the supernatant liquid can be stored for ever as a phage at 4 ℃.

4) Phage titer of original species should range from 10^9 to 10^{10} per milliliter of suspension. If the titer is low, repeat steps 1) to 3), until it reaches far enough phage titer.

5) High dilution (up to several hundred phages) of the phage must be used to spread plate to test the homogeneity of the original species, and screened with a probe again. Test phage should be identified as "uniform" if at least 99% of them are positive phages.

7.5   HLA gene matching and typing technology

7.5.1   HLA gene matching

Human MHC, also known as HLA complex, is located on the short arm of the 6th chromosome. It is 4 centimogens long and contains approximate 4,000 kb. There are nearly 60 loci on the whole complex, and 278 allelic genes have been officially named. According to the characteristics of the different coding elements, these genes were categorized by class Ⅰ, class Ⅱ and class Ⅲ genes (Figure 7-2)[13].

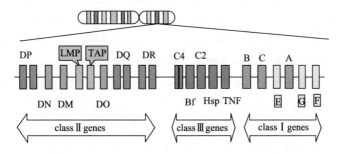

Figure 7-2   Structure of HLA gene(See Colored Figures)

7.5.1.1   Class Ⅰ genes of HLA

Class I genes of HLA are located in the distal end of complex from HCP5p15 of telomere end to MICB and 1. 9 Mb long, including 122 gene loci. It includes 41 functional genes, 12 candidate gene, 3 non-coding genes and 66 pseudogenes. According to distribution function and polymorphism of encoded product, different genes can be divided into classical and non-classical class I genes [14].

1) Classical HLA I genes, also known as HLA-Ia, is the firstly found genes including *HLA-A*, *HLA-B* and *HLA-C*. HLA-Ia genes are highly polymorphic and contain the most allele in whole HLA area. Until July 8, 2006, 805 HLA- Ia genes have been officially named. Each allele may encode the heavy chain (α chain) of HLA I molecule.

2) Non-classical HLA I genes, also known as HLA-Ib, include *HLA-E*, *HLA-F* and *HLA-G*. *HLA-E* is located between *HLA-C* and *HLA-A*, has 5 officially named alleles. *HLA-F* is located at outside of *HLA-C*. *HLA-G* is located at distalis of *HLA-A*, and has 23 officially named alleles.

3) MIC genes, found in 1994, is a family of gene which currently has five members named *MICA*, *MICB*, *MICC*, *MICD* and *MICE*, respectively. *MICA* and *MICB* are functional genes, the others are pseudogene. *MICA* gene is located along the centromere direction of *HLA-B* gene 40 kb away. *MICA* is highly polymorphic, has 61 alleles and is of linkage disequilibrium with *HLA-B*.

HLA I class has largest number of pseudogenes whose functions are unknown, accounting for 54.1% (N=66), such as *HLA-L*, *HLA-K*, *HLA-X*, etc.

7.5.1.2　HLA class Ⅱ genes

HLA class Ⅱ genes are located at centromeric side of HLA complex, containing 0. 9 Mb from DRA to HLA-DPA3. It contains 34 loci, 16 of which are functional genes, 3 of which are candidate gene, and 15 of which are pseudogenes. Except BRD2 (formerly RING3), other than the functional genes all show immunological function. Class Ⅱ region contains at least six sub- region including DR, DQ, DP, DOA, DOB and DM.

HLA-DR sub-region has one DRA gene and 9 DRB genes. DRA gene is non-polymorphic, and its product is the chain of DR molecules. Nine DRB genes were named *DRB1-DRB9* in which *DRB1*, *DRB3*, *DRB4* and *DRB5* are functional genes(Figure 7-3).

DRB1 is most abundant in polymorphism among class Ⅱ region, and contains 527 alleles. DR molecules are composed by β chain encoded by *DRB1* and α chain encoded by *DQA* gene. According to the antigen specificity detected by serological method, these DR molecules are named DRl-DR18. The β chains encoded by *DRB5* gene, *DRB3* and *DRB4* gene compose DR molecules with α chains respectively, and these DR molecules are named DR51, DR52 and DR53 according to their antigen specificity. Therefore, number of DRB genes vary with the difference of individual haplotypes and can be divided into five groups.

DQ sub-region located between the *DRB1* gene and *DOB* genes, including 2 *DQA* genes and 3 *DQB* genes. *DQA1* and *DQB1* are functional gene, encoding α and β chain of *DQ* molecules, respectively. *DQA2*, *DQB2* and *DQB3* all are pseudogenes. Differing from *DR* gene, *DQA1* and *DQB1* genes are highly polymorphic.

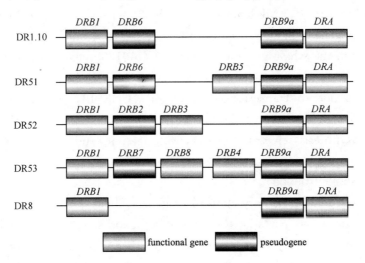

Figure 7-3　Combinations of genes that form DR in different haplotypes

DP sub-region is located near the centromere direction inside DQA, including 2 pairs of *DPA* and *DPB* genes. *DPA1* and *DPB1* are functional genes, encoding α and β chain of DP, respectively, and *DPA2* and *DPB2* are pseudogenes.

DR, DQ and DP genes belong to classical HLA Ⅱ genes, and the encoded molecules are similar in distribution and function and all highly polymorphic.

DM sub-region located between the HLA-DQA and PSMB8 and is compos by two locus, DMA and DMB, which encode α and p chain of DM molecules, respectively. DM gene is polymorphic. DM heterodimer molecule is present in a specific organelle called MHC Ⅱ class compartment (M Ⅱ C), and plays an important and unique role in presentation and processing of exogenous antigen.

TAP and PSMB region: there are a pair of genes, *TAP1 / TAF2*, locating between DMB and DQB2, associating with antigen processing, and aother pair of proteasome sub-units, beta type, (PSMB) gene, *PSMB9* and *PSMB8* (formerly called low molecular weight peptide gene *LMP7* and *LMP2*). TAP is expressed in the endoplasmic reticulum, and responsible for peptide transport into ER lumen. PSMB genes encode cell cytosol proteasome components, which hydrolyze endogenous antigen into small peptides of 10 amino acid residues.

DOA and DOB genes encode α and p chain of DO molecules. Exact function of DO is not known and may be negative regulation of DM function.

7.5.1.3 HLA class III genes region

These genes are located between HLA Ⅰ and Ⅱ in area from *PPIP9* to *BTNL2*, containing 62 loci with 0.9 Mb length, of which 58 genes are functional genes, and two genes are pseudogenes. There is one gene per 14,516 bp, and it is the gene region of highest density in the whole genome. Except *LOCA01250* gene, other genes were discovered before 1999. Some expressing products of class III genes play important role, such as participation in transcription regulation (CREBL1, BTKl9, PBX2, RDBP, SKW2L, BAT1 and VARS2), housekeeper (DOM3Z, NEV1, AGPAT1, CLIC1 and CSNK2B), biosynthesis, electron transfer, water activity and mediating protein-protein interaction.

(1) Complement genes

Complement gene C2 is very close to Bf gene. There is only 421 bp distance between them. These two genes have similar intron-exon structure, probably generated by duplication of the same set of genes. C2 and Bf genes both are polymorphic. At least nine type of C2 molecule can be detected based on isoelectric analysis and C2C is the most common one. Bf is also polymorphic based of the electrophoretic heterogeneity and more than 20 types have been found. Phenotypes and genes frequency of C2 and the Bf show significant regional and ethnic differences. C4 genes, including C4A and C4B, encode CAA and CAB complement proteins with high homology (only four amino acid residues is different between 1,101 and 1,106). The length of C4 is variable. C4A is usually long genes (22 kb) and C4B genes have long (22 kb) and short (16 kb) forms. Presence of invalid gene is high frequency in C4 and is commonly expressed as Q0 (quantity zero). Expression of C4 gene may be associated with autoimmunity, and disease susceptibility. Number of C4 gene in different HLA haplotypes, possibly due to unequal exchange between homologous chromosomes during meiosis.

(2) 21- hydroxylase gene

There are two 21-hydroxylase genes (*CYP21A* and *CYP21B*) which locate at the C4A and C4B gene, 3 kb from the 3′ end of them. The two genes show about 97% homology to C4 gene. CYP21B gene encodes the adrenal 21-hydroxylase.

(3) Heat shock protein (HSP) genes

In type Ⅲ region, there are three HSP genes, *HSPML*, *HSPMA* and *HSPAlB*, encoding heat shock proteins of 70 kDa molecular weight, 1-L, IA and 1B molecules, respectively. These heat shock regulated genes encode chaperone in protein synthesis, folding, composition and degradation.

(4) *TNF*, *LTA* and *LTB* genes

TNF genes encode TNF-α which is produced by monocytes/macrophages. *LTA* and *LTB* (lymphotoxin, LT) encode lymphotoxin α and lymphotoxin β in T cells.

7.5.2 Matching methods for HLA-Ⅱ genes

7.5.2.1 Restriction fragment length polymorphism analysis

(1) RFLP

RFLP is a technique to exploit variations in homologous DNA sequences. In RFLP analysis, the DNA is digested by restriction enzymes, separated by gel electrophoresis and hybridized with cDNA for classification using autoradiography.

RFLP-based DNA typing technique was firstly used by the Bidwell in typing HLA-DR and DQ in 1988. Lu and others[15] successfully used RFLP technology for HLA-DR5 typing. However, The operation of RFLP is complicated and time-consuming (more than 10 days) [16, 17].

(2) PCR-RFLP

PCR-RFLP is a technique based on polymorphism of restriction fragment of PCR products using primers designed by sequence of polymorphic HLA loci ends. This technology was firstly reported by Maeda[18], and now have been used in genotyping of HLA-DQA1, DQB1, DPB1, DRB1, HLA-B44 and HLA-C. The technique greatly improves the sensitivity and accuracy of HLA typing method, and shortens the detection time. But it is difficult to spread due to the complicated operation and classification procedures.

7.5.2.2 Conformational analysis

(1) PCR-single strand conformation polymorphism analysis (PCR-SSCP)

PCR products can generate two complementary single strands after denatured. If there is mutation, two single strands have different conformation, even if only one base is different, which will be discriminated by gel electrophoresis. This technique was firstly reported by Orita[19], and has been successfully used for HLA-A, HLA-DRB1, HLA-DQB1, HLA-DPA1, DPB1 and HLA-DQ4 typing. PCR-SSCP is simple and sensitive method to analyze multiple samples simultaneously, but can't determine the location and nature of mutations. If the PCR products are less than 200 bp, the detection rate will decrease, because the product is too short to cause detectable conformational changes by PAGE. In addition, many factors can affect its reproducibility.

(2) Heteroduplex analysis (HA)

After denatured, wild and mutant DNA all will release single-stranded DNA. They form hybrid double-stranded heteroduplex during renaturation, which can be distinguished from normal double-stranded DNA by polyacrylamide gel electrophoresis.

Heteroduplex can be analyzed using universal heteroduplex of main generator (UHG) introduced in 1994 by Clay for HLA-DPB1 typing in bone marrow transplants.

Arguello and others[20] developed a new complementary strand analysis method (complementary strand analysis, CSA) for typing HLA-I class, and this method has been successfully applied in HLA-A, HLA-B and HLA-C genetyping. This method had similar resolution with PCR-SSCP. Its most distinguishing characteristic is easy to operate.

7.5.2.3 Genetyping based on DNA sequencing

Sequencing HLA alleles subtype is the most direct and accurate method, but the method requires expensive equipment, which is difficult to achieve in many laboratory. DNA sequencing technology for HLA typing includes sequence-based typing (SBT) and PCR-sequencing. This technique was successfully used for typing HLA-DRB1, HLA-DQB1, DQA1[21], HLA-DPB1[22], HLA-I[23], HLA-B15, HLA-C and HLA-B[24,25].

7.5.2.4 Sequence specific primer polymerase chain reaction (PCR-SSP)

PCR-SSP is a widely used genotyping method, which was first used by Olerup and Zetterquist in typing HLA-DRB1×04, HLA-DRB1×07 and HLA-DRB1×09[26]. Based on nucleotide sequences of HLA polymorphism and the known DNA sequence, a series of allele-specific sequence primer can be used in specific PCR reaction amplification, resulting in amplified products corresponding to specific bands. For homozygote and heterozygote, one and two amplicons will be generated, respectively. Specificity of PCR-SSP technique is accurate to distinguish one base difference.

The most distinguishing characteristic of PCR-SSP typing is simple and less dependent on the experimental equipment. PCR-SSP is usually used for single locus genotyping, such as DRB1 × 04. A few studies also used it simultaneously for HLA-I and HLA-II gene typing. The shortcomings of PCR-SSP technology is requiring a large number of primers, producing false positives, and requiring a larger amount of DNA samples, which brings inconvenience to the clinical application. The latest technology of PCR-SSP genotyping include fully-automated and trace sample PCR-SSP typing technique.

7.5.2.5 PCR-sequence specific oligonucleotide probing (PCR-SSOP)

Firstly, the polymorphism of the HLA region will be amplified by PCR and the products will be labelled with isotope or non-isotope during the course. Basing on the principle of base pairing, a series of oligonucleotide probes will be designed and fixed on membrane to hybridize with PCR products for autoradiography analysis. PCR-SSOP typing is the most commonly used and highest recognized DNA typing technique. Uniform standards for experimental conditions and probes for PCR-SSOP were developed in the 11th International Histocompatibility Workshop (IHWC).

PCR-SSOP technique is of high sensitivity and specificity and requiring fewer samples. The technique was successfully used in HLA-DRB1, HLA-DQB1, HLA-DPA1 and DPB1 genotyping[27]. used PCR-SSOP for HLA-I gene typing. It was also successes in typing of HLA-B35 and B35 HLA-A, HLA-B and HLA-C. PCR-SSOP include normal phase SSOP and reverse phase SSOP. The former is implemented through fixing PCR product on membrane to hybridize with labelled probe. The latter is implemented through fixing specific probe on the membrane to combine with the labelled PCR product, and more widely used.

The carriers of PCR-SSOP include membrane and micro-titration plates and as for complicated and numerous HLA allelic genes, they lack integration. In addition, the elution conditions are complicated and difficult to standardize and automatize.

References

[1]   Marinos C, Dalakas, MD. Molecular Immunology and Genetics of Inflammatory Muscle Diseases. Arch Neurol, 1008, 55, 1500 1512

[2]   Fu WL. Modern molecular bacteriology. Shantou:Shantou University Press, 2000:109

[3]   Campbell MJ, Zelenitz AD and others. Use of familiy specific leader region primers for PCR amplification of the human heavy chain variable region gene repertoire. Molecular immunology, 1992, 29(2): 193-203

[4]   Nie QH. Infectious diarrhea. Beijing: The people's medical publishing house, 2000: 752

[5]   Wang CB. Practical experimental techniques of Biochemistry and Molecular Biology. Wuhan:Hubei science and technology publishing house, 2003: 152

[6]   Fu GL. Inspection techniques of molecular biology. Beijing:The people's medical publishing house, 2003: 166

[7]   Zhao JW, Sun YM. Modern Food Testing Technology. Beijing:China Light Industry Press, 2005: 269

[8]   Ding DQ, Tomita Y, Yamamoto A and others. Large-scale screening of intracellular protein localization in living fission yeast cells by the use of a GFP-fusion genomic DNA library. Genes to Cells, 2001, 5(3): 169-190

[9]   Liu Y, Yang HY. The key technology in situ hybridization and its application. Jiangxi Journal of Medical Laboratory Sciences, 2006, 24(6): 560-562

[10] Ding JF and others. Genetic analysis and bio-chip technology. Wuhan:Hubei science and technology publishing house, 2004: 3

[11] Li XY. Molecular biology of cardiovascular disease. Wuhan:People's Medical Publishing House, 2000: 34

[12] Peng XX. Hand Book in Plant Molecular Biology. Beijing: Chemical Industry Press, 2005: 18

[13] Pietras DF and others. Construction of a small Mus musculus repetitive DNA library: identification of a new satedllite sequence in Mus musculus. Nucleic Acids Research, 1983, 11(20): 6965-6983

[14] He L, Wei MT, Zhang KJ. The progress to study the relationship between HLA gene polymorphism and infectious diseases. Acta Academiae Medicinae CPAF, 2005, 14(6): 483-486

[15] Lu PH, Zhou GY, Fu SL and others. RFLP analysis of 15 HLA-DR5 homozygote typing cell and comparison in cell typing . Chinese Journal of Immunology,1990, 6(2), 74-79

[16] Li CT. Progress in human leucocyte antigen typing technology. Journal of Forensic Medicine, 2004, 20(2): 120-123

[17] Tan JM. Progress in DNA matching of HLA anfigen. Foreign Medicine Sciences (Urology and Nephrology Foreign Medical Sciences), 1994, 14(5): 211-218

[18] Maeda M,Murayama N, Ishii H and others. A simple and rapid method for HLA-DQA1 genotyping by digestion of PCR-amplified DNA with allele specific restriction endonucleases. Tissue Antigens, 1989,34(5),290-298

[19] Orita M, Suzuki Y, Sekiya T and others. Rapid and sensitive detection of point mutations and DNA polymorphisms using the polymerase chain reaction. Genomics,1989,5(4) 874-879

[20] Arguello R, Avakian H, Goldman JM and others. A novel method for simultaneous high resolution identification of HLA-A, HLA-B, and HLA-Cw. Proceedings of the National Academy of Sciences of the United States of America,1996,93(20), 10961-10965

[21] Santamaria P, Boyce-janino MT, Lindstrom AL and others. HLA class II typing: direct sequencing of DRB, DQB, and DQA genes. Human immunology,1992,33(2):69-81

[22] Rozemuller EH, Bouwens AG, Bast BE and others. Assignment of HLA-DPB alleles by computerized matching based upon sequence data. Human immunology,1993,37(4):207~212

[23] Santamaria P, Lindstrom AL, Boyce-Jacino MT and others. HLA class I sequence-based typing Human immunology,1993,37(1),39-50

[24] Domena JD, Little AM, Arnett KL and others. A small test of a sequence-based typing method: definition of the B*1520 allele. Tissue Antigens,1994,44(4):217-224

[25] Petersdorf EW, Longton GM, Anasetti C and others. The significance of HLA-DRB1 matching on clinical outcome after HLA-A,B, DR identical unrelated donor marrow transplantation. Blood,1995,86(4):1606~1613

[26] Olerup O, Zetterquist H. HLA-DR typing by PCR amplification with sequence-specific primers (PCR-SSP) in 2 hours: an alternative to serological DR typing in clinical practice including donor-recipient matching in cadaveric transplantation. Tissue Antigens,1992, 39(5): 225-235

[27] Allsopp CE, Hill AV, Kwiatkowski D and others. Sequence analysis of HLA-Bw53, a common West African allele, suggests an origin by gene conversion of HLA-B35. nHuman immunology, 1991,30(2): 105-109   ·

Chapter 8　Immunoblotting Technique

8.1　Introduction

The immunobloting test (IBT) is also called enzyme-linked immunoelectro transfer blot (EITB) and western blot since it is similar to the southern blot developed earlier for nucleic acid detection[1].

The TBT is often carried out in three steps[2]. The first step is gel electrophoresis, most of which is SDS- polyacrylamide gel electrophoresis (SDS-PAGE). Proteins, covered with the negatively charged SDS, will migrate from cathode to anode in gel electrophoresis. Smaller proteins migrate faster through the gel and the proteins are thus separated according to their molecular weight. The result is invisible only when the electrophoresis lands are stained. The second step is electro transfer, in which the lands separated in the gel are transferred into nitrocellulose membrane. The procedure is often implemented within 45 minutes at low potential (100 volt) and high current (1-2 ampere). The protein lands are still invisible in this stage. The third step is enzyme-linked immuno location, in which proteins imprinted on the nitrocellulose membrane will react with a specific antibody and an enzyme-labelled secondary antibody in succession. Then a substrate of enzyme is added and some insoluble and visible compounds are generated, which will stain the protein land. This method integrates the high resolution of sodium dodecyl sulphate polyacrylamide gel electrophoresis (SDS-PAGE) and the high specificity and sensitivity of enzyme-linked immunoassay (ELISA). It has a lot of applications including analysis of antigen composition and immunoreaction activity and diseases diagnosis, i. e. HIV. Diseases diagnosis kits can be developed in the way as followed. The antigen is transferred onto the nitrocellulose membrane and then the membrane is cut into small strip. With some enzyme-labelled antibody and substrate solutions, the strip can be used conveniently and the result also can be presented by the visible land in lab.

8.1.1　SDS-PAGE

The polyacrylamide gel is three-dimensional networks which forms in the following way. After acrylamide molecules, the monomer, are initiated by ammonium persulphate and TEMED (N, N, N′, N′-tetramethyl ehylenediamine), they are polymerized into a long chain and then cross-linked by a difunctional crosslinker, bisacrylamide. The mechanical strength, elasticity and adhesive strength of the gel depend on the total concentration of the gel. The higher the gel concentration is, the higher the mechanical strength is and the smaller the average pore size is. Hypothesize that T means the total concentration of acrylamide and bisacrylamide, and C means the percentage of bisacrylamide divided by acrylamide and bisacrylamide in the gel, namely the degree of crosslinking. When the degree of crosslinking is 5% after the total concentration of the gel is set, the gel pore size is smallest. But the gel pore size will become bigger when the C is more or less than 5%[3].

When added to protein, the anionic detergent, SDS, can destroy the hydrogen and hydrophobic bond of protein, bind to protein molecule and form a kind of complex. The reaction generates a plenty of negative electric charges on protein molecule far surpassing the initial electric charges of it, and masks the quantity of electric charge differences between various proteins. With the addition of mercaptoethanol in protein samples, the mobility ratios of proteins will be independent of their shapes. As a result, the mobility ratios of proteins will depend on their molecule weight.

8.1.1.1 Colleting protein samples

Samples, e. g. adherent cells, suspend cells or tissues, can be broken down by appropriate lysis buffer. Some subcelluar fractions, e. g. nuclear proteins, cytoplasm proteins and mitochondrial proteins, can be dissolved and cell total protein including phosphate proteins can be extracted by SDS lysis buffer. Besides SDS lysis buffer, nonionic detergent and hypotonic lysis buffer are also usually selected. Commercial protein extraction kit is also alternative for sample extraction.

8.1.1.2 Determining protein concentration

The protein concentration of each sample should be determined for keeping each sample injected in the same concentration level. Appropriate analytical method should be selected according to the lysis buffer adopted because the tolerances of detergents and reductants in different method vary considerably. Commercial lysis buffer kit and protein test kit are more facile methods. The samples should be diluted if the protein concentrations of the samples are too high. After the dilution concentration is determined, it can be converted into the sample concentration.

8.1.1.3 Electrophoresis

After gel for SDS-PAGE is prepared, the protein sample collected mixed with some condensed sample buffer, e. g. 2 or 5 times condensed sample buffer. 5 times condensed sample buffer is more favourable for reducing sample volume and increasing sample quantity than 2 times. It should be pointed out that acrylamide and bisacrylamide are of neuro toxicity and dermal toxicity, and TEMED is also destructive to mucosa, upper respiratory tract and derma.

8.1.1.4　Checking the SDS-PAGE

Should SDS-PAGE be checked firstly and then Western Blot be carried out? The answer is that knowing the result of SDS-PAGE is beneficial for next transfer protein into membrane.

Coomassie Brilliant Blue staining is a rapid, facile and cheapest method for protein in PAGE and can detect 1 μg band. Silver staining is more complicated, but more sensitive and can detect 2-5 ng band. However, most people will skip the staining procedure since Coomassie Brilliant Blue and silver staining both are irreversible processes of binding with protein and will interfere with the following experiment. One solution is preparing two PAGE plates for electrophoresis, one of which is for staining and the other for transfer. If we want to obtain the electrophoresis result before transferring with one PAGE plate, we can choose SYPRO Tangerine, a kind of fluorescent metal dye for protein staining. This staining is the as sensitive as silver staining and can detect 4-8 ng protein. Meanwhile, the protein gel staining procedure is fast and simple, does not need organic solvents and can be performed in saline or PBS solutions. However, this staining reagent is very expensive. Another cheap method is utilizing Ponceau-S to stain transfer membrane directly. Ponceau-S can be washed out from the protein and does not interfere with Western Blot. This staining method is as sensitive as Coomassie Brilliant Blue.

8. 1. 2　Transfer

Generally there are two transfer methods: capillary blotting and electro-blotting. Electro-blotting is usually applied rather than capillary blotting because the later is of low efficiency. In electro-blotting, the membrane is placed on the top of the gel, and a stack of filter papers placed on top of them. The entire stack is placed in a buffer solution and proteins can be pulled from the gel into transfer membrane by electric current.

There are several transfer membranes. The best common membranes are nitrocellulose (NC) membrane and polyvinylidene fluoride (PVDF) membrane. Nylon and DEAE cellulose membranes are also alternatives. The membrane is selected according to the following points. The first point is the ability of binding target protein (the quantity of protein that a unit area of the membrane can bind) and pore size of the membrane. The second point is whether it is suitable for colouration and of high signal to noise. Meanwhile, we should consider the other requirement, e. g. protein sequencing or mass spectrometry analysis[4].

Table 8-1 Character of several membranes

	NC membrane	Nylon membrane	PVDF membrane
Sensitivity and resolution	High	High	High
Background	Low	Slightly high	Low
Binding ability	80-110 $\mu g/cm^2$	>400 $\mu g/cm^2$	125-200 $\mu g/cm^2$ (suitable for protein containing)
Texture	Dry membrane is fragile	Soft and solid	High mechanical strength
Solvent tolerance	Intolerant	Intolerant	Tolerant
Operation requirement	Wet with buffer and avoid air bubble	Wet with buffer	Wet with methanol
Detection method	Utilize conventional dye or radioactive detection	Can't utilize anionic dye	Utilize conventional dye, ECL or rapid immunoassay
Application scope	0. 45 μm, molecular weight (Mw) above 20 kDa; 0. 2 μm, protein Mw below 20 kDa; 0. 1 μm, protein Mw below 7 kDa	Small molecular weight protein with low concentration, acidic protein, glycoprotein and proteoglycan (use for nucleic acid detection)	Glycoprotein and protein sequencing
Price	Slightly cheap	Cheap	Slightly expensive

8.1.2.1 NC membrane

NC membrane is a transfer media used most widely and has a lot of advantages such as having high protein binding ability, suitable for conventional colouration and staining, fluorescence, chemiluminescence and isotope detection, of high signal to noise with low background. The pretreatment of it is simple as follows: soaked in pure water to exclude air bubble and balanced in electrophoresis buffer. Blocking and washing are also easy. Protein transferred to NC membrane can be preserved in an appropriate condition for a long time. However, NC membrane is fragile and not suitable for the application needing repeating washing. It is important to select appropriate pore size of NC membrane. Generally, 0. 45 μm membrane is suitable for protein of 20 kDa Mw, 0. 2 μm membrane for protein of below 20 kDa and 0. 1 μm membrane for protein of below 7 kDa. Another point should be pointed out that moderate detergent is preferred in blocking such as 0. 05% or 0. 3% Tween 20 since a part of protein binding with NC membrane will be replaced by detergent. Some common suppliers of the membrane include Schleicher & Schull, Millipore, PALL, Invitrogen, GE (Amersham) and Santa Cruz.

8.1.2.2 Nylon membrane

Usually nylon membrane is applied to nucleic acid transfer, and occasionally employed for protein blotting. Nylon membrane has some advantages such as it being soft, solid and with high mechanical strength and binding ability. It also has drawbacks such as high background, that protein often leaks out from the membrane when transfer buffer contains SDS and can't be stained directly. However, some products, for instance, nylon membrane supplied by Pierce, has low background and high signal to noise and 200 $\mu g/cm^2$ protein binding capacity.

8.1.2.3 PVDF membrane

PVDF membrane as transfer membrane, which has advantages over NC membrane in protein binding capacity, mechanical strength and solvent tolerance, was commercialized by Millipore company at first in 1985. Protein binding capacity of it is 100-200 $\mu g/cm^2$, whereas NC membrane is 80-100 $\mu g/cm^2$. Moreover, the binding strength of PVDF membrane is 6 times higher than that of NC membrane. Comparative study on PVDF and NC membrane in HIV serological test shows that PVDF membrane has higher capacity of intercepting HIV antigen and can improve the performance of detecting glycosylated capsula antigen. Another advantage of PVDF membrane is of high mechanical strength and solvent tolerance, which is preferable for versatile staining and immunoassay. Lane replica in a gel can be used for many purposes, such as N-terminal sequencing, proteopepsis, peptide separation and immunoassay after Coomassie Brilliant Blue stain and being cut down. PVDF membrane is the only choice especially for N-terminal sequencing because it can be rinsed time after time. PVDF membrane is also a better choice for some high hydrophobic protein. Many detection methods can be implemented in PVDF membrane, e. g. common colourations and stains, chemiluminence and isotope detection except fluorescence. Issue that needs attention in particular is that PVDF membrane needs pretreatment with methanol for less than 15 seconds followed by balance with buffer. Now there are two sizes of PVDF membranes, 0. 45 μm and 0. 2 μm, and the latter is better for binding low Mw protein and of higher background than the former. Some common suppliers of the membrane include Millipore, Bio-Rad, Pierce, Pall Life Science, Invitrogen and Whatman.

8. 1. 3 Enzyme immunoassay

8.1.3.1 Blocking

To some extent the sensitivity of Western blot depends on blocking. Skim milk is utilized most commonly and that is also an economical prescription. However, skim milk is not suitable for biotin-avidin system and alkaline phosphatase (AP) based detection method since this blocking reagent possibly contains trace biotin and alkaline phosphotase which will bring about some background.

Blocking reagent, 6% casein containing 1% polyvinylpyrrolidone and 10 mmol/L EDTA in buffer, is preferable for AP-based detection method. The reagent should be heated in 65 ℃ for 1 h and added with fresh 0. 05% sodium azide in order to inactivate alkaline phosphotase. Nowadays horseradish peroxidase (HRP) is applied more widely to immunoassay. Generally blocking reagent can't contain sodium azide for HRP-based detection method since sodium azide can inactivate HRP. A blocking reagent should be resolved in Tis buffer if it is employed in AP-based coloration method.

8.1.3.2 Antibody

Generally monoclonal antibody can be utilized in western blot, but the conformation of the protein's recognition site probably will change and can't be recognized again by its antibody after by denature in SDS-PAGE. Polyclonal antibody is usually better than monoclonal antibody although specificity of the former is lower than that of the latter. Primary antibody from rabbit or micc is often selected in western blot since the succeeding regents are mostly designed against rabbit and mice and they have a wider choice. Primary antibodies labeled with biotin or fluorescence are also selected for Western blot besides unlabelled antibody, which help to decrease the background and unspecific binding. However, there no longer exits cascade signal amplification from secondary antibody in Western Blot and the detection signal will be weakened in this way. Thus, very bright fluorescence label is needed in this detection mode. The dilution of primary antibody is usually determined through preliminary experiment, in which 2 to 3 dilutions near the dilution suggested in antibody specification need to be tried. If chemiluminece is selected in western blot, higher dilutions perhaps are needed for experiment.

Secondary reagents, which help to amplify signal, are chosen according to label and specie of the primary antibody. As for unlabelled primary antibody, labelled secondary antibody corresponding to primary antibody is usually chosen, and the alternative is labelled protein A, which comes from staphylococcus aureus and contains high affinity with IgG protein from human being, rabbit and guinea pig but low affinity with IgG protein from wister rat, goat and chicken. Moreover, the label of secondary antibody also needs consideration according to detection method.

8.1.3.3 Label

Enzyme label is usually applied in Western blot. Different substrates can match with corresponding enzyme-catalyzing reaction, which will implement different detection methods such as chemiluminescence and colour detection. Chemiluminescence detection method is very sensitive, whose sensitivity can reach pg level and is higher than isotope detection method. However, colour detection method is easy and cost-efficient with direct colour development.

The most common enzyme labels are horseradish peroxidase (HRP) and alkaline phosphatase (AP). In addition, glucose oxidase and β-galactosidase are also often utilized as enzyme labels. HRP is the most widely used as enzyme labels since it has high enzymatic specific activity, low molecular weight (40 kDa), stability and many substrates.

The substrates of HRP include DAB, 4-CN, CN/DAB, AEC and TMB, etc, which can be majorly categorized into chemiluminescent and chromogenic types. Beyond that, substrates such as OPD and ABTS, which are suitable for ELISA, will be transferred into soluble products. In Western blot, colour development from substrate, which is different with enzymatic chemiluminescence, is because of its losing electron and producing colour change on the effect of HRP and hydrogen peroxide. DAB (3, 3-diamonobenzidine), which is one of the most common substrate, can be transferred into insoluble brown product on the effect of HRP and hydrogen peroxide. Western blot is sensitive and specific through DAB colouration. However, DAB also has some weaknesses such as toxicity, needing fresh solution, that colour from DAB will fade after hours and should be taken a picture. Some commercial substrate reagents are mixture of DAB and 4-CN, which can form good black precipitation image since they hold advantages of high sensitivity from DAB, low background and aptness to imaging from 4-CN. Moreover, some DAB reagents can reach 20 pg detection limit through signal enhancement by metal ions. TMB is a safe substrate and obtains high signal to noise which can be realized through blocking appropriately and washing for times. Another substrate of HRP, AEC, is similar to DAB except that its sensitivity is lower than that of DAB.

Chemiluminescence substrates can produce strong signal, which can be sustained for a long time, on the effect of enzyme. This method, which is totally different with colour producing reaction that will produce insoluble products precipitating on membrane, is suitable in particular for the membrane that needs detection for different target protein. Nowadays the sensitivity of chemiluminescence can reach femtogram in Western blot.

Alkaline phosphatase (AP) is also a common enzyme for colour producing reaction, and has been applied to detect immobilized antigen in Western bolt and nucleic acid long before. The substrates of AP are more stable than those of HRP since they do not need freshly prepared components such as unstable H_2O_2. Thus, kits of AP can be preserved for longer time than HRP. However, there often exists high background in experiment with AP in Western blot. It is explained in the laboratory manual "Molecular Cloning 3" that high background is produced by AP in the blocking reagents, defatted milk or bovine serum albumin. Thus, the manual recommends that the suitable blocking reagent for AP consists of 6% casein, 1% polyvinylpyrrolidone and 10 mmol EDTA. In addition, the blocking reagent should be heated at 65 ℃ for an hour in order to inactivate the trace AP in them. Although some samples contain high concentration of AP, the endogenous AP in them can be inhibited in experiment; however, the point is actually easy to be ignored.

The product from colour producing reaction of AP is bluish violet and easy to be imagined. Now chemiluminescence substrates of HRP have been continuously improved and the sensitivities of them are becoming higher. As a result, HRP is nearly equal with AP in chemiluminescence detection. But the sensitivity of AP is higher than HRP in colour producing reaction. Substrates of AP can be categorized into two types according to their reaction. Most common substrates of AP are BCIP and INT.

Table 8-2　Character of AP

	BCIP	NBT	BCIP/NBT	BCIP/INT
Name	5-Bromo-4-chloro-3-indolyl phosphate, toluidine salt	Nitro-blue tetrazolium chloride	—	
Effect	Form precipitate after react with O_2	Oxidation reaction	BCIP can react with O_2 and NBT is also electron receptor	Oxidation reaction
Sensitivity	100 pg	100 pg	30 pg	Average level
Colour of precipitate	Violet	Bluish violet	Bluish violet	Brownish red
Application	Color producing reaction of AP	AP or glucose oxidase	AP, hybridization in situ or immunol histochemistry	AP or immunol histochemistry

Some particular substrate such as Fast Red which is supplied by Pierce Company, is designed for Dot blot, Western blot and immunol histochemistry and can produce shiny red precipitate. The most interesting advantage of Fast Red is that it can implement double-staining together with substrate of HRP since the shiny red precipitate produced from Fast Red and insoluble precipitate from substrate of HRP can form striking comparison on NC or nylon membrane. Thus, the two substrates can carry out colour development twice and detect different target protein.

Chemiluminescence substrates of AP can produce strong signal rapidly and obtain sensitive and impressive result. They are not like colour producing substrates which need taking half an hour to accomplish colour development. AP chemiluminescence kit such as Western Breeze supplied by Invitrogen Company can realize femtogram detection limit and chemiluminescence of it can last for 5 days. Other kits such as Lumi-Phos Wb supplied by Pierce Company can realize picogram detection limit on nylon membrane and its chemiluminescence can last for 8 hours.

8. 1. 4   Notices of Western blot

1) Western blot only can detect the relative content of a target protein. Thus, the technique only can confirm whether the target protein exists and that the protein content in a kind of cell is higher or lower than in another one under given conditions, but can't confirm how much target protein a kind of cell contains.

2) In order to compare the relative contents of a target protein in different kinds of cell or the same kind of cells under different conditions, total protein content of samples should be equal.

3) Appropriate gel content of SDS-PAGE should be selected in order to obtain good resolution of target proteins.

4) The cathode and anode should be checked carefully whether they are jointed right before electrophoresis and transfer.

5) Nonspecific binding sites should be blocked carefully in immunoassay.

6) When antigen-antibody reaction is carried out in a sealed plastic bag, bubbles in the bag must be extruded. Otherwise, the antigen-antibody will bind unevenly.

7) The processing should be carried out by wearing gloves and as gently as possible. Transfer membrane should be kept in solution without any drying and shouldn't be scratched.

8.2   Molecular imprinting technique (MIT)

Molecular imprinting technique (MIT), as an emerging technology, has aroused extensive interect. It can be described as a technology of making a lock which can recognize a molecular key. Namely, it is a technology to prepare a polymer (Molecular imprinting polymers, MIPs) complementary to a special target molecule in space structure and binding sites [6].

The origin of MIPs can be traced back to the supposition about the lock-and-key by Fischer [7], the antigen-antibody binding by Pauling [8], clonal selection theory by Burnet [9] and specific adsorption conception by Dickey [10]. However, the establishment of modern MIT should be attributed to Mosbach and Wulff groups, whose pioneering works improved MIP enormously. In 1972, Wulff, an outrunner in MIP, and his coworkers in Germany reported their work on the successful preparation of a MIP by covalent approach recognizing amino acid and sacchrides selectively for the first time. Since then MIT has gradually become known it has achieved vigorous development since Mosbach and others published a paper in *Nature* about a theophylline-imprinted polymer produced by non-covalent approach in 1993[14]. Then MIT has achieved an enormous breakthrough and more rapid development. In 1997, Society for Molecular Imprinting

(SMI) in Lund University was established by Mosbach and others The objective of SMI was to develop and apply MIP technology in sensor, artificial antibody, chromatogram separation fields.

Within this chapter, special attention is paid to the basic principles, synthesis and application of MIPs. The problems that MIPs are facing will be also discussed.

8.2.1 Overview of molecular imprinting

8.2.1.1 Principle and method

Currently, the preparation of MIPs often involves three steps (as shown in Figure 8-1).

1) The host-guest complexes are formed through covalent or non-covalent interactions between the template and functional monomers in organic solvent (nonpolar or weak polar organic solvents, also called porogenic agent).

2) Cross-linkers are added, and copolymerized with host molecules by free radical polymerization, polymer of rigid structure are produced around template molecules.

3) The template molecules are removed from the polymer by a certain solvent. Consequently, the space in the polymer originally occupied by the template molecules are left as a stereo cavity which can match the size, structure of the template. Besides that, these cavities possess functional groups provided by functional monomer which are precisely arranged in cavity and complementary to the template molecule. The space structure and properties of these cavities depend on the template molecule and the functional monomer, thus MIPs can bind efficiently and selectively the template molecule (or its analogs).

The three main featuses promoting the rapid development of MIT are as follows:

1) Predetermination. MIPs can be prepared according to different purposes, thus meet the needs of different kinds of test.

2) Specific recognition. MIPs, are prepared in accordance with imprinting molecules, therefore, can recognize specifically them.

3) Practicability. MIPs not only maintain the advantages of natural recognition system, such as enzyme-substrate, antigen- antibody and receptor- hormone, but also have unique and excellent stability against adverse circumstances and of long service life.

Molecularly imprinting technique are divided into two approaches: covalent imprinting and non-covalent imprinting.

(1) Covalent Imprinting

Covalent imprinting is also called preorganized method which was proposed primarily by Wulff and his coworkers in Germany [11, 12]. In this method, the complex is formed through a reversible covalent binding between the template and the functional monomer,

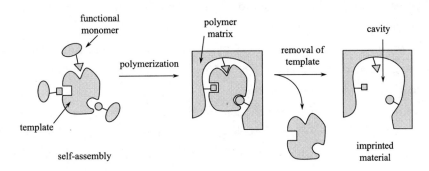

Figure 8-1   Schematic diagram of molecular imprinting process

and then cross-linkers are added in this complex to form a highly crosslinked polymers with porous network. After template is removed by cleaving covalent bond, the MIP obtained can rebind the template exactly through the residual cavities complementary to the removed template.

In order to achieve covalent imprinting successfully, a key point is to select covalent structure between template and functional monomer. The covalent structure should be sufficiently stable and reversible to survive polymerization, and the bond must be cleaved under mild conditions, and after cleavage, have a positive interaction with the analyte.

In order to combine with target guest (or release it) quickly, a fast reaction speed is needed in the formation and decomposition of a functionalized structure. At present, a few covalent bonds which can satisfy the above thermodynamics and dynamics require-ments have been found, and there included boric acid vinegar, acetal, ketal, Schiff base and disulfide bond.

Up to the present, the covalent imprinting method has been applied to prepare all kinds of polymers which can recognize some template molecules selectively, such as sac-charides and their derivatives, glyceric acid and its derivatives, amino acid and its deriva-tives, mandelic acid, aromatic ketone, dialdehyde, propionaldehyde, transferrin, united coenzyme and steroid alcohols, etc.

(2) Non-covalent imprinting method

Non-covalent imprinting method was also known as self-assembly approach, repor-ted firstly by Mosbach and his coworkers [13]. In this method, template molecule and functional monomer self-assembly arrange beforehand, and then multiple forces are formed by non-covalent interactions, such as electrostatic force, hydrogen bonding, di-pole force, ion, metals-chelating, charge transfer, hydrophobic interaction and van der Waals attraction.

Now, this method has been extensively used to prepare MIPs such as dyes, amino acid derivatives, polypeptide, nucleotide, adrenergic inhibitor, theophylline, diamine and vitamin, etc.

As shown in Table 8-3, both covalent and non-covalent imprinting methods have their advantages and disadvantages. Choice of these methods depends on the need and their application.

Beside the above two methods, another imprinting, also called the "sacrificial spacer" method, is the combination of the covalent and non-covalent approaches. In addition, the preparation of MIPs utilizing the coordination of functional monomers and metal ions have been also developed, and are mainly used to separate metal ions and protein molecules. Because of the strong metal coordination, MIPs obtained by this method can recognize the target molecule in polar solvent. The greatest advantages of this technique is that it can flexibly adjust the affinity to imprinting molecule by simply changing metal ion species.

Table 8-3   Comparison of covalent and non-covalent imprinting methods

	Covalent imprinting	Non-covalent imprinting
Advantages	Binding groups are precisely arranged in space; Template molecule can be removed with high percentage; Binding sites keep uniform.	Molecular imprinting systems are diversified and template can be easily removed from the polymer; Recognition is close to natural macromolecule recognition system; The preparation is relatively easy and moderate.
Disadvantages	The rate of association and dissociation is relatively slow; Of low recognition ability and preparation is complicated	The imprinting process is complicated; The presence of nonspecific binding may result in low selectivity.

8.2.1.2   Recognition mechanism of MIPs

Many studies on molecule imprinting mechanism have been reported, and it is generally considered that recognition of imprinting molecule is mainly determined by the following factors.

8. 2. 1. 2. 1   Printing reaction

Printing reaction means the selectively reaction between the functional groups on functional monomer and the functional groups on imprinting molecule, including reaction of reversible covalent bonds and forming non-covalent interaction (e. g. hydrogen bonding, electrostatic interaction, metal coordination-bond, π-π force, hydrophobic force and van der Waals (vdW) force). Each factor that affects these reactions will effect the recognition of MIPs to imprinted molecule. There are mainly four factors: ① inhibitors of functional group; ② the space orientation of functional groups; ③ electrostatic repulsion and sterically exclusion; ④ solvent.

8. 2. 1. 2. 2　Perfect match between imprinted molecule and spatial structure of cavities on polymer

With the perfect match between the configuration and conformation of imprinted molecule and the spatial structure of cavities on polymer, the functional group on the imprinted polymer can keep thoroughly close to imprinting molecule, thus facilitating specific binding between them. The structure and shape of the spatial cavities on polymer are not completely rigid and invariable. This is because polymer will swell in solvent, leading to change in size and shape of cavities, thus changing the selectivity to template to some extent. The specific inter-reaction between functional groups and imprinting molecule plays an important role in selective affinity interaction between polymer and template. In addition, recognizing templates of polymer with the same functional groups are mainly attributed to the size and shape of cavities. In general, functional groups and the cavities of polymer together determine the selectivity of polymers.

8. 2. 1. 2. 3　The process of MIPs recognizing imprinting molecules

It's very important that imprinting molecule can enter into the polymer's cavities correctly, whether MIPs is used for catalysis, separation or other purposes. In general, the less the binding sites are, the faster the combination between polymer and template is, but the specificity of the selection is relatively low. In contrast, more binding sites may enhance the specificity of MIPs, but the combination speed will be affected accordingly. Thus, it requires taking the two factors into consideration during the molecularly imprinting. In fact, the binding constant of two binding sites is usually bigger than that of single binding site. However, as for two binding sites, the combination is slower. Different cavities in the same polymer have different recognition ability, due to the different matching levels between template and cavity as well as functional group. In addition, non-specific binding can also be found in the functional group of polymer cavity. Thus, non-specific and specific adsorption will simultaneous occur in the recognition process, lowering the selectivity of imprinting molecule and causing tailing peak in chromatographic analysis[13].

8.2.1.3　Preparation of MIPs

8. 2. 1. 3. 1　Raw materials of MIPs

All the chemicals for MIPs preparation are as follows: molecule template, functional monomers, crosslinking agent, and solvents.

(1) Molecule template

In theory, all molecules with appropriate functional monomers can be employed as template. Due to the number restriction of functional monomers, templates with corresponding MIPs are mainly compounds with strong polar-group (hydroxyl, carboxyl, amino etc), such as carbohydrates, amino acid and its derivatives, carboxylic acid, steroids, vitamin, protein and some biological macromolecules. Strong polar group and functional monomers can form stable molecular composite, thus benefiting the preparation of polymer with high efficiency.

(2) Functional monomers

The guiding principle in the choice of functional monomer is the structure of templates molecules. Thus, for templates containing basic groups, acidic functional monomers (e. g. trifluoromethylacrylic acid, methacrylic acid) are preferred, whereas basic functional monomers (e. g. 4-vinylpyridine) are used for target templates containing acidic groups. Other neutral templates usually choose neutral or acidic functional monomer (e. g. acrylamide, methyl methacrylate, methyl methacrylate, etc).

For some specific templates, the dual-functional or multi-functional monomers can be considered because multiple functional monomer can form multiple binding sites. For example, Toshifumi [17] utilized methacrylate and 2-trifluoromethyl acrylic acid as dual-functional monomers to produce MIPs. Zheng [18] used sulfamethazine and sulpha-oxazole as the template molecules to prepare MIPs, chose acrylic (AA), 4-vinyl pyridine (4-Vp), the mixture of acrylic acid (AA) and 4-vinyl pyridine (4-Vp) as functional monomer, and obtained better imprinting effect. The multi-functional monomers have the advantage of good specificity and strong interaction between template molecules and imprinted polymer.

In addition, much efforts on functional monomers synthesis have been made, and some new monomers have been synthesized and applied from time to time. Figure 8-2 shows some conventional functional monomers.

(3) Cross-linker

In the process of imprinting, the fundamental role of cross-linker is to fix the guest-binding sites firmly in the desired structure. To maintain hardness (especially MIPs used in chromatography phase) and form stable binding sites of MIPs, high degree of cross- linker (70%- 90%) often is required. Due to the large amount of cross-linkers in polymerization solution, the cross-linked monomer determines the hydrophobicity of imprinted polymer to a high extent. Ethylene glycol dimethacrylate (EGDMA) is the most widely used cross-linker, which can produce MIPs with high selectivity and low non-specific adsorption. Typically, the reference ratio of the template molecule, functional monomer and cross-linker is 1 : 4 : 20.

(4) Solvent

The role of solvents is to dissolve homogeneously the agents for polymerization (including the template molecule, functional monomer, cross- linker and initiator). In addition, they can also provide porous structures for imprinted polymers, thus also called porogen. In the preparation of MIPs, the amount and characteristics of solvents are criti-

acrylic acid　　　　　　methacrylic acid　　　　methyl methacrytate

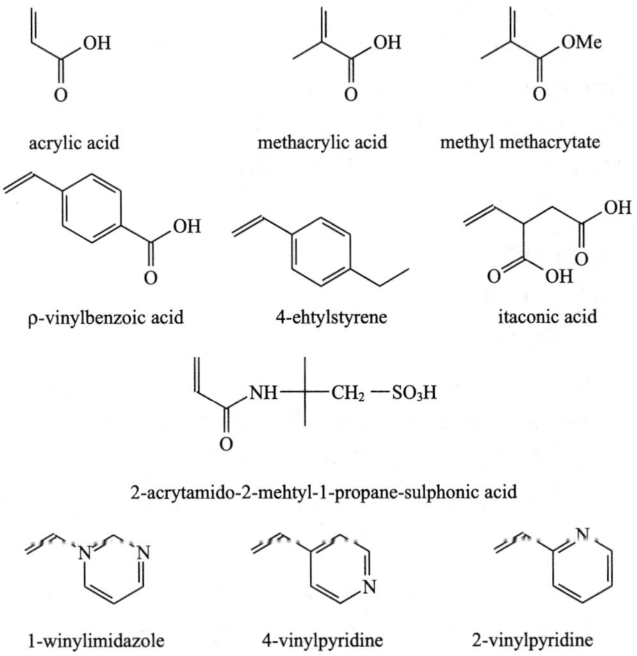

p-vinylbenzoic acid　　　　4-ehtylstyrene　　　　itaconic acid

2-acrytamido-2-mehtyl-1-propane-sulphonic acid

1-winylimidazole　　　　4-vinylpyridine　　　　2-vinylpyridine

Figure 8-2　Structural formula of monomer

cal to the polymer morphology and pore volume [21]. Therefore, as for the selection of solvents, minimizing should be considered about the effect on the binding between imprinted molecules and polymer, and, meanwhile it also should be noticed to simplify the post-process (e. g. extraction). In the non-covalent molecular imprinting, since functional monomer and template molecule form complex by weak non-covalent interactions, the polar solvent will affect their interaction. Thus solvent of low dielectric constant such as toluene, chloroform, dichloromethane should be chosen as possible.

　　Eshter[22] utilized toluene and acetonitrile as the solvent to prepare MIPs against triazine herbicide respectively, and compared their extraction properties for the herbicides. It was found that the recovery was about 90% when toluene of low dielectric constant was adopted as the solvent, while the recovery was about 37.5%. when adopting acetonitrile. To meet the solubility of template molecules, acetonitrile[23,24] and tetrahydrofuran [25] are often applied. In addition, it also has been reported[26] the best recognition performance could be attained when the solvent used during molecule recognition is kept identical with that in polymerization,. This was possibly because of meeting the most suitable imprinting conditions.

　　(5) Initiation

　　Polymerization of MIPs can usually be initiated by thermal, light or radiant rays, such as 2, 2-azobis (isobutyronitrile) (AIBN) and 2, 2′-azobis (2,4-dimethylvaleronitrile) (ADVN). Low-temperature light is the most common initiator with the advantages as follows :① it can imprint compounds with poor thermal stability; ② it can stabilize the complex forming between template molecule and functional monomer;③ it can improve the physical property of MIPs, and then enhance their selectivity. Thermal initiation is also widely used to prepare MIPs, but only applied to the substances of thermal stability.

8. 2. 1. 3. 2　Preparation methods of MIPs

(1) Bulk polymerization

Early MIPs are mostly prepared by bulk polymerization method. In a general polymerization, the template molecules, functional monomers, cross linker and initiator are dissolved in inert solvents in certain ratio. After sealed in a vacuum ampere glass, polymerization is induced and block-shaped polymer is obtained, subsequently crushed, ground and sieved. At last, after the template molecules are removed, the MIP particles are obtained.

Because the bulk polymerization is simple, it is still one of the main methods for MIPs preparation. But this method has many shortcomings, such as troublesome post-treatment and irregular particles, etc. , resulting in low separation ability and hardness in large scale production.

(2) In-situ polymerization

In-situ polymerization is a method that involves mixing template molecule, functional monomer, cross linker, initiator and solvent in a certain ratio, and then polymerization directly in a specific container (liquid column, capillary column). Polymer obtained by this method does not need grinding, sieving, sedimentation and other complicated steps, and can be directly used in analysis.

Zhang and others[27] synthesized a series of MIPs column by this technique to separate the strychnine and its structural analogues, and meanwhile applied them to enrich and detect strychnine in body fluids. In addition, this method also has been successfully used to separate enantiomers of nateglinide , cinchonine[28] , etc.

The degree of polymerization determines the performance of chromatographic column. Long reaction time may affect the compactness of polymer, thus leading to poor flow mechanical property of chromatographic column, namely, low flux and high column pressure. Whereas, short reaction time will reduce the chromatographic column efficiency, due to few selective binding-sites in MIPs.

(3) Precipitation polymerization

Precipitation polymerization is a simple preparation method for spherical polymer, also called non-homogeneous solution polymerization. In this polymerizing, the monomer, cross linker and initiator are dissolved in a dispersed phase to form the homogeneous mixture. Then, initiator in mixture is decomposed to produce free radical in initial stage of the reaction, producing linear oligomers with branches. After separation from the media by nucleus-forming reaction in cross-linked process, these oligomers gathered into particles of each other. By capturing other oligomers and monomers, particles obtained in this method grow into spherical polymer with uniform diameter. However, uniform diameter only can be obtained in low viscous solvent due to a high correlation between viscosity and precipitation polymerization. This is probably because, the monomer and oligomers have the better fluidity in low viscous solvent, and then avoid polymerization of each other. However, only precipitation polymerization reaction in acetonitrile has been reported so far.

Ye and others[23] successfully prepared estradiol imprinted polymer in acetonitrile, and the average size of this spherical polymer was 0.03 μm. It was also found that with the increasing of reactants concentration, the micro-morphology of spherical polymer gradually changed from microspheres to blocks. These microspheres obtained in this method have been used to analysis radioactive ligand by immunoassay binding, and was better than traditional bulk polymerization.

Jiang and others [29] prepared anti-aconitase imprinted polymer microspheres in ace-tonitrile by precipitation polymerization method, and then evaluated its binding characteristics through adsorption test under shaking. The results showed that recognition characteristics of MIP microspheres were superior to those block-shape MIPs and noni-mprinted polymers.

In addition, MIPs prepared by precipitation polymerization have been characterized and explored recently. For example, methyl testosterone printed polymer microsphere was prepared precipitation polymerization method, and it was proved that not only the method was simple, but also MIP achieved high recognition ability and selectivity.

(4) Swelling polymerization

Swelling polymerization, also known as multi-step swelling suspension polymeriza-tion or seed swelling suspension polymerization, often involves two steps: ①small size microspheres synthesized by with or without soap emulsion polymerization are adopted as swelling seeds; ② after seeds are repeatedly swelled in emulsion, MIPs are obtained by adding reduction agent as well as photo-initiator or thermal-initiation.

Liu and others[21] synthesized polystyrene seeds firstly, and then obtained MIP par-ticles with uniform size by ultrasonic polymerization and a two-step swelling, and these particles showed specific adsorption to theophylline. Hganikaa and others[30, 31] prepared polymer microspheres with better monodisperse by a multi-step swelling polymerization.

The swelling polymerization has advantage that the reaction can be carried out in polar solvents, thus the MIP obtained can also be used in polar media and meets the re-quirements of simulation enzyme for bilogical environment. Meanwhile, the MIPs are highly mono-disperse and of regular structure. In addition, the recognition sites are in the MIPs microsphere surface which benefits that the binding and elution of imprinting molecules can quickly be brought into equilibrium. However, this method has the disadvantage of complicated preparation and long preparation period because of its requiring a multi-step swelling process.

(5) Surface Imprinting

The surface polymerization is a new approach recently developed and can mainly be divided into two techniques. They have been used to recognize metal ions and other substances. To recognize metal ions, the silicone resin serves as a matrix and then complexes between the template, metal ion, and functional monomer are formed and introduced onto the silica surface by condensation polymerization. After removing the metal ions, the surface of silica gel or polymer leave binding-sites which can combine these metal ions. Another method of surface imprinting is also called matrix modification method. Template molecule is firstly linked to the silica surface by the reaction (such as the coupling reaction). After polymerization by adding the functional monomer and cross-linker, hydrofluoric acid is used to erode the silicone, and then MIPs are finally obtained[31, 32].

Surface imprinting method has the following advantages: ① polymerization and recognition-all can be implemented in aqueous phase; ② polymer has great surface area and has uniform particles. After removing silica, many pore diameters and channels are left, which is favourable for polymer to combine the template molecules in solution. ③ binding sites are located in the polymer surface. On the one hand, it is easy to elute the template molecules, on the other, the template molecules need not enter their rigid structure, which greatly accelerates the dynamic processes of the combination.

This method also has some limitations. The matrix is mainly confined to the silica gel and it is difficult to remove the matrix after reaction. Hydrofluoric acid is usually used to erode the skeleton structure of the polymer and yet will produce some adverse effects.

8.2.1.4　Applications of MIPs

Molecular imprinting technique has now been used extensively in many fields, including environmental and military field, pharmaceutical and food industry, etc. MIPs have the characteristics of simpl preparation, low cost, durability and wide application and thus are also known as universal molecule recognition materials. These advantages make it possible to replace the natural or other materials with molecule recognition property obtained by other methods, such as antibody.

In the past, the application of MIPs focused on the following aspects: ① solid phase extraction; ② combination simulation of antibody and receptor; ③ mimic enzyme catalysis; ④ sensor material. These aspects will remain priority for a long time.

8.2.1.4.1   Solid-phase extraction

MIP can selectively combine a compound or other compounds structurally similar in a mixture, thus it is very suitable for the purification and enrichment of samples before chromatography analysis. Conventional SPE, such as reversed-phase, normal-phase and ion exchange, make use of the interactions between substance analyzed and adsorbent to extract. Due to their non-specific interactions, the conditions of elution and extraction must be carefully determined based on their physical-chemical characteristics. However, the problem will be simplified when MIPs are applied to solid phase extraction.

Recently, some scholars have used molecular imprinting technology to extract active ingredients from medicinal herbs. The separation and analysis of medicinal herbs is rather difficult due to their complex components, especially for those with less amount of active ingredient. However, the active ingredients can be easily obtained by molecularly imprinted technology. Hu and others [33] prepared MIPs with aescin as template and successful separated aescin and several other compounds coexisting in the medicinal herbs by optimizing the elution conditions.

In addition, great efforts have been made on the solid-phase extraction column packed with MIPs which are used to detect foods, bio-pesticides in the environment, pesticides and veterinary drugs. Zhu and others[34] have prepared a monocrotophos imprinted polymer as solid phase extractant to detect and enrich monocrotophos, mevinphos, phosphorus and oxygen amine dimethoate in water and soil, and then improved the detection ratio.

8.2.1.4.2   Simulation of antibody and receptor

Ligand-binding analysis has been used to detect trace substances in the clinical blood. A receptor (usually antibody), which can combine selectively with analyzed substance, is often needed in this method. However, because of their inconvenience in operation and high cost, MIPs have been called artificial antibodies and receptors and are useful supplements to biological antibodies in theory.

Valatkis and others [13] prepared MIPs with theophylline and sedative drug for the first time in 1993, and they appeared bossessing amazing specific recognition ability. By competitive radioimmunoassay, it was found that MIPs didn't recognize compounds which are similar to the template molecule on structure, but has similar cross-reactivity with monoclonal antibodies.

MIPs, as artificial antibodies and receptor, have the following advantages: ① simple preparation; ② better stability, suitable for harsh environments such as high temperature, high pressure, acid, alkali and organic solvents; ③ a useful supplement to biological antibodies and receptor rarely found in nature; ④ low cost; ⑤ does not need a large number of animals; ⑥ less susceptible to biological degradation. Therefore, MIPs, as ideal substitution of biological antibodies and receptors, have shown promising development prospects.

8. 2. 1. 4. 3 Mimic enzymes catalysis

Artificial simulation of enzyme is the most challenging application and study in molecular imprinting. Since it is difficult for natural enzyme to be extracted, recovered and reused, and is of high cost and poor tolerance, chemists have been trying to prepare chemical catalyst similar to enzyme, but until now no substantial progresses have been made. However, molecular imprinting technique brought about a new look to the field of artificial simulation enzyme.

Compared with natural enzymes, mimetic enzymes of molecular imprinting can be designed according to different substrates and reactions. Thus, they have many advantages such as high selectivity, stable structure and performance, better environmental compatibility, long life time, low cost, and that it is apt to preservation and mass production[35-37]. Additionally, some studies also found that mimetic enzymes of molecular imprinting can catalyse the Diels-Alder reaction between tetrachlorothiophene and maleic anhydride, and yet natural enzymes has not been found to catalyze this reaction[38].

8. 2. 1. 4. 4 Sensor materials

Because of their higher selectivity for template molecule, MIPs are attractive recognition elements in biosensor.

Percival and others [39] coated the electrode with MIP material, and then developed an acoustic sensor to screen a hormonal veterinary drug, nortestosterone. Chun and others [40] developed an electrochemical sensor for bilirubin which attained the sensitivity $1.344 \pm 0.38 \ \mu A/mg$.

Ma [41] filled clenbuterol MIP particles into chemiluminescence circulation pool and established a clenbuterol imprinting flow injection analysis method through reaction between formaldehyde/potassium permanganate luminous system and clenbuterol adsorbed. Zhou and others[42] used a MIP recognition element to fabricate chemiluminescence sensor, and detected albuterol in urine with a detection limit of $0.016 \ \mu g/L$.

Compared with biological sensors that have been studied extensively recently, MIPs based sensor exhibited many advantages such as having good resistance to high temperature, acid, alkali and organic solvent, hard to biodegrade, and that it is apt to be reused and preserved, etc. Moreover, MIP based sensor can be synthesized by standard chemical method. Therefore, MIPs are expected to become an ideal substitutes of biological materials.

In recent years, scientists have begun to explore the some new applications of MIPs, in which the screening of combinatorial chemistry library, trace element enrichment, selective catalyst and by-products separation have made some improvement[43].

8.2.2   Methyl testosterone and molecular imprinting technique

Food safety is closely related to peoples life and has significant effects on economy. In recent years, with the increasingly occurring food-borne pathogen, food security problem has becomes a focus. Therefore, the detection of food pollutants and hazards will be very important. Good inspection system requires a large number of rapid, sensitive and accurate analysis methods.

8.2.2.1   Methyl testosterone

Methyl testosterone (MT) which is a chemical anabolic steroid, as a growth promoting agent, has been widely applied in animal husbandry and aquaculture in the last century. However, because of slow metabolism and high accumulation, the known negative effects occur when MT gets into humans through food chain, such as liver toxicity, carcass poisoning and carcinogenic toxicity etc. Structure of methyl testosterone and its analogues are shown in Figure 8-3.

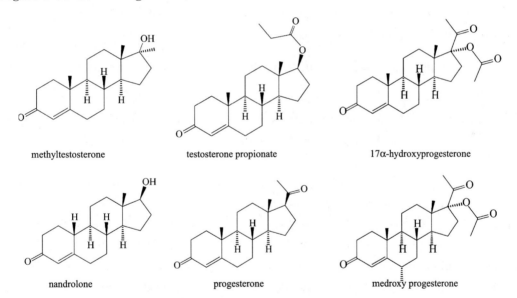

| methyltestosterone | testosterone propionate | 17α-hydroxyprogesterone |
| nandrolone | progesterone | medroxy progesterone |

Figure 8-3   Structure of MT and its analogues

8.2.2.2   Detection and molecular imprinting technique of MT

The EU began to prohibit application of steroid hormone back in 1988, for example in raiseing livestock, and MT has also been banned as growth promoting agents since 1999 in China. In order to prevent application of MT illegally, the most important key points are to control the production and application of veterinary drug from the source and establish rapid and accurate detection methods. However, detection methods of MT have been studied extensively to date, including HPLC [44, 45], GC-MS [46, 47], LC-MS [48], ELISA [49] etc. Due to complex and long time pretreatment in these methods, it is urgent to provide a simple, quick, high sensitivity analytical procedure for MT.

Molecularly imprinted polymers with specific recognition sites have been widely studied. Compared with other artificial receptors, these polymers have the advantage of simple synthesis, design and preparation, low cost and good resistance to acid, alkali and high temperature [12]. and play important roles in food and drug detection, industrial production and environmental monitoring, and medical diagnosis, and have broad application prospects [50-53]. For example, Haupt and others has synthesized a molecularly imprinted polymer (MIP), templated with methyltestosterone, for the cleanup of hydrolyzed urine samples for subsequent testosterone quantification by LC-MS/MS. A concentration of 2 ng/mL testosterone could be quantified after a single step extraction on the MIP. The limit of detection and quantification with the criteria of a signal-to-noise ratio of 3 and 5 were 0. 3 and 2 ng/mL, respectively. Hence, the polymers can offer a more specific extraction procedure, resulting in increased sensitivity with limits of detection 10 times lower than the ones achieved by the standard SPE C18 sorbents employed in official testing laboratories.

8. 2. 3　The problem and outlook of molecularly imprinting technique

Although some progress in molecularly imprinting technique have been made and MIP is expected to be applied in the fields of the separation, catalysis and sensor, many problems constraining the development of these materials still need be discussed and resolved.

1) The preparations of MIPs have been mainly completed on the non-water system in the past years, whereas the practical recognition system and procedure are mostly undergone in water phase. Thus, preparation and recognition in the aqueous phase is an urgent problem to be resolved. Many natural molecule recognition systems, such as enzyme and substrate, antigen and antibody, donor and receptor, the identifications of them all are implemented in an aqueous environment. This is the main weakness of imprinted polymer mimic enzymes and antibody compared with natural enzyme and biological antibody. Many researchers are making efforts on this problem, and have made some breakthrough progresses. It is expected in the near future that the molecularly imprinting technique and its recognition process will be implemented in the water phase rather than in the organic phase, so as to improve the affinity and specificity of molecularly imprinted polymer which are similar to the characters of natural molecular recognition system.

2) At present, we still lack quantitative and systematic research in molecular imprinting technique, especially about the thermodynamics and kinetics characteristics of

imprinting at molecular level. Sellergen and others[54] studied the thermodynamics and kinetics property of enantiomers in mass transfer on a MIP. Chen and others [55] discussed the effect of heat treatment of MIP on the dynamics and mass transfer dynamics. In short, few studies have been reported on this aspect, and the relationship between binding- sites, polymer shape and the mechanism of mass transfer still remains unclear, and thus further research is needed about how to well explain molecularly imprinted process and recognition process at the molecular level.

3) Now the functional monomers and crosslinkers which can be used in the preparation of MIP are very limited. The functional monomer is so limited that it cannot satisfy some certain requirements, and only highly crosslinked rigid polymers can be produced through the available cross linkers, thus the imprinting polymers have a great disadvantage that they are not suitable for biological macromolecules (e. g. protein) of big volume. Therefore, an important prerequisite of maintaining the development of molecular imprinting technique is to create various valid functional monomers and cross linkers.

4) The preparation method of MIPs mainly relies on bulk polymerization and solution polymerization to date, and usually the products by bulk polymerization are blocky, need to be ground and sieved before use. Although this polymerization process is simple, there are still some disadvantages, such as troublesome post-treatment procedure, too long time, severe loss in grinding and screening process, irregular shape and the poor dispersibility. In contrast, MIP microspheres can be used directly after template molecule are eluted due to their regular shape, large specific surface area, high adsorption capacity, controllable particle diameter and simple post-treatment procedure. Therefore, the research about preparation and application of MIP microshpheres has been becoming a hotspot in MIP.

5) The application scope of MIP needs to be broadened. Most studies currently template with amino acids, dyestuff, metal ions and drugs, and yet rarely with peptide, protein and enzyme. In addition, the template leakage problem of MIP is also a problem demanding prompt solution [56, 57].

Although there still exist many problems, many polymerization systems for molecular imprinting have been presented with the rapid development of biological technology, electronic technology and modern detecting technology. In addition, as the synthesis, characterization and theoretical system of molecularly imprinted polymer have been becoming perfect, we can prepare molecularly imprinted polymers with various physical and chemical properties according to actual needs.

8.3 Applications of immunobloting test

Tan and others investigated the application potential of immunobloting test (IBT) in detecting the syphilis antibody in the serum of patient contracted syphilis [58]. Comparative studies on the immunobloting test, fluorescence treponema pallidum absorbent test (FTA-ABS), and enzyme-linked immunosorbent assay (ELISA) for the serodiagnosis of syphilis serum samples were implemented, and it was found that the IBT was two times more sensitive than the FTA-ABS and ELISA method. The positive ratio obtained through IBT was 100%, and slightly higher than those obtained through FTA-ABS and ELISA.

Cao and others applied IBT method to detect helicobacter pylori antibody in serum and completed the detection of 120 samples. The results showed that the IBT was a reliable non-invasive serological diagnosis method [59]. Compared with other methods, this detection is highly specific and sensitive, and doesn't need expensive instruments. Thus it is suitable for patients intolerant to gastroscopy, and in addition, can be applied in the epidemiology survey and screening of helicobacter pylori [60-62].

References

[1] Liu Y, Gou S. Experimental technology of animal immunology. Changchun: Jilin science press, 1989: 49

[2] Wabg Q. Experimental methodology of clinical medicine. Beijing: Science press, 2002: 721

[3] Lv B. Receptor. Hefei: Anhui science press, 2000:364

[4] Hiatt EE, Hill NS, Bouton JH. Tall fescue endophyte detection: commercial immunoblot test kit compared with microscopic analysis. Crop Science Society of America, 1999, 39: 796-799

[5] Vincent C, Simon M, Sebbag M and others. Immunoblotting detection of autoantibodies to human epidermis filaggrin: a new diagnostic test for rheumatoid arthritis. J Rheumatol. 1998, 25(5): 835-837

[6] Ramstrom O, Ansell RJ. Molecular imprinting technology: challenges and prospects for the future. Chirality, 1998, 10: 195-209

[7] Fischer E. Synthese der mannose und lavulose. Ber Dtsch Chen Ges,1890, 23: 799-805

[8] Pauling L. A theory of the structure and process of formation of antibodies. J Am Chem Soc, 1940, 62: 2643-2657

[9] Burnet FM. A modification of Jeme's theory of antibody production using the concept of clonal selection. Australian Journal of Science, 1957, 20: 67-69

[10] Dichey FH. The preparation of specific adsobents. J Proc Natl Acad Sci USA, 1949, 35: 277-299

[11] Wulff G, Sarhan A, Zabrocki K. Enzyme-analongue built polymers and their use for the resolution of racemates. Tetrahedron Lett, 1973, 4329-4332

[12] Wulff G, Sarhan A. The use of polymers with enzyme-analogus structures for the resolution of racemates. Angew Chem Int Ed Engl, 1972, 11: 341

[13] Vlatakis G, Andersson LI, Muller R and others. Drug assay using antibody mimics made by molecular impringting. Nature, 1993, 361: 645-647

[14] Jiang Z, Wu H, Molecularly impringting technique. Beijing: Chemistry industry press, 2003: 1-10

[15] Khasawneh MA,Vallano PT, Remcho VT. Affinity screening by packed capillary high performance liquid chromatography using moceluar imprinted sorbents II. Covalent imprinted polymers J Chromatogr A, 2001, 922: 87-97

[16] Vallano PT, Remcho VT. Highly selective separations by capillary electrochromato-graphy: electrochromatography:molecular imprint polymer sorbents. J Chromatogr A, 2000, 887: 125-135

[17]　Toshifumi T, Daigo F and others. Combinatorial molecular impirinting: an approach to synthetic polymer. Receptors Anal Chem, 1999, 71: 285-289

[18]　Zheng N, Li YZ, Chang WB. Sulfonamide imprinted polymers using co-functional. Anal Chim Acta, 2002, 452 (2): 277-283

[19]　Tanabe K, Takeuchi T, Matsui J and others. Recognition of barbiturates in molecularly imprinted co-polymers using multip le hydrogen bonding. J. Chem. Soc. Chem. Commn, 1995, 2303- 2304

[20]　Vidyasankra S, Ru M, Arnold FH. Molecularly imprinted ligand-exchange adsorbents for the chiral separation of underivatized amino acids. J Chromatogr A, 1997, 775(1-2): 51-63

[21]　Liu J, Luo G, Shen J. Molecularly imprinted polymer and its applicaton. Journal of functional polynmers, 1998, 11(4): 561-565

[22]　Esther T, Antonio ME, Pilar F and others. Molecular recognition ina propazine imprinted polymer and its application to the determination of triazines in environmental samples. Anal Chem, 2001, 73(21): 5133-5141

[23]　Ye L, Mosbach K. Molecualrly imprinted microspheres as antibody binding mimics. React Funct Polym, 2001, 48(1-3): 149-157

[24]　Piacham T, Josell A, Arwin H and others. Molecularly imprinted Polymer thin films on quartz crystal microbalance using a surface bound photo-radical initiator. Anal Chim Acta, 2005, 536(1-2): 191-196

[25]　Xie HC, Zhu LL,Luo HP and others. Direct extraction of specific pharmacophoric flavonoids from gingko leaves using a molecularly imprinted polymer for quercetin. J Chromatogr A, 2001, 934(1): 1-11

[26]　Imma F, Francesca L, Antal T and others. Selective trace enrichment of chlorotriazine pesticides from natural waters and sediment samples using terbuthylazine moleeularly imprinted polymers. Anal Chem, 2000, 72(16): 3934-3941

[27]　Zhang J, He L, Fu Q. Prepariton of strychnine molecularly impriinted monolithic columns. Chinese journal of anlytical chemistry, 2005, 33(2): 113-116

[28]　Huang X, Zhou H, Mao X and others. Preparation of molecularly impriinted chiral monolithic column and seperation of diastereomers. Chromatography, 2002, 20(5): 436-438

[29]　Naka K, Kaetsu I, Yamamoto Y. Preparation of microphere by radiation-induced polymerization. I. Mechanism for the formation of Monodisperse poly(diethylene glycol dimethacrylate) miceosphrere. J Polym Sci PartA: PolymChem,1991, 29(6): 1197-1202

[30]　Haginaka J, Takehira H, Hosoya K and others. Uniform-sized molecularly imprinted polymer for(S)-naproxen selectively modified with hydrophilic external layer. J Chromatogr A, 1999, 849(2): 331-339

[31]　Haginaka J, Kagawa C. Uniformly sized molecularly imprinted polymer for d-chlorpheniramine: evaluation of retention and molecular recognition properties in an aqueous mobile phase. J Chromatogr A, 2002, 948(1-2): 77-84

[32]　Katz A, Davis ME. Molecular imprinting of bulk microporous silica. Nature, 2000, 403: 286-289

[33]　Hu S, Li L, He XW. Solid-phase extraction of escuetin from the ash bark of Chinese traditional medicine by using molecularly imprinted polymer. J Chromatogra A, 2005, 1062(1): 31-37

[34]　Zhu XL, Yang J, Su QD and others. Selective solid-phase extraction using molecularly imprinted polymer for the analysis of polar organophosphorus pesticides in water and soil samples. J Chromatogr A, 2005, 1092: 161-169

[35]　Ohkubo K, Sagawa T, Sawakuma K. Shape-and stereo-selective esterase activities of cross-linked polymers imprinted with a transition-state analogue for the hydrolysis of amino acid esters. J Moleeular Caltalysis A: Chemical, 2003, 165: 1-7

[36]　Visnjevski A, Schomacker, Yilmaz E. Catalysis of a Diels-Alder cycloaddition with differently fabricate molecularly imprinted polymer catalysis com Diels-Alder. Munications, 2005, 6: 601-606

[37]　Ollver B. Chemical reaction engineering using molecularly imprinted polymeric catalysts. Anal Chim Acta,

2001, 435(1): 197-207

[38] Liu XC, Mosbaeh K. Studies towards a tailor-made catalyst for the Diels-Alder reaction using the technique of molecular imprinting. Macromol Rapid Commum, 1997, 18(7): 609-615

[39] Percival CJ, Braithwaite A, Newton MI and others. Molecular imprinted polymer coated QCM for the detection of Nandrolone. The Analyst, 2002, 127(8): 1024-1026

[40] Chun YH, Mei JS, Yeng SC and others. A portable potentioa for the bilirubin-specific sensor prepared from molecular imprinting. Biosensors and Bioelectronics, 2007, 22: 1694-1699

[41] Ma J. Detection of clenbuterol by molecularly imprinting chemiluminescent method. Journal of China University of Mining & Technology, 2005, 4: 36-41

[42] Zhou H, Zhang Z, He D and others. Flow through chemiluminescence sensor using molecular imprinted polymer as recognition elements for detection of salbutemol. Sensors and Actuators B, 2005, 107(2): 798-804

[43] Yang H, Zhou W, Ying X. Mimic detectiong kit for organic phosphorous pesticides. Chinese patent: 1773287, 2006-05-17

[44] Marwah A, Marwah P, Lardy H. Development and validation of a high performance liquid chromatography assay for 17α-methyltestosterone in fish feed. Journal of Chromatography A, 2005, 824: 107-115

[45] Jiang J, Lin H, Fu X and others. Preliminary validation of high performance liquid chromatography method for detection of methyltestosterone residue in carp muscle. Journal of Ocean University of China, 2005, 4(3): 248-251

[46] Andrew RM, Craig JS, Allen MS. A stereochemical examination of the equine metabolism of 17 alpha-methyltestosterone. Analytica Chimica Acta, 2007, 581(26): 2663-2670

[47] Hasegawa H, Shinohara Y, Hashimoto T and others. Prediction of measurement uncertainty in isotope dilution gas chromatography/mass spectrometry. Journal of Chromatography A, 2006, 1136 (2): 0021-9673

[48] George K, Georgios T, Ioannis NP. Determination of anabolic steroids in muscle tissue by liquid chromatography-tandem mass spectrometry. Journal of Agricultural and Food Chemistry, 2007, 55(21): 0021-8561

[49] Lu HH, Conneely G, Crowe MA. Screening for testosterone, methyltestosterone, 19-nortestosterone residues and their metabolites in bovine urine with enzyme-linked immunosorbent assay (ELISA). Analytica Chimica Acta, 2006, 570(1): 0003-2670

[50] Jenkins AL, Bae SY. Molecularly imprinted polymers for chemical agent detection in multiple water matrices. Anal Chim Acta, 2005, 542(1): 32-37

[51] Mathieu LN, Anne SL, Benoit G and others. Selective removal 17β-estradiol at trace concentration using a molecularly imprinted polymer. Water Research, 2007, 41: 2825-2831

[52] Baggiani C, Baravalle P, Giraudi G and others. Molecularly imprinted solid-phase extraction method for the high-performance liquid chromatographic analysis of fungicide pyrimethanil in wine. Journal of Chromatography A, 2007, 1141: 158-164

[53] Huang Z, Tang Y. Pretreatment of cephalosporin in biological samples by molecularly imprinting technique. Chinese jounal of analytical chemistry, 2005, 33(10): 1424-1426

[54] Yan L, Huang Z, Pu S and others. Preparation of chloramphenicol molecularly imprinted polymer. Journal of Nanchang Institute of Aeronautical Technology, 2005, 19(3): 1-5

[55] Sellergen B. Noncovalent molecular imprinting: antibody-like molecular recognition in polymeric network materials. Trends in Analytical chemistry, 1997

[56] Chen Y, Kele M, Sajonz P. Influence of thermal annealing on the thermodynamic and mass transfer kinetic properties of D- and L-phenylalanine anilide on imprinted polymeric stationary phases. Anal Chem, 1999, 71: 928-938

[57] Cormark PAG, Mosbach K. Molecular imprinting: recent developments and the road ahead. React Funct Polym, 1999, 1(13): 115-124

[58]　Tang A, He M and others. Clinical value and methodology evolutiation of immunobloting for detecting IgG in the serum of patient contracted syphilis. Modern Laboratory Medicine, 2002, 17(3): 19-21

[59]　Zeng Y. Evolutiation of immunobloting for detection of helicobacter pylori antibody in serum. The Journal of Medical Theory and Practice, 2008, 21(1): 88-90

[60]　Xu Z, Lin X, Yao G.. Evlutication of immunobloting test for auto-antibody of diabetes patient's serum. Laboratory Medicine, 2008, 26(2): 197-197

[61]　Zhao J, Wang X, Ding H and others. Immunobloting test for specific IgE of allergen. Journal of Modern Laboratory Medicine, 2008

[62]　Tse Sum Bui B, Merlier F, Haupt K. Toward the use of a molecularly imprinted polymer in doping analysis: selective preconcentration and analysis of testosterone and epitestosterone in human urine. Analytical Chemistry, 2010, 82, 4420-4427

Chapter 9   Immunoelectron Microscopy

9.1 Overview

Great contribution has been made by immunocytochemical technique on research of immunoreaction at cellular level. However, for the limitation of optical resolution, it could not be used to study immune response of cell ultrastructure. In 1959, Singer applied ferritin with high electron density to label antibody, which made it possible to explore antigen-antibody reaction of cell ultrastructure. Since then many related technologies developed on this basis, such as hybrid antibody technology, ferritin and anti-ferritin complex technology, PA-ferritin labeling technology, inmunoenzyme technology, colloidal gold technology, etc. Electron microscopy immunocytochemistry (hereinafter abbreviated as immunoelectron microscopy, IEM) is different from immunocytochemistry and conventional electron microscopy, mainly in the following aspects.

9.1.1   Tissue fixation and sampling

The ultrastructure and antigenicity of tissue should be well-preserved. Therefore, the fixative used in this respect should not be too strong. Common fixatives[2] for immunoelectron microscopy are paraformaldehyde-glutaraldehyde mixture and periodate-lysine- paraformaldehyde mixture (called PLP for short). And Bouin's solution, Zamboni's solution or 4% paraformaldehyde solution (preparation method see appendix) are also adopted. In many literatures, PLP solution are recommended to apply for the immunoelectron microscopy for its excellent fixing effect on carbohydrate-rich tissue. Since most tissue antigens are composed of protein and carbohydrate and the epitopes are located in the protein part, the antigencity of tissue will be stabilized if carbohydrate are fixed selectively. Periodic acid in PLP solution can oxidize carbohydrates into aldehyde which can bring about intermolecular and intramolecular connect by the effect of lysine. In this way, the tissue antigens will be stabilized. However, paraformaldehyde and glutaraldehyde are economical, simple and efficient as fixatives compared to expensive lysine. In sampling, the immunoelectron microscopy is more rapidly, finer than immunocytochemical light microscopy technique.

9.1.2   Immune staining

It can be divided into three kinds: staining before embedding, staining after embedding and immune staining of ultrathin sections [3].

9.1.2.1   Staining before embedding

Immune staining will be carried out first, and then the immunoreactive sites is taken out from positive part and cut into small pieces under a dissecting microscope. At last, the pieces are fixed, dehydrated and embedded by osmium tetroxide according to conventional pretreatment approach in electron microscopy. If the range of specific immune response is too small, a second embedding can be carried out for accurate orientation. The detailed steps are as follows: the tissue is placed between two layers of thermanox plastic film during first embedding, epoxy resin held within the folder like a sandwich bread will undergo polymerization at high temperature, and then the required parts are removed for second embedding under a dissecting microscope. The middle part of tissue is more desirable for stained before embedding than the surface and the deep part. These are because the structure of the surface part are hardly kept integrated due to mechanical finishing, and the immune response of the deep part is weak or even undetected because antibody can't penetrate deeply. Semi-thin sections should be preceded by ultrathin sections in order to find out the immunoreactive sites. Sometimes semi-thin sections can be observed under phase contrast microscope without staining (i. e. PAP staining) and the parts of the immune response are black spots. On semi-thin section stained by the HE or toluidine blue, the parts of immune response are brownish yellow. According to this, positioning for ultrathin sections can benefit enhancement of the positive detection rate. To avoid confusion between electron microscopy lead, uranium staining reaction and the immune reaction, the ultrathin sections from the continuous phase can be picked up by two copper networks respectively. One sample is for staining observation, and the other for control observation with uranium staining or without staining.

The advantages of staining before embedding are as follows: ① Slices avoids the procedure of osmium tetroxide fixation, dehydration and resin embedding before staining, antigen is not destroyed and good immune reaction is easily achieved. ② The immunoreactive sites may be positioned for ultrathin sections and the detection rate under electron microscopy will be improved especially for tissues with fewer amounts of antigen. But certain ultrastructural damage usually appears after a set of steps in the immune staining.

9.1.2.2   Staining after embedding

The staining after embedding is implemented as follows: after fixation, dehydration and resin embedding, tissue is cut into ultra-thin slices, and then perform immunohistochemical staining. For the ultrathin sections posted on the network are the object of immune staining, it is also called on grid staining. Some points should be noted: ① There are different views on the osmium tetroxide application after fixation. In general, it is better without osmium tetroxide or to allow osmium tetroxide affect as short as possible. Theoretically, osmium tetroxide would protect antigen, but in practice it has been proved that osmium tetroxide will significantly reduce the antigenic activity. ② During on grid staining, nickel or gold network should be chosen since copper can easily react with chemicals. ③ During whole immunohistochemical process, the network should be kept wet since the drying will reduce antibody activity. This method holds some advantages: simple, that ultrastructure is well preserved, positive structures are highly repeatable and multiple immune staining can be done on the same section. However, the activity of antigen may be weakened or even lost during the processing of electron microscopy biological sample; epoxy group in epoxy resin may react with tissue components and the properties of antigens may change during the polymerization; tissue embedded in epoxy resin is difficult for immune reaction. Therefore, researchers in immunohistochemistry had tried to reduce or remove the embedding medium by various approaches such as saturated benzene, ethanol, NaOH saturated solution or ethoxy sodium solution, which obtained some effect to some extent. Nowadays the approach, etching with H_2O_2 before immune staining to remove osmium and enhance resin penetration, is widely applied.

9.1.2.3 Ultra-thin frozen section

According to the Tokuyasu method, the tissue was placed in 2.3 mol/L sucrose solution, frozen in liquid nitrogen, then sliced in a frozen section machine. The slice thickness will be slightly thicker than conventional resin sections. The ultrathin frozen section is directly immunostained without steps such as fixation, dehydration and embedding, etc. , so antigenicity is well preserved and has advantages of the aforementioned two staining methods.

9.1.3 Embedding

9.1.3.1 Resin embedding

Epoxy resin embedding method is now widely used in China. Tissue can be not only directly embed after dehydration, but also embedded in situ by attaching small pieces of tissue or semi-thin sections to the slide, inverting a gelatin capsule filled with epoxy on the slice for polymerization and hardening.

9.1.3.2 Embedding at low temperature

The antigenicity of tissue may be lost fully or partly during conventional resin embedding for processes, for example, polymerization at high temperature. Thus, low techniques are developed in many laboratories in immunoelectron microscopy, such as cryogenic technologies and ultra-thin frozen section, etc. The low temperature embedding is easier to implement than ultra-thin frozen section for the latter need to be equipped with frozen thin slicer and is more difficult. Study on low-temperature embedding medium started in the 60's of the 20th century. In the 80's, the wide applications of immunocytochemical technology in the electron microscope level opened up vast areas of experimental study on low-temperature embedding medium. Since then numerous applications have been reported in China. Ethylene compounds are the most common low temperature embedding medium, such as glycolmethacrylate (GMA), Lowicryls, LR White, LR Gold, etc. They are provided by manufacturers such as Polysciences INC, Reichert-Jung, and LKB series at present. Common low temperature embedding mediums and their application are briefly introduced as follows.

1) Lowicryls are chemicals of acrylate and methyl acrylate, including products such as K4M, HM20, K11M, KM23 (polysciences INC), which are characterized by maintaining low viscosity at low temperatures (K4M: -35 ℃; HM20: -70 ℃; K11M, HM23: $-80 \sim -60$ ℃). In addition they have the ability of polymerization in the light irradiation (UV, wavelength 360 nm) while the light polymerization is independent of temperature. K4M and are K11M are hydrophilic and particularly suitable for the application of immunohistochemistry, because they can not only keep the structure and antigenicity of tissue well, but also reduce the background staining. HM20 and HM23 are hydrophobic, can produce high contrast images and suitable for making sections under scanning electron microscopy, transmission electron microscopy and dark-field observation. All of these types of low-temperature embedding medium are suitable for frozen technology.

2) LR White and LR gold are a transparent resin composed of mixed acrylic monomers with a very low viscosity (8 cps) and high hydrophilicity. Therefore, they have strong penetrability, which benefits antibody (or antigen) and immune chemicals to penetrate through the LR resins into tissue binding sites. They have achieved good results when applied in light microscopy immunocytochemistry (semi-thin sections) and electron microscopic level. Specimens can keep antigenicity well when dehydrated to 70% ethanol. The suppliers of LR white and LR gold are Poly-sciences INC, etc. LR gold, an embedding medium of light induced polymerization at low-temperature, is especially suitable for immunocytochemistry to keep tissue antigen and antibody activity to the maximum extent. The optimum temperature for photopolymerization is at -25 ℃, and the polymer shows gold from which the name is. LR White can polymerize in both hot and cold conditions. Thermal polymerization is at 60 ℃ for 24h, while cold polymerization is at -25 ℃ and needs adding accelerator to adjust the proportion in preparation. Processing and immunohistochemical staining of biological samples is similar to conventional resin sections.

3) GMA is abbreviation of 2-hydroxyethyl methacrylate (HEMA). As far back as

the 1960s, electron microscopy workers tried to use it as a biological embedding medium in light and electron microscopy. As an embedding medium for electron microscopy, GMA has the advantage of high electronic density and good image contrast. However, it also has three main drawbacks: firstly tissue embedded after polymerization is crisp, not easy to trim and slice; secondly, the polymerization easily leads to tissue damage, for example, artificial organelle swelling; thirdly, the polymer lacks stability. The embedding medium cannot afford electron beam bombardment while heat will sublime it and result in tissue collapse and deformation. So it was afterwards replaced by epoxy resin which has good stability after polymerization and can withstand the electron beam bombardment, and, meanwhile it can achieve good image contrast and high resolution. But the embedding medium based on epoxy resin is not satisfactory with semi-thin section staining. Yet GMA-embedded sections is superior to epoxy resin in terms of staining effect. Thus electron microscopy scholars such as Leduc and Bernhard tried to increase the proportion of plasticizer such as methyl acrylate and a small amount of water. Besides, they added an appropriate amount of plasticizer such as polyethylene glycol 400 (PEG400) to change its hardness and also added appropriate amount of ethylene dimethacrylate to enhance its stability against electron beam. bombardment. In order to avoid tissue structure damage caused by excessive polymerization at high temperature, low temperature initiator-benzoyl peroxide (benzoyl peroxide) was adopted at the temperature range $-30 \sim -10$ ℃. Through on going preparation improvement, GMA is widely used in semi-thin sections (1-3 μm) of the light microscope, especially the histochemical research and immunocytochemistry at the electron microscopy level.

Low temperature embedding medium, now commonly used in embedding staining of ferritin or colloidal gold immunoelectron microscopy can help detect a variety of antigen which can't be detected when the embedding medium is epoxy resin.

In short, no matter what kind of immunoelectron microscopy, all of them are facing the contradictions of the preservation of fine structure and the preservation of the antigen activity of tissue. The fixatives such as glutaraldehyde and osmium tetroxide benefit the preservation of microstructure, but damage the antigenic activity. H_2O_2 can increase the resin penetration, but damage the microstructure and produces pores in reaction site. These two aspects should be noted during processing biological samples. In addition, a thorough cleaning should be noted in each immune staining, otherwise non-specific products and some pollutants will affect the display and observation of specific reaction products. Some scholars suggest spraying along nickel mesh by plastic bottle with conical spray head is better for cleanout than the usual washing method in glass water currently used. One should be careful not to touch the network when drying the residual water droplets by dry filter paper after rinsing. Filter can be cut into triangles to suck up water droplets by using cutting-edge to contact with water droplets. Double distilled water and special containers must be applied during the whole process and special containers should be dedicated.

9.2   Ferritin immune technology

9. 2. 1   Basic principles

Ferritin is a protein containing iron ion with high electron density. Antibody and ferritin can combine together into a two-molecule complex by bifunctional agents with low molecular weight. The complex retains the original activity of antibody. Moreover, ferritin containing dense iron ion core which can be seen in the electron microscope. Thus, the ultrastructure of antigen or antibody in tissue cell can be identified and located with the trace precipitation caused by specific binding of antigen and this complex[4].

9. 2. 2   Extraction and purification of ferritin

Healthy horse spleen (fresh or frozen) is treated as follows: remove the lymph nodes and adipose tissue outside the spleen, add distilled water to wet weight by ratio of 1 : 1. 5 or 1 : 2, homogenize them by tissue homogenizer and then heat in a water bath to 75-85 ℃ to denature proteins except ferritin; after cooling, filtrate the homogenized tissue with 2 to 4 layers of gauze and keep the supernatant after centrifugation; after adding 35 g ammonium sulphate per 100 mL supernatant, stir the solution at 4 ℃ overnight to precipitate ferritin; discard supernatant after centrifuged at 3,000-10,000 r/min for 20 min and scrape and wash the precipitate with distilled water to dialysis bag in order to remove ammonium sulphate; then add cadmium sulphate at 4-5 g per 100 mL to crystallize ferritin at room temperature or 4 ℃; dissolve the ferritin crystal with 2% ammonium sulphate (pH 5. 58) and then crystallize it by cadmium sulphate again; repeat the crystallization for six times and then obtain purified ferritin. This ferritin can be stored in the half-saturated ammonium sulphate solution for 1 to 2 years and can be applied immediately after dialyzing with distilled water. The negative staining can be used in electron microscopy to understand the integrity and purity of ferritin[5].

Commercial ferritin is 1%-2% solution (pH 5. 85) diluted by 2% ammonium sulphate solution. In order to obtain satisfatory results, the commercial ferritin should be purified before application. Purification can be carried out as follows, adjust pH value to 5. 85 by 0. 1 mol NaOH or 0. 1mol HCl, and then add 20% cadmium sulphate solution to reach the final concentration to 5% in the ferritin solution. Afterwards, the solution is placed in the refrigerator at 4 ℃ for 2 h (or overnight) to completely crystallize ferritin, and then the supernatant is removed after centrifugation for 1h (2,000 r/min). Then the ferritin crystal is dissolved and centrifuged to remove insoluble particles. Add 5%

cadmium sulphate to the supernatant until the ferritin crystal show brownish red under the microscope. The crystals one dissolved in 2% ammonium sulphate solution and precipitated with 50% ammonium sulphate for three times. The third precipitate is dissolved in a small amount of distilled water, dialyzed with water firstly and then with 0.05 mol/L phosphate buffer (pH 7.5) for 12-24 h. The purified ferritin solution is ultracentrifuged at 100,000 r/min for 2 h, 3/4 colurless supernatant is removed and precipitation (containing iron protein) is kept at 4 ℃ overnight until completely dissolved. Then it is filtered with a microporous filter (pore size 0.25-0.45 μm). The product can still be effective after preservaction for 1 to 2 years.

9.2.3 Combination of ferritin and immunoglobulin

They are generally combined together by bifunctional reagents, such as dimethyl phenylene diisocyanate (abbreviated as XC), 2,4-toluene diisocyanate (abbreviated as TC), O-anisidine (abbreviated as BDD), O-fluoride-dinitrodibenzohalonium alum (abbreviated as FNPS) and glutaraldehyde. In recent years, it is generally considered that glutaraldehyde is a good conjugation agent because of little effect on antibody activity and high yields of labelled antibody. It is divided into one-step and two-step methods[6], which are outlined below.

9.2.3.1 One-step method

15 mg of ferritin and 3 mg of globulin are dissolved in 0.9 mL 0.1 mol/L phosphate buffer (pH 7.0), then 0.1 mL freshly prepared glutaraldehyde solution is added to a final concentration of 0.005% to 0.05%. Then 0.02% sodium azide (NaN_3) is added for antisepsis. The mixture is placed without stirring at 37 ℃ for 24 h and then 0.01 mol/L lysine solution is added to terminate the conjugate reaction.

9.2.3.2 Two-step method

1) Diluted glutaraldehyde is added into 0.1 mol/L phosphate buffer (pH 7.0) containing 50-80 mg ferrtin and finally 1 mL of 0.05%-0.15% glutaraldehyde is obtained.

2) After setting at 37 ℃ for 2 h, the solution is filtered through dextran G-25 column to remove unbound glutaraldehyde. To avoid unnecessary dilution, only the middle part of diluted peak are collected. Then globulin is added (ratio of ferritin to globulin is 5 : 1) to achieve 15 mg/mL of ferritin and globulin 3 mg/mL finally.

3) After the mixture is set without stirring at 37 ℃ for 12 h, 0.01 mol/L lysine are added to stop crosslinking. Two-step reactions are required to be carried out in 0.02% NaN_3 under preservative conditions.

9. 2. 4   Preparation of electron microscope specimens

9.2.4.1   Fixation

Aldehyde and osmium tetroxide are applied for double fixation in this specimen preparation similar to conventional electron microscopy in order to maintain cell and tissue ultrastructure. It was reported that antigen activity would not be affected when 4% formaldehyde solution in pH 7. 2 phosphate buffer was used for fixation at 0 ℃ and treated with osmium tetroxide again. Potassium permanganate can lead to the activity loss of most of antigens, so it should not be used generally. In addition, one of the characteristics of ferritin-labelled antibody is of large molecular weight. Therefore, as for the localization of cell surface antigens, sample can be directly put into the fixative. Otherwise, appropriate measures must be taken to break the cell membrane in order to enhance the permeability of cells to labelled antibodies. It often takes the following approach [7].

1) Freeze-thaw method: small pieces of tissue or cells after fixed are frozen and thawed once to rupture the cell membrane, so that the labelled antibody can enter the cell.

2) Frozen section method: After getting fixed, the tissuse are rapidly frozen and cut into 10-15 μm thick slices, and then the melted slices are soaked in ferritin labelled antibody solution.

3) It has been reported that the tissue was immersed into 4×10^{-3} mol/L digitalis saponin solution for 1-2 min after getting fixed by glutaraldehyde, which could enhance the permeability of cell membranes and help labelled antibody solution enter the cell.

9.2.4.2   Immune reaction processing

The common method is staining before embedding, which can be divided into direct and indirect methods. Before staining, the tissue will be cut into about 10-20 μm thick slices.

(1) Direct method

The slices are soaked in the labelled antibody solution at room temperature or 37 ℃ for 1-2 h or longer time. They are rinsed by a buffer (cold buffer is suggested) to remove unbound labelled antibody, and then conventional double fixation and electron microscopy embedding is carried out.

(2) Indirect method

1) the tissue sections are immersed in primary antibody solution at room temperature for 30-60 min.

2) Stirring and washing with a deal of cold buffer for at least 3 times and 30 min

each time

3) Immersed in ferritin labelled antibody solution at room temperature for 30 min.

4) after being washed with buffer like ②, natural precipitation or centrifugation can be used to separate tissue sections.

5) The small piece of centrifugation are fixed with osmium tetroxide for 30 min.

6) after dehydrated, embedded, sectioned and stained, the slices can be observed on conventional electron microscoe.

9.3　Other immunoelectron microscopy

9.3.1　Colloidal gold labelled method for immunoelectron microscopy

Colloidal gold method was proposed by Faulk and Taylor in 1971[8] and used in immunoelectron microscopy at first. Colloidal gold labelled method is implemented by its adsorption with antibodies for its negative nature in alkaline environment. Since colloidal gold solution seems bright cherry red in light microscope, slice does not need additional staining after gold labelled antibody reacts with antigen on slice [9]. Since gold particles have high electron density and are clearly visible under electron microscope, colloidal gold labeling method for immunoelectron microscopy has been successfully applied in various fields of biology in recent year, made encouraging progress and fixed some problems that had remained in the past. Since 1980s there is a trend to replace PAP technology of immunoelectron microscopy by this method. Colloidal gold labelled antibody technology has many advantages in application at electron microscopy level. Firstly, it is not as cumbersome as PAP method, and has less impact on the ultrastructure without damage by H_2O_2. Secondly, the gold particles with high electron density are clearly visible and are easy to be distinguished with other immune products under the electron microscope. Therefore, colloidal gold labelled method can combine with PAP method to locate the ultrastructure by dual or multiple staining. In addition, it is a powerful tool for studying neurotransmitters coexisting in synaptic vesicles by using different diameter gold particles to label different antibodies. The number of gold particles in antigen-antibody reaction site can be used for quantitative chemical study of immune cells roughly. Gold labelled antibody can also be added in culture medium to mark the antigens in cultured cells. Gold label has been reported about studying the cell skeleton successfully. Since gold label has a strong ability of exciting electrons, it can be applied not only to observe the ultra-thin sections under transmission electron microscopy, but also to analyze the surface antigen and receptor position of cells under scanning electron microscopy.

9.3.1.1   Immuno-gold staining at electron microscopy level

Immune methods applied at the electron microscopy level can be divided into staining before embedding and staining after embedding. Because the method of staining before embedding has poor permeability to cell membrane, it is generally used to mark antigen on cell surface. For penetrating the cell membrane, it should be aided by freeze-thaw method or by adding Triton X-100, saponin or other active agents which will enhance destruction of cell ultrastructure. Thus, staining after embedding method is commonly used now.

In an ideal immuno-gold staining section, the background is clean without residual gold or other inorganic particles, and gold particles are concentrated in the reaction site of antigen and antibody. To obtain ideal immuno-gold staining sections, many factors should be paid attention to and mainly included: ① the antibody should be of high specificity and affinity; ② the tissue to be detected should be of high antigen concentration; ③ the cleanness of rinse fluid and various vessels applied during whole process, and the effect of rinsing; ④ all solution should be filtered though micropore membrane filter with 0. 2-0. 45 μm pore size and all vessels should be cleaned and for special usage. The whole procedure should be carried out in a wet box so that network can be kept moist.

9.3.1.2   Colloidal gold labelled protein A technique

Colloidal gold labelled protein A technique (PAg) is widely used at electron microscope level, because of its high specificity and sensitivity, simpleness and slight background staining, etc Immune specificity of protein A (PA) has been introduced in chapter 5. The preparation of PAg complex is simple. As secondary antibody, it has no species specificity, which can avoid preparing specific immunoglobulin for different specie of animal. PAg complex rarely has non-specific interaction with the embedding medium and cells composition. Non-covalent binding between protein A and gold particles affects neither the activity of protein A, nor high stability of it. PAg complex molecule is small and can penetrate the tissue easily.

9.3.1.3   Colloidal gold dual-labelled technique (usually is PA-colloidal gold)

1) Single-sided method. Two different antibodies are labelled with different diameter gold particles. Firstly stain with one antibody by indirect method and then with another antibody after washing.

2) Two-sided method. Antibodies labelled with different diameters gold particles are immunostained at both sides of the nickel network respectively. This method can avoid not only interference between two kinds of gold labelled antibodies, but also space exclusion. Here space exclusion means that the second antibody cannot have appropriate

space to combine its corresponding antigen after the first antibody combines with its corresponding antigen and occupies most of space.

3) Staining method antigen-antibody from different specimen. For example, the first antibodies are antiserum of human and rabbit such as human anti-HLA A and, rabbit anti-HLA B, respectively. The second antibodies were anti-IgG of human and anti-IgG of rabbit labelled with different diameter gold particles. Because of different species, two anti-sera won't interfere with each other. The primary antibodies and secondary antibodies can be blended for operation respectively, and consequently four steps can be reduced to two steps.

4) Gold-labelled antigen detection method (GLAD method). This method was firstly proposed by Larson in 1977. The specific antibody can react with tissue antigen first and then with labelled pure antigen, based on which Larson proposed the application of colloidal gold in the double or multiple antigens localization at electron microscopic level. The principle of GLAD is as follows: if bivalent IgG molecules are excessively added to tissue sections with antigen, only one antigen binding site of IgG molecule which has two antigen binding sites can combine with antigen and the other binding site will react with gold-labelled antigen. In double labelling method, two different antigens are labelled with different diameter gold particles, respectively. And the localizations of these two antigens at tissue section can be shown after the two antigens react with corresponding primary antibody. The GLAD method has advantage of high specificity. This is because the site of the antigen is shown thorough two options and labelled antigens can only bind to the corresponding specific antibody, not to nonspecific IgG, However, this method needs pure antigen as the same as the one to be checked and labelling different antigens separately. Since the requirements can't be achieved in a common laboratory, the GLAD method has not been widely used so far [10].

9.3.2　Scanning immunoelectron microscopy

9.3.2.1　Labels

Labels used for scanning electron microscope should be in the resolution range of the microscope and have good locating capability to cell or tissue antigens. Labels should be chosen according to research purpose. The large-sized labels are suitable for large-sized cells, for example, to identify positive cells (labelled cells) and negative cells (unlabelled cells), whereas, the small-sized and easily recognizable labels are suitable for locating receptors[11].

Common labels are particulate and can be classified according to their characteristics as follows.

1) Proteins such as hemocyanin and ferritin.

2) Pathogens such as tobacco mosaic virus, southern bean mosaic virus, bacterio-phage T4, *E. coli* F2, phage, and so on.

3) Metal particles: such as colloidal gold and silver particles in immuno-gold stai-ning, immuno- silver staining and isotope autoradiography.

Metal particles are the labels most widely applied and colloidal gold is the most common in them. The diameter of commercial available colloidal gold ranges from 3 nm to 150 nm. Gold particles of 30-60 nm are usually adopted in scanning immunoelectron microscopy. Gold is a heavy metal itself and has a strong secondary electron emission, so there is no need spraying metal film when colloidal gold is adopted as a label. This is why colloidal gold label is superior to others in scanning immunoelectron microscopy. Immuno-gold-silver staining can enhance the density of metal particles on the surface of cells or tissue, because gold and silver particles with high electron density will be shown as clear-shaped particles and are easy to be identified and located under electron micro-scope. As for pathogen labels, the purpose to locate is mainly achieved by utilizing its special shape and structure. For examples, the shape of bacteriophage T4 is like star racket, its head is like hexagonal star with about 100 nm diameter, and the tail length is about 100 nm. In addition, the head of the virus connects with its neck. Tobacco mosaic virus is a baculovirus of 15 nm×30 nm size and southern bean mosaic virus is round par-ticles with 25 nm diameter. These pathogens are very easy to be identified by their typi-cal shapes. Ferritin has high electron density because of containing a dense iron core, so it can achieve the purpose of label. Hemocyanin is multi-molecular polymer extracted from the conch mollusks and its shape is 35 nm×50 nm cylindrical. It is mostly applied for virus research. Sometimes it is also applied for the cell membrane receptor location because the glycosyl part of hemocyanin can combine with lectin.

9.3.2.2   Immune labelling methods

Immune labelling of metal is based on the same principle as section immuno-stai-ning. After combining the label with antibody, the antigenic site will be shown through direct or indirect method. Colloidal gold adsorpted with Protein A can combine with Fc segments of IgG molecule and it can react with biotin after adsorpted with avidin. In im-muno-gold-silver staining development of silver solution can be carried out after labelling with colloidal gold. Viruses (including phage) labels are mostly used through non-la-belled antibody method which is called bridging method. This method utilizes two spe-cific primary antibodies against an antigen and a label respectively (e. g. rabbit anti-A antigen and rabbit anti-HRP) and the secondary antibody prepared against IgG from the former animal serum (e. g. goat anti-rabbit IgG). By using the secondary antibody as the bridge, the specific antibodies against the antigen and the label combine together. Its basic principles are similar to PAP method. Pathogen immune labels can display by using their morphological characteristics or antigen-antibody agglutination which takes place after a virus reacts with a specific antibody of viral antigen. The localization can be completed under electron microscope after concentrated and negative staining.

9.3.3 Freeze-etching immunoelectron microscopy

Freeze-etching, also known as freeze replica or freeze fracture, is an important method in research on membrane structures [12]. The main steps are as follows: firstly the sample is frozen in liquid nitrogen and then placed in a vacuum sputtering instrument to cut off. There are organelles and also frozen water on the section. After heated to sublimate ice and evaporate the water, the organelle membrane structure will be exposed. And this step is called freeze-etching. If without etching it is called frozen cut. Exposed membrane structures are projected with platinum-carbon ion and then sprayed the carbon again for reinforcing. Consequently, in the cut sample surface a complex film forms. That means the three-dimensional structure of cell section is printed in the complex film. The sample is removed from the vacuum and then the tissue under complex film is etched down. The complex film is fished to the copper network and observed under transmission electron microscope.

As for the molecular structure of biofilm, fluid mosaic model is currently recognized and confirmed by frozen replica electron microscopy (Figure 9-1). The model is also called lipid-globular protein mosaic model. According to this model theory, biofilm is a fluid and plastic lipid bilayer membrane mounted with protein molecules. Each molecule on bilayer membrane has two poles. One end is hydrophilic pole towards the surface of the inner and outer membrane, while the other end is hydrophobic pole towards the center line of membrane. The two rows of molecules which are arranged opposite to each other form the structure basis of the biofilm. Protein molecules can be categorized into the mounted and the external. The former account for about 3/4 of total protein and seem approximately spherical and mount in the lipid molecules layers within different depth, while the latter are mostly attached to the cytoplasmic surface of cell membrane. After freeze fracturing, the horizontal section of biofilm mostly occurs at hydrophobic pole of membrane unit. The hydrophilic part of membrane, named PS (the surface adjacent to the cytoplasm, nucleoplasm or mitochondrial matrix) and ES (the surface adjacent to the extracellular space or intracellular space such as endoplasmic lumen, nucleoplasm space, lumen between the inner and outer membrane of mitochondria and lumen of other bubble, etc.). The hydrophobic part of membrane, also known as the fracturing surface, is called EF (the side facing the cytoplasm, nucleoplasm or mitochondrial matrix) and PF (the side facing the extracellular or intracellular space). From the early 1970s, freeze etching immunoelectron microscopy has come into vogue. But because immuno labelling should be carried out before freeze etching, only the outer surface of cells

(ES) cab be labelled. Since 1980s, a fracture-labelled cell chemical method started to establish. In this method, both sides of the central section (EF and PF), each surface of organelles membrane, cytoplasm and nucleus can also be labelled after fracturing the cell membrane, which pave a way for a wide application of this method. This method can also be used in quantitative statistics of antigen and receptor molecules.

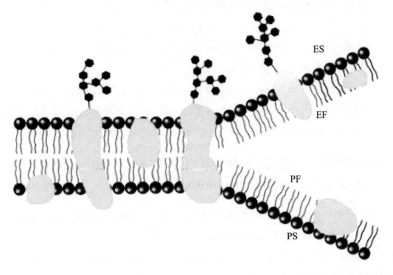

Figure 9-1　Lipid mosaic model and frozen fracture surface of biofilm molecules(See Colored Figures)

9.4　Situ hybridization for electron microscopy

Over the past 20 years since Gall and Pardue established situ hybridization technique, this technique has provided extremely valuable information for gene localization and expression, gene evolution, developmental biology, oncology, microbiology, virology, medical genetics and genetic analysis, and played a role that other techniques cannot replace. In recent years, applications of this technique gradually expanded and improvements have taken place technically in two major aspects [13]. On the one hand, a set of amplification methods are applied to enhance the sensitivity, on the other hand, its resolution was improved towards electron microscope level. At first, Jacob[14] hybridized RNA probe labelled with ^{3}H and DNA and observed them successfully at electron microscopy level. Since then many scientists have done a lot of work in this area, but all work was based on the applications of isotope-labelled DNA probe. Until 1986, Binder[15] utilized biotin-labelled rDNA probe and UI probe in analysis of the drosophila ovary. He implemented situ hybridization on a Lowicryl-K4M low-temperature embedded sections with good preservation of the morphological structure of the tissue, which obtained high signal-to-noise ratio with protein A and colloidal gold complex as display system. In 1986, Webster and others [16] used biotin-labelled probe to show the distribution of intracellular mRNA. Singer and others [17] successfully applied dual-labelled situ hybridization not only to detect the mRNA combined with the matrix in whole mount extracted cells but also to show the protein expressed by the mRNA. Jiao and others [18] underwent situ hybridization electron microscopy on whole mount extracted cells by the biotin labelled adenovirus genome probe. They observed the colloidal gold particles combined with nuclear matrix as cluster or beaded string and successfully proved the close combination of adenovirus DNA and nuclear matrix. Xiang and others [19] showed the POMC mRNA localization in the rough endoplasmic reticulum of neurons by using staining before embedding method with digoxin labeled probe and HRP. In short, situ hybridization at electron microscopy level is being improved unceasingly. Here are some typical situ hybridization techniques at electron microscope level.

9. 4. 1　Application of isotope-labelled cRNA probe with situ hybridization electron microscope in chromosome

1) The whole chromosome were stretched on the gold network, fixed in 70% alcohol vapour for 4-15 h and then dried in air.

2) DNA denaturation by alkali. The gold network was placed in $2 \times SSC$ (adjusted to pH 12 with 0. 1mol NaOH) at room temperature for 2 min.

3) Dehydration. The gold network was washed with 70% and 95% alcohol respectively for 3 times, and then dried in the air.

4) Hybridization. The gold network was covered on hybridization droplet in a petri dish. Every droplet is of about 10-15 μL. The small dishes are placed in hybridization buffer in a large plastic box, kept moist and incubated at 60-65 ℃ for 4-24 h.

Preparation of hybridization solution: $6 \times SSC$ or $6 \times TNS$ ($1 \times TNS =$ 0. 1mol/L NaCl, 0. 01 mol/L Tris, pH 6. 8, containing 30,000-50,000 cpm/10μL ^{3}H cRNA).

5) Washing after hybridization. After removed from the hybridization solution, the gold network is immediately washed with a large volume of $2 \times SSC$ solution for several times. After washed with RNA enzyme solution (20 μg/mL) at room temperature, the network is washed again with $2 \times SSC$ solution at room temperature. At last, it is dehydrated with gradient alcohol (70%- 95%) and then dried in air.

6) Immersed in latex film. It is applied L-4 nuclear emulsion, diluted with water for 4 times. After immersed, the gold network was fixed on glass slides with tape and sealed in the black plastic box at 4 ℃. The exposure depends on isotope type and the content of target mRNA. Different exposurse can be compared and at last an exposure is determined by development degree.

7) Imaging. Cassette is warmed at room temperature for at least 30 min to prevent the emulsion film shrinkage after being taken out of the cold room or refrigerator. Then

it is put in the developer (for example, Kodak Microdol X) to image at 18-24 ℃ for 3-5 min. After washing with water, it is put in 15% Kodak rapid fixer to fix for 5 min, and then dried in the air.

8) Observed under electron microscope.

All solutions should be freshly prepared as far as possible.

9. 4. 2   Application of biotin-labelled DNA probe with situ hybridization electron microscopy

9.4.2.1   Basic principles

Through base matching provided that a DNA chain serves as a template, a biotin-labelled deoxy nucleotide triphosphate (for example, Bio-dUTP) will inset into a gap in the DNA chain by applying DNA probes nick translation method. The biotin-labelled DNA probes can undergo situ hybridization with cell or tissue. And there are two kinds of display method. One method is through HRP. It is similar to staining before embedding in immunoelectron microscopy. Digoxin-HRP colour development method are applied to situ hybridization and immune cell staining of thick slices at light microscope level in order to reduce the amount of reagents, such as H_2O_2 and Triton X-10, which are easy to damage cells and tissue structure in the embedding process. After colour development, the positive parts undergo fixation after osmium tetroxide, dehydration, embedding, sectioning and observation with conventional procedures of electron microscopy. This method is not easy to achieve accurate localization since the product in immune reaction has a very high electron density by using HRP, is of different sizes and for non-specific products in reaction often cover the background of ultrastructure. Now biotin -PA colloidal gold label method is more widely used in electron microscope localization of situ hybridization and has achieved satisfactory results. Its basic principle is as follows: after biotin-labelled DNA probes situ hybridize with the cells or tissue, anti-biotin antibody-PA-gold complex can label the hybrids and consequently complete ultrastructural localization. This method is similar to the staining after embedding in immunoelectron microscopy. As for the embedding medium, successful application of epoxy resin is reported. However, most satisfying results are from the low temperature hydrophilic embedding medium-Lowicryl-K4M. LR-white also achieves good result [20].

9.4.2.2   Application of biotin-labelled DNA probe-PA-gold electron microscopy hybridization

1) Fixation. It is considered that mixed agent of 4% paraformaldehyde and 0. 1% glutaraldehyde in phosphate buffer (pH 7. 4) is an ideal fixative. A single agent with 4% paraformaldehyde is also recommended because it has been confirmed that the appli-

cation of high concentrations of 4% glutaraldehyde could reduce 60% hybrid rate of samples than paraformaldehyde. The fixed time by paraformaldehyde-glutaraldehyde is from 15 min to 1 h. Then the sample is placed in Ringer's solution, PBS or 5 mol/L phosphate buffer containing 7% sucrose at 4 ℃ overnight. Application of alcohol-acetic acid mixed solution is also reported.

　　2) Dehydration. Next day, the sample is washed with 0. 15 mol / L phosphate buffer (pH 7. 4) for 30 min and then dehydrated as per the following procedure: ① 65% ethylene glycol, 0 ℃, 60 min; ② 80% ethanol, −35 ℃, 120 min; ③ 100% K4M:80% ethanol=1 : 1, −35 ℃, 120 min; ④ 100% K4M:80% ethanol=2 : 1, −35 ℃, 60 min; ⑤ 100% K4M, −35 ℃, overnight.

　　3) Embedding. According to the method proposed by Roth and others[21], tissue is embedded in the capsule containing K4M and polymerization is effected by irradiation from ultraviolet light (wavelength 360 nm) at −35 ℃ for 24-48 h. After removing out the capsule, it is irradiated continuously with ultraviolet light at room at temperature for 24 h in order to harden the tissue for ultrathin sections. Capsules can be stored at room temperature for short-term and also in −70 ℃ refrigerator for nearly 1 year.

　　K4M preparation (medium hardness)

Cross-linker	2. 7 mg
Monomer	17. 3 mg
Initiator	0. 10 mg

The ratio of the compounds can be adjusted according to the hardness. As the amount of cross-linker increases, the hardness of tissue increases.

　　4) Slice. The ultrathin sections are 50-60 nm thick, fished and put on a nickel network coated Formavar and carbon film.

　　5) Pre-treatment and pre-hybridization of tissue. The sample is washed with 0. 1 mol/L glycine (pH 7. 4) containing 0. 2 mol/L Tris buffer for 15 min in order to remove the aldehyde fixative and decrease the impact on hybridization and detection. And then it is washed with 2×SSC for 15min and treated with denaturing solution (70% deionized formamide, 2×SSC) at 65 ℃ for 5-10 min in order to achieve denaturation. The denatured samples are pretreated in pre-hybridization solution (50% deionized formamide, 0. 5 mol/L NaCl, 10mmol/L Tris, 1 mmol/L EDTA, pH 7. 5) at 37 ℃ for 15 min.

　　6) Hybridization. The side containing section of nickel network is covered on the hybridization droplet (about 20 μL), and placed in a wet box at 37 ℃ for 1-2 h. The components of hybridization solution are similar to situ hybridization at light microscopy immunocytochemistry. DNA probe needs to be boiled in boiling water for 2-3 min in order to denature, and then quickly moved to ice bath.

　　One should note to prevent hybridization solution from drying on nickel network during the hybrid process.

7) Rinsing after hybridization.

$2 \times$ SSC solution containing 50% formamide	37 ℃	30 min
$1 \times$ SSC solution containing 50% formamide	room temperature	30 min
$1 \times$ SSC solution without formamide	room temperature	20 min
Sample set in $1 \times$ SSC	4 ℃	overnight
0.01mol/L PBS(pH 7.2)	room temperature	10 min
4% BSA	blocking	15 min

8) Dilute Goat anti-biotin IgG antibody (Sigma) with PBS solution in the ration of 1 : 100 (PBS solution: 2% NaCl, 0.05% KCl, 0.05% KH_2PO_4, 0.278% Na_2HPO_4, adding 300 mmol/L NaCl, 0.5% Triton X-100) and react at room temperature for 2 h.

9) Wash with c PBS containing 0.1% BSA for 3 times and 30 min each time.

10) Dilute rabbit IgG anti-goat IgG bound to 10 nm colloidal gold (Sigma) with a dilution buffer at ratio of 1 : 100 (the dilution buffer: 20 mmol/L Tris-HCl, pH 8.2, 0.9% NaCl, 0.02 mol/L sodium azide, 0.5% Triton X-100 and containing 1% BSA), at room temperature for 1 h.

11) Wash with PBS (pH7.2) containing 1% BSA for 3 times and 30 min each time and then wash with distilled water for 5 min, and then air dry.

12) Stain with uranyl acetate for 20 min and lead citrate for 15 min. Then air dry and then observed under electron microscopy.

The labelling method has some advantages, as follows: ①can realize rapid signal detection; ② can preserve morphology well; ③ SNR (signal/noise ratio) is similar to isotope-labelled situ hybridization electron microscopy. It is widely recognized as the ideal method for application of radioactive situ hybridization at electron microscopy level. Colloidal gold particles with high electron density is clearly distinguished from nonspecific staining and pollution and helps to achieve satisfying ultrastructural localization.

This technique also has some difficulties which include, ① long experimental period; ② when preparing section with hydrophilic low-temperature embedding medium, for the strong hydrophilicity, it is difficult to fish section and the section is brittle; ③the purity of probe and thorough rinsing after hybridization are the key points of successful experiment. For rinsing to excess, easily leads to section breakdown on nickel network, and so it is recommended to utilize plenty of rinsing solution, or rinse with multiple rinse cups, or frequently change the rinse solution, or adopt soft plastic bottle with pen-like nozzle, or increase water pressure, etc. But whether rinsing or washing with water, it should be noted that water flow must be parallel, not perpendicular, to the surface of network, otherwise the rinse is more likely to result in the section breakdown. ④ hybridization time. Binder and others proved specific signal will appear after hybridization for 1 h and be continuously increasing within 5 h. Lawrence and Singer [22] also showed that specificity could achieve maximum value after 3-4 h hybridization. Although hybridization was still continuing after this, the specificity didn't increase. Jiao and others[18] also achieved the same result on whole mount cells. They found that hybridization for 4h was similar to for 20 h, but the former could preserve cell structure better. Their experiments showed that the hybridization time of section was about 9-10 h. Therefore, the shortest and most efficient hybrid time is very important to maintain the morphological structure. Binder considered that the ideal hybridization time of biotin PA-Gold electron microscopy hybridization is 1-2 h. ⑤rinse buffer containing formamide easily leads to non-specific background from gold particles accumulation. Therefore, Binder recommended PBS buffer to rinse for 5×10 min instead of salt solution containing formamide, which was proved to be able to avoid or reduce non-specific staining; ⑥ most researchers found K4M embedding medium was the best in the electron microscope hybridization technology and the SNR (signal/noise ration) could be equal to that of cell smear and stretched chromosome without embedding agent.

Biotin-PA-Gold electron microscopy hybridization has obtained satisfying results in K4M embedding sections, but it is less sensitive than some isotope-labelled technique. Lawarence and Singer [22] compared three detection system as for their SNR. Result showed that the SNR with ^{32}P label, ^{3}H and biotin label was 70, 20, 10, respectively. Webster and others [16] also compared biotin labelled P0-HRP glycoprotein cDNA probe (staining before embedding) with biotin labelled cDNA probes P0-glycoprotein-PA- gold (staining after embedding) in hybridization electron microscopy technique. They observed the rat trigeminal ganglion Schwann cell myelin marker for 15 days with fixatives, PLA solution (4% paraformaldehyde, 15% saturated picric acid (V/V)) and 0.08% glutaraldehyde (in phosphate buffer). The both got color rendering. However, authors still concluded that the application of electron microscopy PA-Gold on K4M embedded sections in the network to hybridization was more ideal for displaying results and morphology preservation.

9. 4. 3　Application of digoxin-labelled rRNA probe in situ hybridization electron microscope

9.4.3.1　Basic principles

After digoxin-labelled DNA probes were applied widely in situ hybridization at light microscopy level and achieved very satisfying results, scientists began to try applying it at electron microscopy level [23]. The basic principles are similar to biotin-labelled DNA probes-PA-Gold electron microscopy hybrid technology. Firstly DNA probes were modified with digoxin, hybridized with cells or tissue and then combined with anti-digoxin antibody conjugated with colloidal gold for specific nucleotide ultrastructural localization of cell or tissue. To enhance the display effect of gold, silver enhancement can be used.

9.4.3.2 Basic Operation

(1) Tissue treatment

1) Fixation. Fixative is mixed solution of 4% paraformaldehyde (PFA) and 0.5% glutaraldehyde (GA) prepared in PBS and stored at 4 ℃. The immersion method can is applied as follows. Rat liver is put in cold fixative (4% PFA without GA) and cut into small squares around 1 mm^3. The dressed tissue is removed into another glass dish filled with 4% PFA and 0.5% GA and fixed for 2 h.

2) Rinsing with PBS (4 ℃) for 3-5 min.

3) Dehydration.

Time	Ethanol concentration	Temperature
2 × 15 min	30%	4 ℃
30 min	30%	4 ℃
30 min	50%	−20 ℃
2 × 30 min	70%, 90%, 96%, 100%	−20 ℃

Note not to cause tissue drying by alcohol evaporation.

4) Infiltration and embedding [24]. For K4M embedding agent is volatile. This procedure should be carried out in a fume hood and operated with gloves. Note not to put K4M embedding medium near the O$_2$. All operation should be conducted at low temperature and preferably at constant low temperature in a freezer.

a. Lowcryl K4M preparation

crosslinker A	2 g
monomers B	13 g
initiator	0.075 g

After the crosslinker A is blended the monomer B, initiator is added into them and mixed well. Then they are placed at −20 ℃.

b. Infiltration is carried out at −20 ℃

Time	Liquid
1h	100% ethanol: K4M (1 ∶ 1, V / V)
2 × 1h	K4M
Overnight	K4M
2 × 3h	K4M

c. Embedding. Firstly capsules are cooled and several droplets of K4M are added into them. Each capsule is put into a tissue, filled with K4M and placed at room temperature for 30 min in order to release bubbles in embedding medium. Then they are placed at 0-4 ℃ and irradiated with UV light of 360 nm wavelength for 5 days. Note that the temperature must be kept constant. Then they are moved to room temperature. As the temperature rises, capsules are hardened.

d. Section. Since the capsule is transparent, embedded liver tissue is easy to be identified. Slice after the tissues are dressed. Since K4M is hydrophilic, it should be noted that the surface of tissue must not be soaked by water during section. The nickel network with 200 meshes is covered with Formvar film and carbon film to fish the sections, and then the sections are dried by air for hybridization.

(2) Probe preparation

The rRNA probe is labelled with Dig-UTP. Note that the probe should not be too long, otherwise it will affect its penetration into tissue. If it is too long, it should be hydrolyzed before usage.

(3) Hybridization

The steps are carried out in the wet box. The hybridization solution is dropped on a wax film, the droplets are covered with the side of nickel network with sections and then hybridization will be implemented at 65 ℃ for at least 3 h. The hybridization solution contains the following agents: 5 × SSC, 0. 1 mg/mL tRNA, Dig-UTP antisense probes 10 ng/μL (1 × SSC containing 150 mmol/L NaCl and 15 mmol/L sodium acetate).

(4) Rinse after hybridization

Rinse at room temperature with 2 × SSC for 3 times and 5 min each time, and then with PBST (PBS containing 0. 1% Tween-20) for 2 times and 10 min each time.

(5) Colour development

1) Blocked by PBST and BT (BT preparation: PBST, 1% BSA) and incubate for 15 min.

2) Anti-digoxin antibody is conjugated with 1 mm diameter gold particles are diluted with PBST and BT in the ratio of 1 : 30 and are incubated for 1 h.

3) Rinse with PBST for 3 times and 5 min each time.

4) Wash by soft plastic pot with pen beak for 6 times and 15 min each time.

5) Silver enhancement method is adopted. Incubate with mixed solution of developing solution and enhancer (1 : 1) in the dark for 4-20 min.

6) Repeat 4).

7) Stain with 2% uranyl acetate (4 min), lead citrate (1 min), rinse and then air drying.

8) Observe under electron microscopy.

References

[1] Yu DH, Cui XL. Auarantine norms of entry and exit animal of China. Beijing: China agriculture press, 1997: 160

[2] Liu SX. Practical techniques of biological tissue. Beijing: Science press, 2004: 159

[3] Chen FC. Department of basic medical, Bethune medical university edit. Histological experimental techniques. Changchun: Bethune medical university press, 1980: 210

[4]　Hong T. Biomedical ultrastructure and electron microscopy. Beijing: Science press. 1980: 197

[5]　Yang YJ. Practical biomedical electron microscopy. Shanghai: Second military medical university press. 2003: 125

[6]　Cai WQ, Wang BY(ed). Practical immunocytochemistry. Chengdu: Sichuan science and technology press. 1988: 164

[7]　Cai WQ. Modern practical experiment techniques of cellular and molecular biology. Beijing: Pelople's medical press. 2003: 246

[8]　Faulk WP, Taylor GM. An immunocolloid method for the electron microscope. Immunochemistry, 1971, 8(11): 1081-1083

[9]　Geuze HJ, Slot JW, Strous GJ, Lodish HF, Schwartz AL. Intracellular site of asialoglycoprotein receptor-ligand uncoupling: double-label immunoelectron microscopy during receptor-mediated endocytosis. Cell, 1983, 32(1): 277-287

[10]　Zhu LP, Chen XQ. Common experimental methods of immunology. Beijing: Pelople's medical press. 2000: 424

[11]　Ba DN. Modern immunological techniques and application. Beijing: Beijing medical university, Peking union medical university joint press, 1998: 542

[12]　Li HY. Virology tests. Beijing: People's medical press, 2006: 158

[13]　Shen GX, Zhou RL. Experimental techniques of modern immunology. Wuhan: Hubei science and technology press, 1998: 465

[14]　Jacob JT, Birnstiel ML. Molecular hybridization of 3H-labelled ribosomal RNA with DNA in ultrathin section prepared for electron microscope. Biochem Biophys Acata, 1971, 228: 761

[15]　Binder M, Tourmente S, Roth J and others. In situ hybridisation of the eletron microscope level: Localization of transcripts on ultrathin sections of Lowicryl K4M-embedded tissue using biotinylated probes and PA-Gold Complexes. J. Cell Biology, 1986, 102: 1646-1653

[16]　Webster HF, Lamperth L, Favilla JT and others. Use of a biotinylated probe and in situ hybridization for light and electron microscopic localization of Po mRNA in myelin-forming Schwann cells. Histochemistry, 1987, 86: 441-444

[17]　Singer RH, Langevin GL, Lawrence JB. Ultrastructural visualization of cytoskeletal mRNAs and their associated proteins using double-label in situ hybridization. J Cell Biol, 1989, 108: 2343-2353

[18]　Jiao RJ, Yu WD, Ding MX and others. Localization of adenovirus DNA in its host cell using electron microscopic in situ hybridization. Electron microscopy, 1992, 2: 81-84

[19]　Xiang ZH, Cai WQ, Meng L and others. Distribution and coexistence of somatostatin mRNA and neuropeptide Y or oxytocin in the rat forebrain: combination of immunocytochemistry and insite hybridization histochemisry with digoxigenin-labelled cRNA. Chinese journal of chemistry and cytochemistry, 1994, 2: 123-127

[20]　Ben CE, Li SG. Histochemistry. Beijing: People's medical press, 2001, 891

[21]　Roth J, Bendayan M, Carlemalm E and others. Enhancement of structural preservation and immunocytochemical staining in low temperature embedded pancreatic tissue. J Histochem. Cytochem, 1981, 29: 663-667

[22]　Lawrence JB and Singer RH. Quantitative analysis of in situ hybridisation methods for the detection of actin gene expression. Nucleic Acids Res, 1985, 13: 1777-1799

[23]　Wang Q. Experimental methods of modern medicine. Beijing: People's medical press, 1997, 101

[24]　Carlsmalarm E, Villger W. Low temperature embedding. In: Tehcniques in immunocytochemistry (Eds Bullock GR and Petruoz P). Acad Press, 1989, 4: 29-44

Chapter 10　Pesticide Immunoassay

10.1　Organophosphorus pesticides

10.1.1　Overview

Organophosphorus pesticide is one of three backbones in current chemical pesticides. Ever since 1940s when the first series of organophosphorus pesticides were successfully developed and became widespread, they have under gone evolution for more than half a century. In the recent two decades, organophosphorus pesticides have been the mainstay in the pesticide industry, predominating among all types of pesticides in no matter quantity, output or market share.

Things are similar in China. According to statistics, a total of more than 150 kinds of pesticides are being produced and used in our country, falling into different categories, with a yearly output of more than 200,000 tons (as active ingredients). About ten kinds of them are produced at a yearly output of approximately 10,000 tons, half of which being organophosphorus varieties. For example, yearly output is more than 10,000 tons for methamidophos, dichlorvos, trichlorphon, methyl parathion and folimat, and is more than or approximately 5,000 tons for glyphosate, monocrotophos, phoxim and isocarbophos[1].

10.1.1.1　Classification, structure and properties

(1) Classification and structure

Organophosphorus pesticides containing phosphoryl ($P = O$) or thio-phosphoryl ($P = S$) group are the most widely used now. Usually, the organophosphorus pesticides can be classificated in two ways. On the basis of their application, they can be distingushied as insecticides, acaricides, nematicides, herbicides, fungicides, rodenticides and plant growth regulating agents. This classification is intuitive, but it does not reflect the structural characteristics of pesticides. The other classification mode is based on different groups connected to central phosphorus atom.

①Phosphate Esters

$$
\begin{array}{c}
R^1O \diagdown \quad \diagup\!\!\!\diagup O \\
P \\
R^2O \diagup \quad \diagdown OR^3
\end{array}
$$

② Phosphorothioate

$$\begin{array}{cc} R^1O & S \\ & \diagdown \diagup \\ & P \\ & \diagup \diagdown \\ R^2O & OR^3 \end{array} \qquad \begin{array}{cc} R^1O & O \\ & \diagdown \diagup \\ & P \\ & \diagup \diagdown \\ R^2O & SR^3 \end{array} \qquad \begin{array}{cc} R^1O & S \\ & \diagdown \diagup \\ & P \\ & \diagup \diagdown \\ R^2O & SR^3 \end{array}$$

③ Phosphamide Ester and Thio-phosphamic Ester(X=O、S)

$$\begin{array}{cc} R^1O & X \\ & \diagdown \diagup \\ & P \\ & \diagup \diagdown \\ R^2O & NR^3R^4 \end{array}$$

④ Phosphoryl Halide Ester(X=Cl、F)

$$\begin{array}{cc} R^1O & O \\ & \diagdown \diagup \\ & P \\ & \diagup \diagdown \\ R^2O & X \end{array}$$

⑤ Pyrophosphate

$$\begin{array}{cc} R^1O \;\; O & O \;\; OR^3 \\ \diagdown \;\; \| & \| \;\; \diagup \\ P-O-P \\ \diagup & \diagdown \\ R^2O & OR^4 \end{array}$$

⑥ Phosphonate and Sub-phosphonate

$$\begin{array}{cc} \diagdown | & \\ C \;\; O & \diagdown | \\ \diagdown \diagup & C \;\; O \\ P & \diagup \diagup \\ \diagup \diagdown & P \\ C \;\; OR & \diagup \diagdown \\ \diagup | & -O \;\; OR \end{array}$$

(2) Major organophosphorus pesticides

① Dichlorvos

Dichlorbos is colourless or amber liquid with an aromatic odour, which is slightly soluble in water but well miscible with most organic solvents. Dichlorbos belongs to middle-toxic pesticides, which can be used as a contact, fumigation, and stomach-poison insecticide. Dichlorbos is used extensively for disinsection in grain, potato or vegetable production and store. Therefore, residues may occur because of the excessive application. Poisoning will occur when people are exposed to high dose of dichlorvos from food. The activity of cholinesterase is soon inhibited resulting in acetylcholine accumulation in the body and then, symptoms of poisoning appear. In the case of mild poisoning, whole blood cholinesterase activity is lowered to 50% -70%, and many poisoning signs such as headache, dizziness, nausea, vomiting, chest tightness, weakness, blurred vision, and miosis and so on will occur. If moderate poisoning takes place, whole blood cholinesterase activity can be reduced to 30%-50% and poisoning symptoms such as sweating, tears, salivation, nasal secretion, bronchoconstriction, increased secretion of digestive juice, miosis, muscle twitching, abdominal pain, diarrhea, blood, instability and so on will appera. For severe poisoning, whole blood cholinesterase activity can be completely lost and saliva and digestive juices sharply increase. Severious muscle weakness may also occur, and the pupils can be minified to small as a pinpoint while difficulty in breathing, pulmonary edema, incontinence, coma, seizures, edema would also appear. Finally, severe poisoning patients will die of respiratory muscle paralysis. Dichlorbos is unstable and is easy to be decomposed, resulting in short residual time. Generally, chronic poisoning occurs less.

② Parathion

Parathion, also known as 1605, is a colourless, odourless liquid or white needle-like crystal. It is insoluble in water but easily dissolved in organic solvents. Parathion is an effective, highly toxic, broad-spectrum insecticide killing through contact, stomach poisoning and fumigant actions. It is used against insects for many crops (such as cereals, potatoes, fruits and vegetables) and also used for control of pests indoors. Poisoning will occur if contaminated food by parathion is digested carelessly. Generally, 10-20 mg parathion will result in adult death and a few milligrams of parathion will cause death of a child.

Parathion is a cholinesterase inhibitor as other organophosphate insecticides. When parathion is taken into body, oxidation reaction takes place resulting paraoxon, which is a strong inhibitor for cholinesterase. Herein, the clinical course of parathion poisoning is long. The main symptoms of poisoning include headaches, nausea, vomiting, sweating, salivation, weakness, excitement, disorientation, language barriers, hallucinations, etc. Long-term exposure of parathion may cause chronic poisoning with main syptoms such as headache, fatigue, loss of appetite, muscle tremors, pupillary constriction and so on.

③ Methyl- parathion

Methyl-parathion, also known as methyl-1605, is colourless and odourless crystal. It is slightly soluble in water but soluble in aromatics. Methyl- parathion belongs to highly effictive and high toxic organophosphate insecticides and its residue period is shorter compared with parathion. Food may be contaminated with methyl-parathion due to abuse in agricultural production, and then food poisoning takes place. The main symptoms include headache, nausea, vomiting, sweating, salivation, weakness, excitement, hallucinations, etc. Long-term exposure to methyl-parathion may cause chronic poisoning, resulting in continuously lowering cholinesterase activity and different degrees of autonomic dysregulation, such as the vagus nerve excitability. Several poisoned people suffer from delayed neuronal injury behaving lower limb ataxia, muscle weakness, loss of appetite and so on. In serious poisoning cases, the syptoms above developed with time, often results in lower limbs paralysis.

④ Malathion

Malathion is pale yellow oily liquid. It is slightly soluble in water but soluble in organic solvents and its residue period is short. As a slightly toxic, broad-spectrum organophosphate insecticide, it is used for controlling pests in storage, flies indoors and livestock ectoparasites through contact-toxicity, stomach-toxicity and fumigant actions. In agriculture, Malathion is often used together with other pesticides, resulting in joint toxic effects. For example, stronger toxicity will be produced if kitazine and Malathion are used together, which is far greater than the sum of toxic activity of the two individual pesticides. Acute poisoning will take place if exposed to large dose of Malathion. Malathion is an indirect inhibitor of cholinesterase and the latency is long. The posioning symptoms include salivation, miosis and other symptoms that last for a long time if overdose of malathion is intaken.

⑤ Fethion

Fenthion, a colorless liquid, is insoluble in water but soluble in organic solvents. It is a moderately toxic, highly efficient and broad-spectrum organophosphate insecticide. It is mainly used for prevention pests from vegetables, rice, beans, fruits, cotton with a long residual term. It also has good killing effect on mosquitoes and can be used for mosquito control in malaria endemic areas. Poisoning may occur after exposuse to fenthion. Chronic poisoning from fenthion results in neurasthenia syndrome and vagus nerve excitability. Fenthion is also indirect inhibitor to cholinesterase, resulting in a long latency after poisoning and poisoning symptoms last a long time.

⑥ Demeton

Demeton also known as 1059, is a colourless viscous liquid with garlic odour, which is slightly soluble in water but quite soluble in toluene, ethanol, propylene glycol and other organic solvents. It is a highly toxic pesticide and good at prevention of aphids, spider mite, nematode and other pests. Demeton may enter human body by ingestion, dermal adsorption and inhalation causing acute poisoning. Demoton displays a longer residue period compared with other organophosphates. Till recently, it was not used in vegetables, tobacco and tea production. Demoton can be activated and then it may inhibit the activity of cholinesterase if it was intaken by human. Poisoning symptoms appear due to the accumulation of acetycholine in the body. In serious cases, toxic hepatitis, paroxysmal atrial fibrillation and even mental sequelae may occur.

⑦ Dimethoate

Dimethoate, a shiny white needle like crystal, is slightly soluble in water but soluble in organic solvents. It belongs to the class of efficient, low toxicity, low residue pesticids and is widely used for controlling many insects through contact killing and systemic poison, especially pests with sucking and chewing mouthparts. Dimethoate is now widely used in vegetables, fruit trees, and tea, cotton and oil crops planting. After entry into the human body, dimethoate can be oxidized into omethoate, which is more toxic, resulting in poisoning symptoms such as headache, fatigue and nausea. The toxicity enhances a lot when it is used together with other pesticides.

⑧ Monocrotophos

Pure monocrotophos, a white crystal, is soluble in water, alcohol, acetone but insoluble in petroleum. It is a highly toxic pesticide with contact killing and systemic poison. It is mainly used to control pests in cotton, rice and tobacco planting. Monocrotophos can inhibit cholinesterase activity directly when a human is exposed to it, resulting in dizziness, headache, and loss of appetite, nausea, vomiting, abdominal pain, diarrhea, sweating, salivation, miosis, blurred vision, increased respiratory secretions and even pulmonary edema.

⑨ Acephate

Acephate, also known as aimthane, is a white crystal. The solubility is 65 % (m/m) in water and more than 10% in ethanol and acetone. It is unstable in alkaline media. Acephate is of low toxicity, high effectiveness and is a broad-spectrum insecticide through contact killing and systemic posioning. Its residual Period is medium. Acephate is widely used to control pests in vegetables, fruits, rice, cotton, fruit trees, tea, tobacco, maize and wheat. Acephate can inhibit cholinesterase activity if it is taken into body.

⑩ Fenitrothion

Fenitrothion, also known as fenitrothion, is a pale yellow oily liquid with garlic odour. It is insoluble in water but soluble in ethanol, ether and other organic solvents. Fenitrothion is a highly effective broad spectrum pesticide with low toxicity. It is mainly used to kill aphids, red spider, rice and other flying insects. Accumulative poisoning appears when people are exposed to it. Fenitrothion can directly inhibit cholinesterase activity, resulting in acetylcholine accumulation in the body and then a series of poisoning symptoms occur. In a mild poisoning case, the symptoms such as headache, dizziness, nausea, vomiting, chest tightness, weakness, blurred vision, and miosis appear. In a moderate poisoning case, more serious symptoms including lacrimation, salivation, nasal secretion, bronchoconstriction occur. In the servere posioning case, the cholinesterase activity is lost completely and secretion of digestive juic and saliva sharply increase. The posioned patients may die to respiratory muscle paralysis[2, 3].

10.1.1.2 Toxicity mechanism

Humans are very sensitive to organophosphate pesticides. For example, parathion's lethal dose for adults is 15-30 mg and that of demeton is 10-20 mg. The obvious characters for organophosphorus pesticide poisoning is the inhibition of cholinesterase activity, which results in nervous system dysfunction and a number of organs controlled by the nervous system (heart, bronchi, intestine, stomach, etc.) can't operate normally. The main dysfunctions are sumarized below.

(1) M-like symptoms (Muscarinic symptoms)

Peripheral M receptor (or M cholinergic receptor) is over-excitement resulting in disorder of effectors. The main symptoms include nausea, vomiting, diarrhea, incontinence, miosis, blurred vision, salivation, sweating, increased heart rate, breathing difficulties and so on. These symptoms usually appear in the mild poisoning case.

(2) N-like symptoms (nicotine-like symptoms)

When peripheral N receptors are over simulated due to exposure to organophosphate, the symptoms including secretion of adrenal medulla, increased and the skeletal muscle excitement will appear. Main clinical signs are blood pressure increasing, heart rate enhancing, muscle tremors and convulsions. In the moderate poisoning case, N-like symptoms and M-like symptoms often appear together.

(3) Central nervous system symptoms

The accumulation of acetylcholine over-excited the central cholinergic receptor, resulting in the disorder of the central function and some symptoms like irritability, delirium, convulsions and so on will appear. In some cases, the activity of cholinergic receptor is inhibited due to over-excitement, which leads to coma, blood pressure decreasing, paralysis of respiratory center and even respiratory arrest to death. In the severe poisoning case, the M-like, N-like and central nervous's symptoms problems may appear at the same time.

The organophosphorus compounds inhibit cholinesterase activity after they come into the body. Normally, acetylcholine is released when the cholinergic nerve is stimulated and the neural impulses will be transferred to the dominated effector organs. At the same time, acetylcholine will be quickly decomposed by acetylcholinesterase (AchE). This mechanism ensures neuropsychological balance and coordination.

AchE has two active parts, an anion (also called as binding site) part and esterlysis part (catalysis reaction site). Under normal conditions, the anion attracts the cationic active center of acetylcholine while the part of esterlysis attracts the acetyl group, and then a complex is formed. Following, the C-O bond of the acetylcholine breaks to form acetylase and sinkaline. The acetylase is unstable and returns to AchE easily by releasing an acetic acid.

Usually, the contents of cholinesterase in the human body is in excess, so the small amount of oganophosphorus pesticides will not cause poisoning. However, large amount of organophosphorus pesticides exposed significantly decrease the cholinesterase activity, leading to a series of clinical symptoms[4, 5].

10.1.1.3　Metabolization

Most organophosphorus pesticides are fat-soluble and may enter into body by the respiratory tract, the gastrointestinal tract and the skin contact. These pesticides may be transferred to tissues and organs all over the body through the blood and lymphatic system. The liver contains the most organophosphorus pesticide residues. The kidney, lungs, and bones closely follow, while the muscle and brain tissue contains the least residues[6].

Organophosphorus pesticides may undergo different types of biotransformation in the human body including oxidation, hydrolysis, de-amine, dealkylation, reduction side chain oxidation and so on.

(1) Oxidation

Desulfidation reaction: the P=S bond of organophosphorus pesticides can be transformed into P=O bond due to the liver microsomal mixed oxidase function. chemicals with P=O bond show higher anti-cholinesterase activity, resulting in the enhanced toxicity.

O-dealkylation reaction: the O-dealkylation reaction will occur due to the liver microsomal mixed oxidase function after the organophosphorus pesticides enter into the body, which eliminates the toxicity of organophosphorus pesticides. For example, methyl-parathion will be rapidly excreted out of the body after methyl group is removed.

S-oxidation reaction: The organophosphorus pesticides containing -C-S-C will undergo oxidation reaction due to the liver microsomal mixed oxidase function. The toxicity of the products including sulfone or sulfoxide compounds is 5-10 times of that of parent pesticides.

(2) Hydrolysis

Hydrolysis is an important biotransformation mode. The toxicity of some organophosphorus pesticides can be reduced under catalysis of hydrolyases. Phosphatase, carboxylesterase and amidohydrolase are the typical hydrolyses. In addition, organophosphorus pesticides may join in disoxidation or combination reaction. For example, the nitro-group from parathion or EPN is transformed into amino group under catalysis of reductase, which reduces a lot the inhibiting of organophosphorus pesticides for AchE. The most important combination reactions are between organophosphorus pesticides and glycuronic acid or glutathione(Figure 10-1). The conjugation shows low bioactivity and is easily discharged from the body.

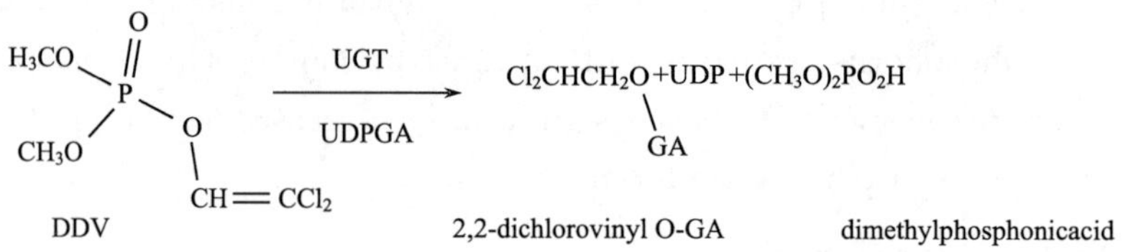

$$H_3CO \diagdown \underset{\underset{O}{|}}{\overset{\overset{O}{\|}}{P}} \diagup O \quad \xrightarrow[\text{UDPGA}]{\text{UGT}} \quad Cl_2CHCH_2O + UDP + (CH_3O)_2PO_2H$$

CH$_3$O ... CH=CCl$_2$

DDV   2,2-dichlorovinyl O-GA   dimethylphosphonicacid

Figure 10-1 Degradation process of dichlorovos in mammals

The metabolic process of organophosphorus pesticides is complex and various trans-
formations have been found. For example, parathion is oxygenized to paraoxon, which
then is detoxicated by phosphoesterase and finally biotransformed into p-nitrophenol,
diethylphosphorodithioate and bis (2-ethylhexyl) phosphoric acid. The p-nitrophenol
may exist alone or further detoxicated by conjugation with glucuronic acid Figure 10-2.
Also some p-nitrophenol undergoes reduction to p-aminophenol and is then discharged
with urine.

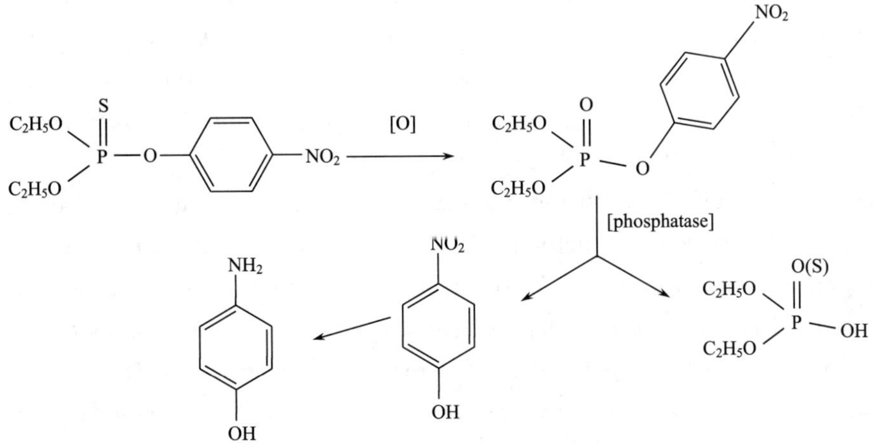

Figure 10-2   Degradation of parathion

10. 1. 2   Immunoassay of organophosphorus pesticide

Due to the selectivity and sensitivity, as well as simplicity of performing immunoassays, many literatures have been found in the immunochemical technologies of organophosphorus pesticide. Rui and others[7] discribed an ELISA method for parathion analysis
with a limit of detection (LOD) 2.0 ng/mL. The antigen was prepared by diazotization.
Parathion is converted to deoxidize parathion firstly and then conjugated to BSA or
OVA. The polyclonal antibody was obtained by immunization of rabbits. The antibody
was specific to parathion and the cross reaction with methamidophos and chlorpyrifos
was less than 0.1%.

Lv and others[8] reported the rapid detection of parathion residues in rice based on
ELISA. Two antigens were developed by diazotization and purified. An immunoassay
was established with polyclonal antibody and good linearity of standard curve was found
with parathion concentration ranging from 39 ng/mL to 5,000 ng/mL. The antibody
showed negligible cross-reactivity with methyl parathion, paraoxon. The spiked rice
samples at different levels (0-5,000 ng/mL) were detected by this immunoassy and
HPLC. A good correlation was found between the proposed method and the reference

chromatographic method. The reproducibility of the ELISA was tested by 3 different operators with various spiked samples. The satisfactory results were found (coefficient of variability was 5. 2%).

Liu and others[9] prepared polyclonal antibody against methamidophos. The dilution titer of the antibody is between 1 ∶ 12, 800 and 1 ∶ 51, 200. The cross-reactivity is less than 2. 8% with most organophosphorus pesticides but 10. 5% with acephate. This proposed method was used to detect the methamidophos residues in environment samples, grain and cabbage. The linear range of inhibition curve is 0. 032-500 μg/mL. The recovery tests in rice and cabbage matrix is 90. 3% and 93. 2% respectively. 31 rice samples and 45 pakchoi samples were detected with this method and the instrument method (GC). The percentages of positive samples were 16. 1% and 46. 7% respectively by ELISA, which was identical to GC results.

Zhang and others[10] prepared methamidophos-BSA conjugates by carbodiimide method and glutaraldehyde method. Various polyclonal antibodies were obtained through immunization of rabbits. A homogeneous radioimmunoassay was established by synthesis of ^{125}I labelled methamidophos- succinic acid- histamine conjugates. The LOD was 0. 03 nmol/L (4. 2 pg/mL) and coefficient of variability was 5. 2%, Also, a solid-phase radioimmunoassay method was established by coating polystyrene tube with purified antibody and the LOD was 0. 05 nmol/L (7 pg/mL). In addition, a direct competitive immunoassay was developed by preparation of methamidophos- horseradish peroxidase. The LOD was 5 nmol/L (700 pg/mL) and coefficient of variability was 12. 6%. Another immunoassay based on biotin-avidin signal amplification system was tested with LOD 1 nmol/L (140 pg/mL). An immunochromatographic strip test based on gold nanoparticle-labeled antibody probes for the detection of methamidophos was also studied [the LOD is 100 nmol/L (14 ng/mL)]. Results showed that enzyme-linked immunosorbent assay was suitable for quantification analysis and the immunochromatographic strip was the best in qualitative analysis.

Han and others[11] developed haptens by modification of methyl parathion with two types of spacer arms. The artificial antigen was prepared using activated ester method and mixed anhydride method. Two ELISA for methyl parathion detection were established with LOD 2. 0 and 6. 1 ng/mL. Two antibodies appeared to prosses high specificity for methyl parathion and no cross-reaction with other orgnaophosphorus pesticides but parathion was found. The tests for water and soil showed that the results were consistent with the GC method. The detection sensitivity of the proposed ELISA method was better than the GC method.

Li[12] prepared hapten nd obtained polyclonal antibody against dimethoate firstly. The affinity constant for antibody and dimethoate was 3. 84 × 10^{10}. Only dimethoate was found to cross reacted with the antibody. An ELISA kit was assemblied (LOD

0. 01 mg/L) with good stability.

Yin[13, 14] synthesized a hapten of methyl paraoxon using hydroxide phenylpropionic acid and p-dimethoxy phosphorus chloride. Three steps including methyl esterification, acylation and hydrolysis were used and the resulting products was O, O- dimethyl-O-(4-propionic acid phenyl) phosphate. The hapten was directly conjugated with BSA and OVA respectively to get complete antigens. Polyclonal antibodies generated from rabbits were highly sensitive for parathion and methyl paraoxon, demonstrating IC50 values of 0. 343 μg/mL and 0. 371 μg/mL respectively in ELISA. The tests in spiked cabbage showed good recoveries (72. 94%-82. 07%). Class selective immunoassay for diethoxy organophosphorus pesticides (such as chlorpyrifos, diazinon, parathion, paraoxon, phoxim) was also studied. Their common parental moiety was $(C_2H_5O)_2P$ (O) or $(C_2H_5O)_2P$ (S). Then diethoxy acetic acid $(C_2H_5O)_2P$ (O) CH_2COOH) was prepared and coupled with carrier proteins by active ester method. The results showed that the antibody obtained recognized paraoxon and pirimiphos with IC_{50} values of 1. 69 μg/mL and 2. 33 μg/mL respectively.

Chen[15] prepared two haptens THBu [O -ethyl, O -(1-phenyl -1H-1, 2, 4 -triazol-proyl) N-(3-carboxypripyl) phosphoramidothioate] andTHHe[O-ethyl O- (1-pheyl-1H-1, 2, 4-triazol-3-proyl) N-(5-carboxypripyl) phosphoramidothioate]. Two polyclonal antibody against triazophos were obtained and a direct competitive ELISA format was used for detection of triazophos in different matrix including water, rice, soybeans (dry), carrots, potatoes, cabbage, onions, eggplant, pears, citrus. The IC_{50} was 62. 0 ng/mL and LOD was 5. 76 ng/mL.

As a small molecule (MW350. 62), Chlorpyrifos should be attached to carrier protein to raise antibodies from animals. Wang and others[16] utilized nucleophilic substitution between the reactive chlorine(site 6) of pyridyl of chlorpyrifos and 3-mercaptopropionic acid to prepare hapten, O, O-diethyl-O[3, 5-dichloro-6- (2-carboxyethyl) thio-2-pyridyl]phosphorothioate. Then, artificial antigens and coating antigen of chlorpyrifos were synthesized by conjugation hapten with carrier protein. Monoclonal antibodies were prepared and the ELISA method for chlorpyrifos detection was established. The result showed that IC_{50} value was 0. 14 μg/mL. Specificity tests indicated that no cross-reaction with fenitrothion, parathion and malathion were found.

Methamidophos, methyl parathion and parathion belong to highly toxic and high residual organophosphate pesticides, which was banned for use in China. Wang[17] designed hapten according to their structural features and obtained polyclonal antibody by immunization of rabbits. A direct competitive ELISA format was used to detect the residues of three pesticides in the vegetables. The result showed that LOD values of methamidophos, parathion and methyl parathion were 0. 01 μg/mL, 1 μg/mL and 0. 01 μg/mL respectively. An ELISA kit based on polyclonal antibody was developed.

The mean intra-plate coefficient of variation is 4. 05 %, inter-plate coefficient of variation is 13. 66 %. The kit could be stored at 4 ℃ for a period of over six months.

　　Biosensor was another way to detect the pesticide rapidly and simply. Based on the selective binding between antigen and antibody, an ultrasensitive piezoelectric immunosensor using an amplification path based on an insoluble biocatalyzed precipitation product was proposed for detection of methyl parathion by Lu and others [18]. This fast and simple immunosensor can be used for rapid detection of methyl parathion residues in field testing and monitoring. The detection scheme was shown in Figure 10-3. The key point of developing the piezoelectric immunosensor is to immobilize the antibody onto the golden electrode of the quartz crystal. Two immobilization methods including protein A (SPA) and self-immobilized membrane (SAMs) method were used and the latter mehod showed better stability. Three systems including direct detection, competitive method and enzyme catalyst aggregation amplifying method was optimized to detect the methyl parathion residues. For the direct method, the linearity scope of detecting methyl parathion is 1. 0-25 μg/mL. The competitive method based on the competition between the immobilized antigen and the dissolved methyl parathion in solution. The good linear response was obtained when methyl parathion ranged from 0. 1 to 10 μg/mL. The piezoelectric immunosensor based on enzyme-catalysis deposition method could significantly amplify the signal with LOD 0. 074 μg/mL. In addition, the regeneration of the immunosensor was discussed. One method was the dissociation of antigen-antibody complex on the surface of the gold electrode using weak acidic or weak basic chemicals while

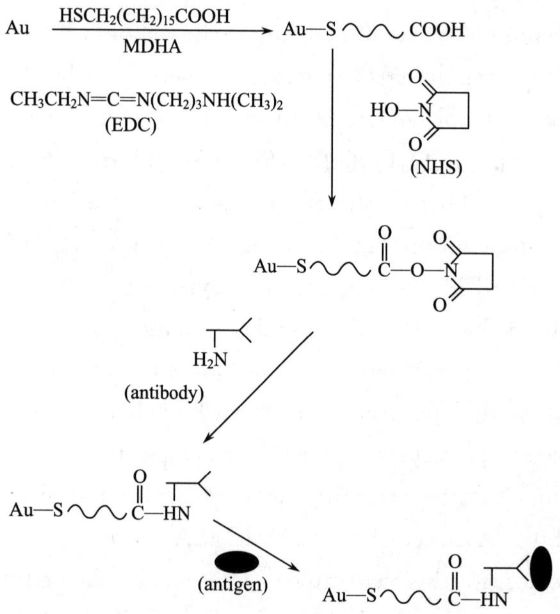

Figure 10-3　Scheme for electrode medication with self-assembly membrane method

another way was to remove the antigen-antibody complex with stong acidic or basic solution. Results showed that the sensor can be regenerated 5-10 times using the former method while more than 10 times using the latter method.

He[19] also developed an enzymatic electrode biosensor, which was prepared by immobilizing acetylcholinesterase with the crosslinking agent named glutaraldehyde. Chlorpyrifos was studied as targeted molecules. The conditions of the preparation of an enzymatic electrode, such as the concentration of glutaraldehyde and enzyme were investigated. The factors affecting enzymatic electrodes including pH and the temperature were optimized. Based on the analysis of the mass transfer of hydrogen ion, it was proved that the agitation operation was negative as to the improvement of sensitivity of the biosensor. The quantification results showed that the linear response to chlorpyrifos was ranging from 1.64 to 20.52 $\mu g/mL$.

10.2　Immunoassy for organochlorine pesticides

10.2.1　Overview

10.2.1.1　Classification, structure and properties

(1) Classification and structure

Organic compounds containing at least one covalently bonded chlorine atom are named organochlorine compounds. Organochlorine pesticides are hydrocarbon compounds containing multiple chlorine substitutions. There are four main types of OCPs: dichlorodiphenylethanes; cyclodienes; chlorinated benzenes; and cyclohexanes. All share a similar pair of carbon rings, one ring being heavily chlorinated.

Organochlorine pesticides are mostly used as insecticides. Specific uses take a wide range of forms, from pellet application in field crops to sprays for seed coating and grain storage. The application of organochlorine pesticides resulted in direct environmental pollution and accumulation in plants, which caused great adverse health effects to human and wildlife.

Many OCPs are persistent organic pollutants (POPs), a class of chemicals known to break down very slowly and bio-accumulate in lipid rich tissue such as body fat. The pollution and harmness from organochlorine pesticides had caused widespread concern. In 1995, 12 kinds of organic substances were recognized as persistent organic pollutants by the United Nations Environment Programme. All of them belong to organochlorine compounds and nine of them including aldrin ($C_{12}H_{12}Cl_6$), dieldrin ($C_{12}H_8Cl_5O$), endrin ($C_{12}H_8Cl_5O$), DDT ($C_{14}H_9Cl_5$), chlordane ($C_{10}H_8Cl_8$), toxaphene ($C_{10}H_{10}Cl_5$), hexachlorobenzene ($C_6H_6Cl_6$), mirex ($C_{10}Cl_{12}$), heptachlor ($C_{10}H_9Cl_7$), were organochlorine pesticides(OCPs). Also, these nine OCPs were listed in the Stockholm Convention

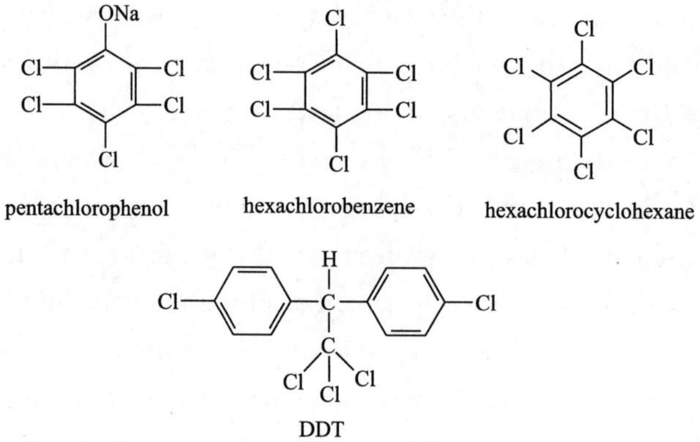

Figure 10-4　The chemical structure of several typical OCPs

on Persistent Organic Pollutants as the first restriction polluss. Sample chemical structures for organochlorines are available here (Figure 10-4).

(2) Properties

OCPs are not particularly volatile, hydrophobic, lipophilic and are extremely stable. They break down slowly and can remain in the environment long after application and in organisms long after exposure. They can travel long distances via wind and deposit on soil and water, so they can be found hundreds or thousands of miles from their point of use. They can also be transported on foods and other products treated with them.

Once in the environment they are subject to global deposition processes and bioaccumulate in the food chain. Wind and rain may move pesticides away from where they were used causing contamination of surface waters, ground-water and/or soil. Diet is the main source of human exposure, primarily through consumption of animal products where OCPs pesticides have bioaccumulated. Many studies have linked organochlorine pesticide exposure with consumption of contaminated animal products, mostly meat, dairy, fish, and marine mammals [20].

① DDT

DDT is an organochlorine, similar in structure to the insecticide methoxychlor and the acaricide dicofol. It is a highly hydrophobic, colourless, crystalline solid with a weak, chemical odour. It is nearly insoluble in water but has a good solubility in most organic solvents, fats, and oils. DDT does not occur naturally, but is produced by the reaction of chloral (CCl_3CHO) with chlorobenzene (C_6H_5Cl) in the presence of sulphune acid, which acts as a catalyst. It exhibits good thermal stability and acid stability; however, it is easy to hydrolyze in alkaline medium. DDT is classified as "moderately hazardous" by the World Health Organization (WHO), based on the rat oral LD50 of 113 mg/kg.

DDT is a persistent organic pollutant that is extremely hydrophobic and strongly absorbed by soil. Depending on conditions, its soil half life can range from 22 days to 30 years. Foods including poultry meat, wool, eggs, milk, fish and some vegetables are prone to build up DDT after exposure. Long-term exposure of DDT contaminated food causes nerve or parenchymal organs toxicity to humans.

Routes of loss and degradation include runoff, volatilization, photolysis and aerobic and anaerobic biodegradation. When applied to aquatic ecosystems it is quickly absorbed by organisms and by soil or it evaporates, leaving little DDT dissolved in the water itself. Its breakdown products and metabolites, DDE [1,1-dichloro-2,2-bis(p-chlorophenyl) ethylene] and DDD [1,1-dichloro-2,2- bis(p-chlorophenyl)ethane], are also highly persistent and have similar chemical and physical properties.

② Chlordane

Chlordane is an organochlorine insecticide, more specifically a chlorinated cyclic hydrocarbon within the cyclodiene group. The major use of chlordane in agriculture was to control insect pests in soil and on plants. The actual technical product is a mixture of various chlorinated hydrocarbons, including isomers of chlordane and other closely related compounds. It is colourless or slightly yellow liquid.

Chlordane is virtually insoluble in water (5 ppm), but highly soluble in organic solvents and oils. This property, coupled with the persistence of chlordane in the environment, gives it a propensity to accumulate in organisms (i. e. , to bioaccumulate), especially in animals at the top of food webs. Because of its insolubility in water, chlordane is relatively immobile in soil, and tends not to leach into surface or ground water. The acute toxicity of chlordane to humans is considered to be high to medium by oral ingestion and hazardous by inhalation. Chlordane causes damage to many organs, including the liver, testicles, blood, and the neural system. It also affects hormone levels and is a suspected mutagen and carcinogen. Chlordane is very toxic to arthropods and to some fish, birds, and mammals.

③ Lindane

Lindane, also known as gamma-hexachlorocyclohexane, (γ-HCH), gammaxene, Gammallin and erroneously known as benzene hexachloride (BHC), is an organochlorine chemical variant of hexachlorocyclohexane that has been used both as an agricultural insecticide and as a pharmaceutical treatment for lice and scabies. In addition to the issue of lindane pollution are concerns related to the other isomers of HCH, namely alpha-HCH and beta-HCH, which are notably more toxic than lindane, but lack its insecticidal properties, and are byproducts of lindane production. Lindane accumulates in adipose tissues as indicated by measurements in humans and animals.

10.2.1.2 Mechanism of toxicity

Organochlorine pesticides (dieldrin, DDT and its compounds, aldrin) have been identified as endocrine disrupting substances. At low concentrations, organochlorine pesticides exhibit relatively low acute toxicity to humans; however, they may mimic human hormones like estrogen, or have other properties that cause long term health effects. At higher concentrations, organochlorine pesticides can be very harmful, causing a range of problems including mood change, headache, nausea, vomiting, dizziness, convulsions, muscle tremors, liver damage, and death[21].

(1) Neurodevelopmental outcomes and neurodegenerative disease

There is clear evidence that exposure to organochlorine pesticides disrupts normal development. In one study, exposure to DDT shortly after birth created a lifelong sensitivity to other pesticide exposures in mice, and permanent behavioral changes upon secondary exposure through food. Prenatal exposure to the organochlorine pesticide chlordane has been linked to reduction of testosterone in adult female rats and behavioral changes in both sexes. Rats prenatally exposed to varying levels of DDT showed behavioral alterations that lasted into adulthood. Organochlorine pesticide exposure is associated with neurodevelopmental health effects in humans. Exposure to organochlorine pesticides has been linked to decreased psychomotor function and mental function, including memory, attention, and verbal skills in children. Children born in agricultural areas where pesticides were applied were found to have lower performance on numerous neurobehavioral assessments when compared to children not born in an agricultural region. There is also some evidence that organochlorine pesticide exposure is associated with the development of autism, although this is based on limited research.

Neurodegenerative diseases such as Parkinson's disease and Alzheimer's disease are more common in people with general pesticide exposures, including organochlorine pesticide exposure.

(2) Endocrine-disrupter

Many organochlorine pesticides are endocrine disrupting chemicals, meaning they have subtle toxic effects on the body's hormonal systems. Endocrine disrupting chemicals often mimic the body's natural hormones, disrupting normal functions and contributing to adverse health effects.

Organochlorine compounds, including pesticides, have been found to alter levels of maternal thyroid hormones during pregnancy. Women with hexachlorobenzene concentrations that ranged from 7. 5-841. 0 ppb (ng/mL) had altered thyroid hormone levels. Another study found that dieldrin exposure was associated with decreased thyroxine levels. Women with decreased thyroid hormone had average dieldrin concentrations of 5,380 ppb (ng/mL).

10.2.1.3 Metabolism and residues

(1) DDT and its analogues

DDT degradated slowly and it could accumulate in plants and animals fat via bioaccumulation of the food chain. Some metabolites were isolated by thin layer chromatography and identified by gas chromatography- mass spectrometry, which included DDE, DDD, DDM [bis-(p-chlorophenyl) methane], DDA [2,2-bis (p-chlorophenyl) acetic acid] and conjugates with natural contents. It was reported that DDT was transferred to DDD in the anaerobic conditions, which may relate with the metabolism of the unsaturated fat. This reaction could be promoted by the carotenoid compounds. DDT residues in silage grass were widely broken down into DDD and DDE [22].

The metabolism of DDT residues in plants was also affected by sunlight. Archer's tests showed that DDT sprayed on *astragalus sinicus* L. no changes were found in darkness while DDD was detected after six days exposure to sunlight. Under the UV irradiation, DDD can be found four days later. However, DDE wasn't found to form in the three conditions above.

A series of tests showed that the effcts of the light on DDT residues degradation was related with the status of pesticide itself and the presence of initiator such as the oxygen and ionization inducer. Miller and other researchers showed that the absence of weak inducer to promote ionization, the role of photolysis was negligible. Photolysis of DDT does not occur unless an inducer which has a low ionization potential, such as diethylaniline is present. The DDT-diethylaniline mixture is stable in the dark, and the induced photolysis is not affected by triplet quenchers. As a pure solid and in hexane solution, DDT readily decomposed when irradiated with ultraviolet light. Principal products identified by gas-liquid and thin-layer chromatography from irradiations of the solid phase were DDD, DDE and 4,4'-dichlorobenzophenone. DDD and hydrochloric acid were identified from irradiated solutions of DDT in hexane. The photodecomposition of DDT in methanol was more complex and more photoproducts were formed. Plimmer has invesitigated the photolysis process and main degradation reactions were summarized as followings:

$$R_2CHCCl_3 \xrightarrow{\text{irradiation}} R_2CH\dot{C}Cl_2 + \dot{C}l$$

$$R_2CH\dot{C}Cl_2 + CH_3OH \longrightarrow R_2CHCHCl_2 + \dot{C}H_2OH$$

$$R_2CHCHCl_2 \xrightarrow{\text{irradiation}} R_2CH\dot{C}HCl + \dot{C}l$$

$$R_2CH\dot{C}HCl_2 + CH_3OH \longrightarrow R_2CHCH_2Cl + \dot{C}H_2OH$$

When O_2 was present,

$$R_2CH\dot{C}Cl_2 + O_2 \longrightarrow R_2CHCCl_2(\dot{O}_2)$$

$$2R_2CH\dot{C}Cl_2(\dot{O}_2) \longrightarrow 2R_2CHCCl_2\dot{O} + O_2$$

$$R_2CHCCl_2\dot{O} \longrightarrow R_2CHCOCl + \dot{C}l$$

$$R_2CHCOCl + CH_3OH \longrightarrow R_2CHCOOCH_3 + HCl$$

$$R_2CHCCl_3 + \dot{C}l \longrightarrow R_2\dot{C}CCl_3 + HCl$$

$$R_2\dot{C}CCl_3 + O_2 \longrightarrow R_2C\dot{O}_2CCl_3$$

$$2R_2C\dot{O}_2CCl_2 \longrightarrow 2R_2\dot{C}OCCl_3 + O_2$$

$$R_2\dot{C}OCCl_3 \longrightarrow R_2CO + \dot{C}Cl_3$$

The reactions above indicated that the presence of oxygen was realted with the category of the photolysis products. Laboratory experimetns showed that dechlorination took place at the site of trichloro ethane at 260 nm illumination while the chlorine atoms on the aryl ring will be displaced under shorter wavelength light irradiation.

DDE will be further decomposed, which has been confirmed. The degradation pathway of DDE in hexane solution under UV irradiation was shown in Figure 10-5. However, the photolysis products from DDE were TDEE and other isomeration substances. Based on the reseaech of Plimmer and others, the degradation process was very complex and more than 30 kinds of catabolites had been found. The photolysis of DDE was related to oxygen. Another degradation product- methoxy DDT was investigated. The degradation products maily were dichloro derivatives after illumination at 310 nm for 100 min.

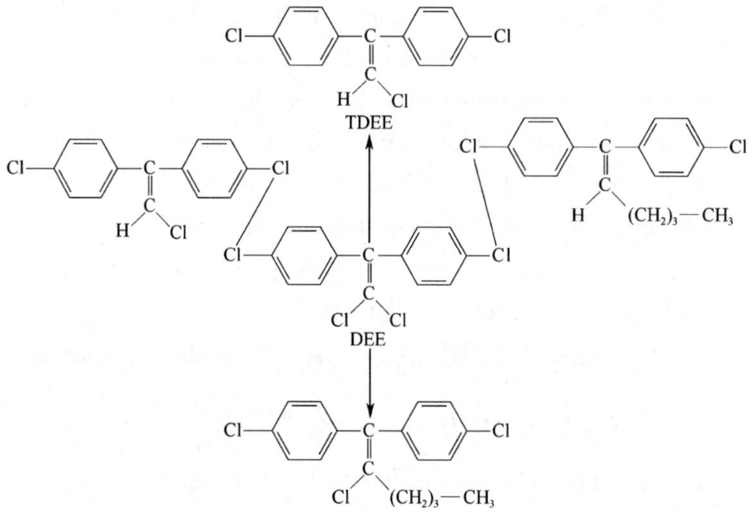

Figure 10-5　Photolysis of DDT in hexane

For algae, such as *cylindrotheca cleosterium*, *skeletonma costatum*, *cyclotella nana*, *thalassiosina fluviatilus*, *dunaliella tertiolecta*, etc., a small part of DDT was observed to be metabolized to DDE. The conversion capacity was widely different, depending on the category of algae.

(2) Hexachlorocyclohexane (BHC)

The main degradation products in plants are some chlorinated phenols (free state or conjugated state), chlorinated benzene and other polar compounds. For example, administration of BHC in lectuce four weeks later, more than 77% residues is BHC itself and free tetrachloro-phenol, pentachlorophenol and their conjugates were also found.

(3) Aldrin and dieldrin

After administration aldrin on the cabbage, six kinds of decomposed substances including dieldrin and other photochemical hydrophilic compounds could be detected. If seeds of carrot and onion were treated with aldrin, only aldrin residues can be found in the carrot plant tissue while aldrin was partly transformed into dieldrin in onion.

(4) Chlordane and heptachlor

When ^{14}C-labelled trans-chlordane was applied to the leaves of cabbage seedlings, three degradation products were found in the plant 4 or 10 weeks later and one of them was identified as dihydroxy-β-dihydro heptachlor. Also, these three products can be found in in carrots if the soil was pretreated with trans-chlordane 12 weeks before. For photochemical decomposion of heptachlor, the photolysis process was proposed by Menzie. When heptachlor was administrated on bean plants in acetone or together with rotenone, series of photochemical keto products can be formed. Heptachlor epoxide I and II has been isolated and identified.

10. 2. 2   Analysis method

Dong [23] determinated DDT and its major metabolite residues in marine samples and seafood with indirect competitive ELISA. DDA[2, 2-bis (4-Chlorophenyl) acetic acid], which is one metabolite of DDT, was coupled with bovine serum albumin and ovalbumin protein by carbodiimide method respectively to produce immunogen and coating antigen. An ELISA was developed based on the polyclonal antibody from immunized mice. The detection limits were 1.56 ng/mL (DDA) and 25 ng/mL (DDT, DDD and DDE) depending on the residues.

Two competitive immunoassays, one based on microtiter plate and the other based on polystyrene tubes were developed by Zhang [24] for the determination of endosulfan residues in agricultural products. The detection limits were $0.8 \pm 0.1 \mu g/kg$ and $1.6 \pm 0.2 \mu g/kg$ respectively. The performance of the two methods was evaluated and three kinds of real samples including tea, tobacco leaf and spinach were detected. 0.5% fish skin gelatin-phosphate buffered saline was the most effective in removing interferences and polyvinyl pyrrolidone was most effective to remove matrix effects in tea analysis. A good correlation was found when the proposed ELISA was compared with a gas chromatography method, which showed the potential use of ELISA in detection of endosulfan residues in agricultural products.

　　Shao[25] prepared hapten of quintozene and coupled it with bovine serum albumin (BSA) or ovalbumin (OVA) to get immunogen and coating antigen. A polyclonal antibody was obtained from immunized rabbits. The IC_{50} value for quintozene was 76. 7 ng/mL based on indirect competitive ELISA. The detection limit was 4. 7 ng/mL (IC_{10} value). Cross-reactivity was tested among some structural analogues: pentachlorophenol 3. 23%, hexachlorobenzene: 12. 22% and less than 2% for 2, 4-D. The impact factors of Indirect ELISA method including pH, ionic strength, and organic solvent resistance were tested. Water samples and soil samples were analysed and good recovery was achieved: 86. 7% for well water, 81. 3% for river water and 78. 9% for soil samples.

　　A surface acoustic wave immunosensor for detection of DDT was developed by Yang[26]. The material 36 °-YX LiTaO₃ was used as base and the surface was modificated with Cr/Au (20 nm/100 nm). The DDT antibody was immobilized to the sensor Au surface via self-assembled monolayer immobilization. The mechanism of this two-channel horizontal shear surface acoustic wave (SH-SAW) DDT immune biosensor was shown in Figure 10-6. The DDT immuosensor manifest high sensitivity and good reproducibility with the detection limit less than $1. 57 \times 10^{-9}$ μg/L and linear range from 6×10^{-9}-29×10^{-9} μg/L.

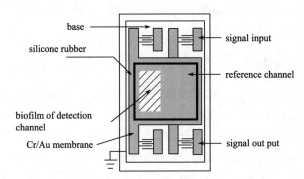

Figure 10-6　Schematic diagram of SAW sensor

10.3　Carbamates

10. 3. 1　Overview

10.3.1.1　Classification, structure and properties

（1）Classification and structure

　　Biological activity of carbamate compounds attracted attention long time ago. In the west coast of Africa, the local people initially used physostigmine juice to make poisonous arrows. Until 100 years from now, the compound was separated and known as physostigmine (physostigmine). Its chemical structure was identified in 1931.

The chemical structure of physostigmine

It is confirmed that carbamate part of physostigmine was active group to cause mio-sis and more research were carried out on simple carbamate developments. The strong insecticidal activities of carbamate derivatives were found in late 1940s. Carbaryl was first synthesized by Union Carbide in 1953 and it was confirmed as a broad-spectrum, low toxicity, high efficient insecticide in 1956. The application of carbaryl effectively promoted the study of carbamate pesticides. The raw materials for carbamates was easi-ly available and the synthesis process was simple. Thus, more than 60 marketed carba-mates appeared in the following several years. Now, the annual yield of carbamates was more than 10,000 tons. In recent ten years, more studies were focused on its toxicity reduction[27]. General formula of carbamate insecticides anhydride is:

$$R_1 \diagdown \atop R_2 \diagup N—\overset{\overset{O}{\|}}{C}—OAr$$

Where, Ar is benzene group, heterocyclic or polycyclic aromatic part. In most ca-ses, R_1 and R_2 are hydrogen or alkyl. Usually, the carbamate insecticides are divided in-to three categories depending on their chemical strucuture.

① Substituted phenols-methyl carbamates

The general formula of these compounds could be expressed as follows:

$$CH_3NH—\overset{\overset{O}{\|}}{C}—OAr$$

Isoprocarb and fenobucarb are the typical presents of this class. Ar group mostly is phenyl, naphthalene or other polycyclic aromatic groups.

② N, N-dimethyl carbamate

Most are N, N-dimethyl carbamic acid derivatives. There is enol unit in its ester group and Ar group here usually is heterocyclic or carbocyclic part as the following for-mula shows. Carbofuran, carbaryl and fenfuracarb belong to this class.

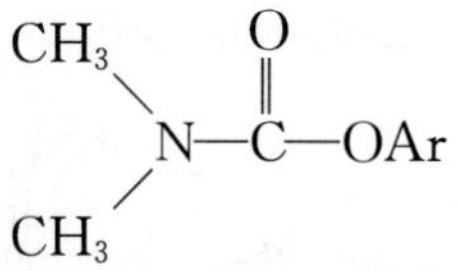

$$CH_3 \diagdown \\ \diagup N—C—OAr \\ CH_3 \qquad O$$

③ N-methyl carbamate proxetil

The oxime ester makes this class carvamtes highly toxic. General formula of these compounds can be expressed as follows.

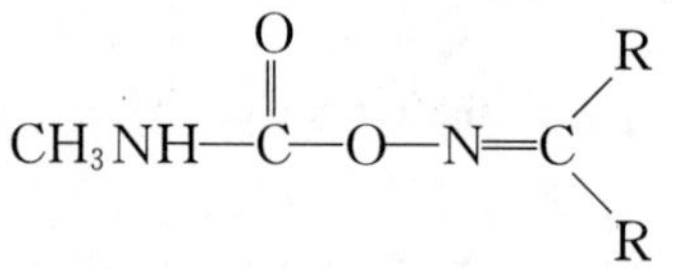

$$CH_3NH—C—O—N{=}C\diagup R \diagdown R \qquad O$$

R in formula can be hydrogen or other alkyl derivatives. Aldicarb, methomyl and thiodicarb belong to this family.

In the 3 categories of carbamates metioned above, the first category is the major commercialized products. Since many highly efficient carbamate pesticides had high toxicity. Scholars dedicated to the study of low toxic carbamates development. A successful case is to connect the nitrogen atom of amide bond to other group through a sulpher atom. Now, many new compunds present such as N-aromatic sulpher-based (or N-alkylthio) carbamate; N, N-sulfur double-carbamate; N-b alkyl ammonium thio carbamate; N-phosphorus amide thio carbamate; N-amino acid ester based compound carbamate; N-alkoxy thio carbamate; N-sulfonyl carbamate; N-aryl (or alkyl) with carbamate dithio-thymine.

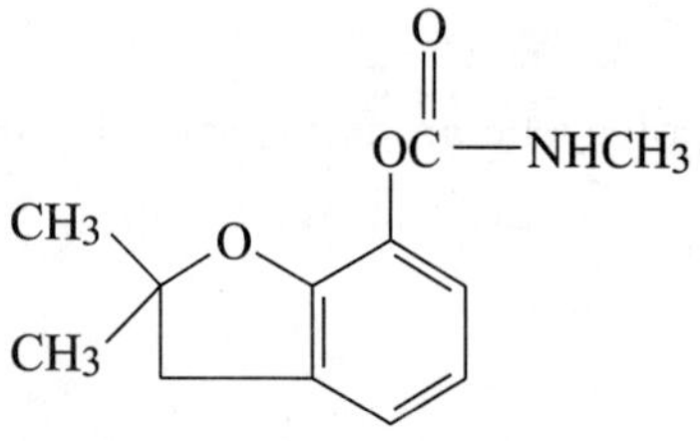

(carbofuran) 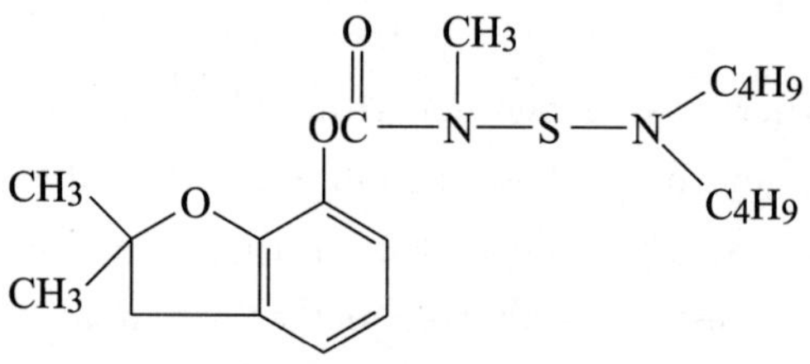　　　　　　　　　　　(carbosulfan)

(2) Properties

The carbamate ester derivatives, used as insecticides (and nematocides), are generally stable and have a low vapour pressure and low water solubility. The carbamate herbicides (and sprout inhibitors) have the general structure $R_1NHC(O)OR_2$, in which R_1 and R_2 are aromatic and/or aliphatic moieties. Carbamate fungicides contain a benzimidazole group. The common varieties in China are: carbaryl, bassa, cartap, carbofuran, pirimicarb, metolcarb, aldicarb, mixed off Granville, isoprocarb, propoxur, methomyl, pirimicarb, and thiodicarb and so on. In general, the vapour pressure of the carbamates is low. From the view for its application in pest control, the performance of carbamate insecides manisfesst the general characteristics as follws.

1) Most of thecarbamate insecide varieties have quick bioactivity, short residual period and good selectivity. They have a good control of planthopper, thrips and other pests but invalid to mites or beatles.

2) Carbamates are not very stable under aquatic conditions and will not persist long in this environment. In general, the toxicity of carbamates for wildlife is low. However, some varieties have very high acute toxicity such as carbofuran, aldicarb. These high toxic caramates are only allowed for use as granules. Aldicarb has good water-solublility and is limited in use for protection of water enviroments.

3) Carbamates in different structures vary a lot in their biological activity. Carbaryl whose moleculor contained naphthalene ring has a broad insecticide spectrum in control of bollworm, twill night mindanao, army worm, cotton leafhopper, cotton thrips, aphids, leaf roller, yellow striped flea beetle and a variety of lepidoptera pests. Carbofuran whose structure contains benzofuran ring manifest wider insecticide spectrum with good control of borer, planthoppers, leafhoppers, rice bud worm, army worms, thrips, gall midge, rice weevil, aphids, nematodes. Carbamates containing substituted phenyl, such as metolcarb and isoprocarb have special toxic effects to most leafhoppers and planthoppers, but no killing effects on other pests.

4) Synergist used for pyrethroid insecticides, such as sesame oil, sesamin, piperonyl butoxide, etc. , have good effect in improving carbamate bioactivity. Also, carbamate insecticides with different structure used in mixture manifest better killing effects on pests and also avoid drug-resistance induced by individual use. Carbamate insecticides can also be used as a synergist for some organic phosphorus insecticides.

10.3.1.2　Mechanism of toxicity

Carbamates are usually easily absorbed through the skin, mucous membranes, and respiratory and gastrointestinal tracts, and are mainly distributed in liver, kidney, fat and muscle tissues. Generally, the metabolites are less toxic than the parent compounds and are mainly excreted rather rapidly in the urine. 70% -80% of the intake drug could be emited in 24 h.

Poisoning mechanism of carbamate pesticide was similar to organophosphorus pesticides. Carbamates inhibit acetylcholinesterase (AChE) in the nervous system by carbamyl reaction of serine in the enzyme active centre. The carbamylation of the enzyme is unstable, and the regeneration of AChE is relatively rapid compared with that from a phosphorylated enzyme. Thus, carbamate pesticides are less dangerous with regard to human exposure than organophosphorus pesticides. The posioning symptoms from carbamates are similar to organophosphorus pesticide, but are relatively lessened. The muscarinic symptoms are typical such as dizziness, headache, fatigue, nausea, vomiting, salivation, sweating and miosis. The AChE activity is slightly inhibited. The symptoms were generally mild, short duration and the poisoned can recover fast. However, pulmonary edema, cerebral edema, coma or respiratory depression may occur if large dose was oral intaken. The delayed peripheral neuropathy symptoms did not occur in carbamate poisoning. Mostly, the poisoning symptoms from carbamates can be seen a few minutes after exposure, last for a few hours and the patients can fully recover within 24 h (except maximum dose poisoning) without after-effects of disability[27].

10.3.1.3　Metabolism of carbamates

Carbamates degrade quickly into CO_2, N_2, H_2O and other simple compounds in the soil due to microbial action. The rate of degradation depends on the carbamate, soil type, temperature and the species of microbes present. The decomposition rate in dry soil is lower than wet soil (Table 10-1). The metabolized products included oxidation products and hydrolysis products and completely was degraded into CO_2, which has been confirmed by degradation experiments of carbofuran labelled with ^{14}C. It showed that the amount of resulted $^{14}CO_2$ is closely related with the moisture content in soil.

Table 10-1　Half-Lives of some carbamate pesticides in soil

Carbamate pesticides	Half-life in soil	Carbamate pesticides	Half-life in soil
Aldicarb	30 days	Desmedipham	30 days
Aldicarb oxygen	20 days	1-Naphthalenyl methyl carbamate	10 days
Bendiocarb	5 days	Furadan	50 days
Benomyl	67days	Methomyl	30 days
Oxamyl	4 days	Propham	10 days
Propoxur	30 days	Chlorpropham	30 days
Propamocarb	30 days		

Almost all microbial fungi or bacteria could be involved in metabolism of carbamates. Hydrolysis and oxidation reactions are the basic decomposition modes of these compounds. The major products from hydrolysis include phenols or carboxylamine. The former can be oxidized to polyhydroxy derivatives, which bind to organic matter for further breakdown while the latter is quickly tansformed into methylamine and CO_2. Carbofuran and carbaryl degrade in this way. The breakdown of carbaryl is shown in Figure 10-7 and Figure 10-8. In plant, the resulting hydroxy compounds bind to glucose while it is transformed into water soluble sulphates or glucuronic acid conjugates. Figure 10-9 and Figure 10-10 are the metabolism processes of arprocarb and ethiofencarb. The carbamate compounds could be completely degraded in the soil but gradually break down into various compounds in other media as shown in the figure.

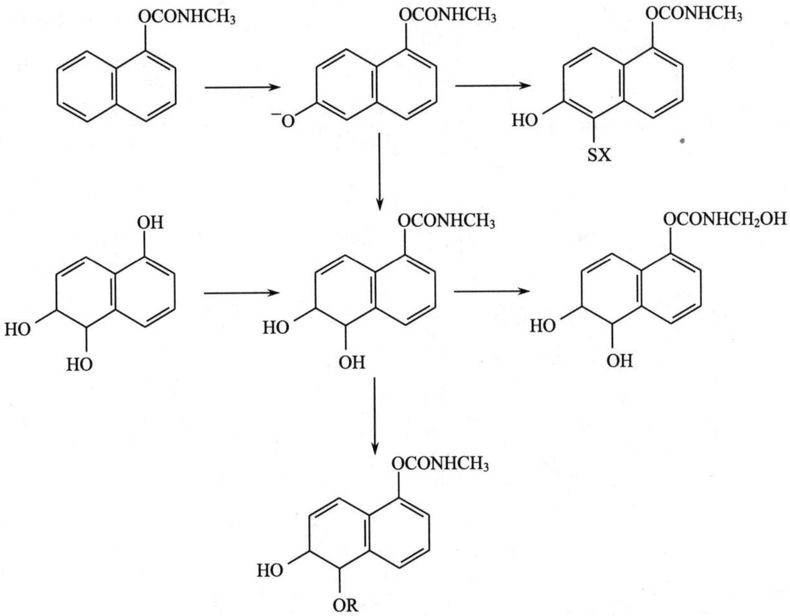

Figure 10-7　Breakdown of carbaryl(I)

R present glucose,glucuronic acid or carbonates

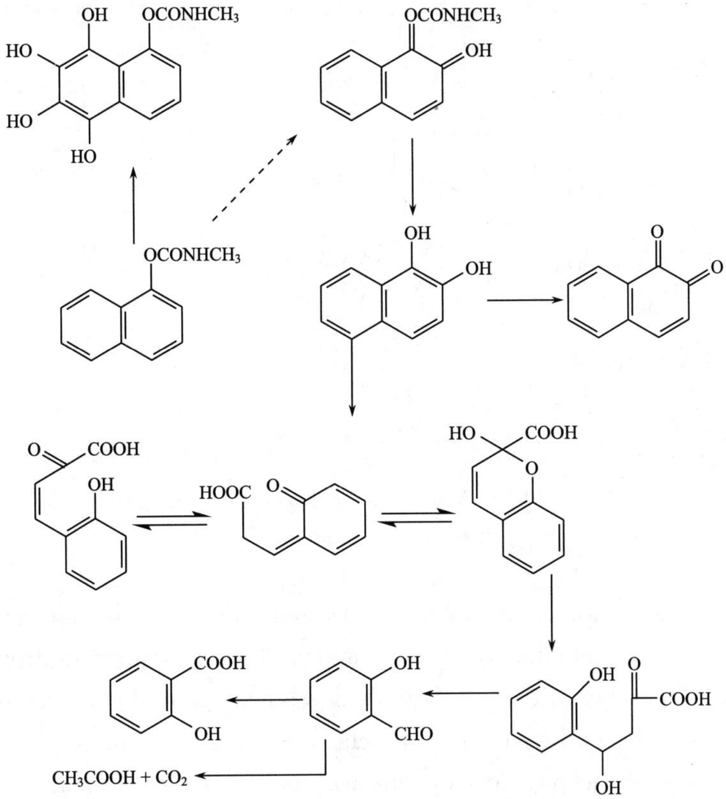

Figure 10-8　Breakdown of carbaryl(II)

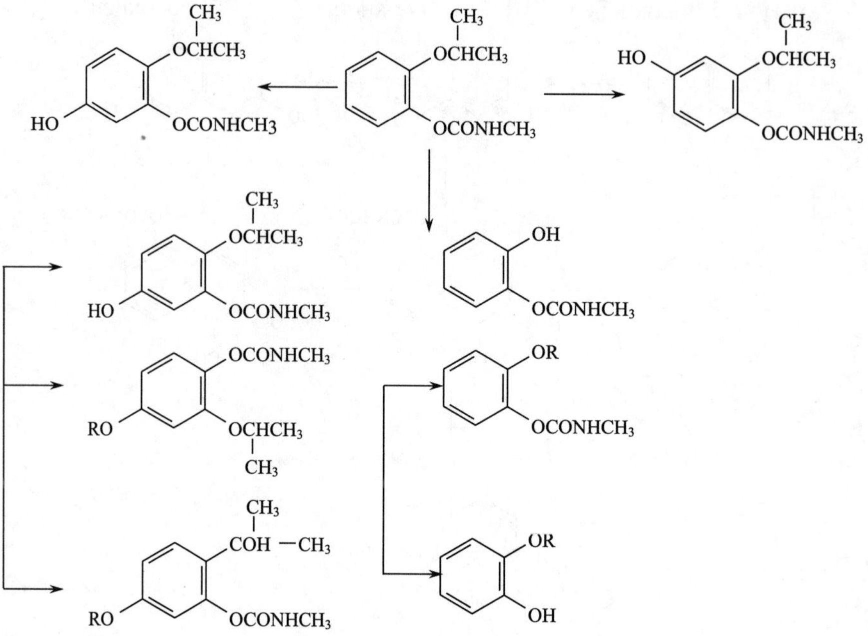

Figure 10-9   Metabolism pathway of arprocarb

R=sulphate or glucuronic acid

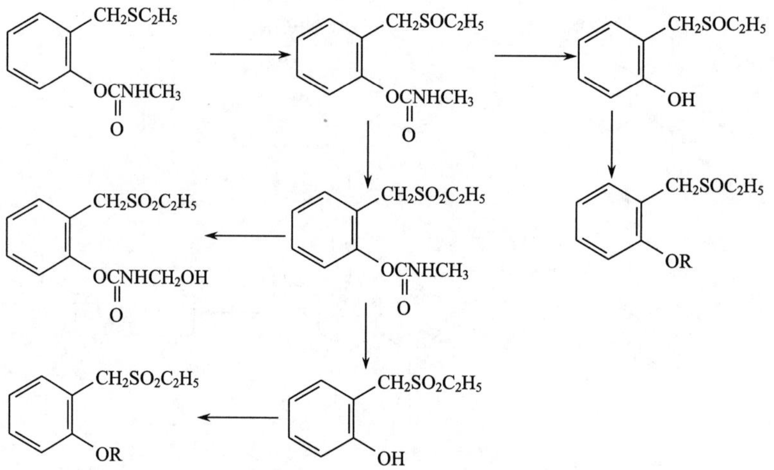

Figure 10-10   Metabolism pathway of ethiofencarb

R=sulfate or glucuronic acid

Metabolic pathway of carbosulfan in rats and drosophil are not same, which is shown in Figure 10-11. Higher toxic compounds will be generated in drosophil, which shows the high selectivity of carbosulfan. Aldicarb belongs to one of the most toxic and stable species in carbamate pesticides, which is toxic to fish, insects, nematodes and mammals. The metabolic products of aldicarb such as sulfoxide and sulfone are also highly toxic. The photolysis of aldicarb includes methylamine, dimethylamine ether, di-cyano-tetramethyl succinic acid, dimethyl urea, thio cyano methyl maleic acid and their derivatives.

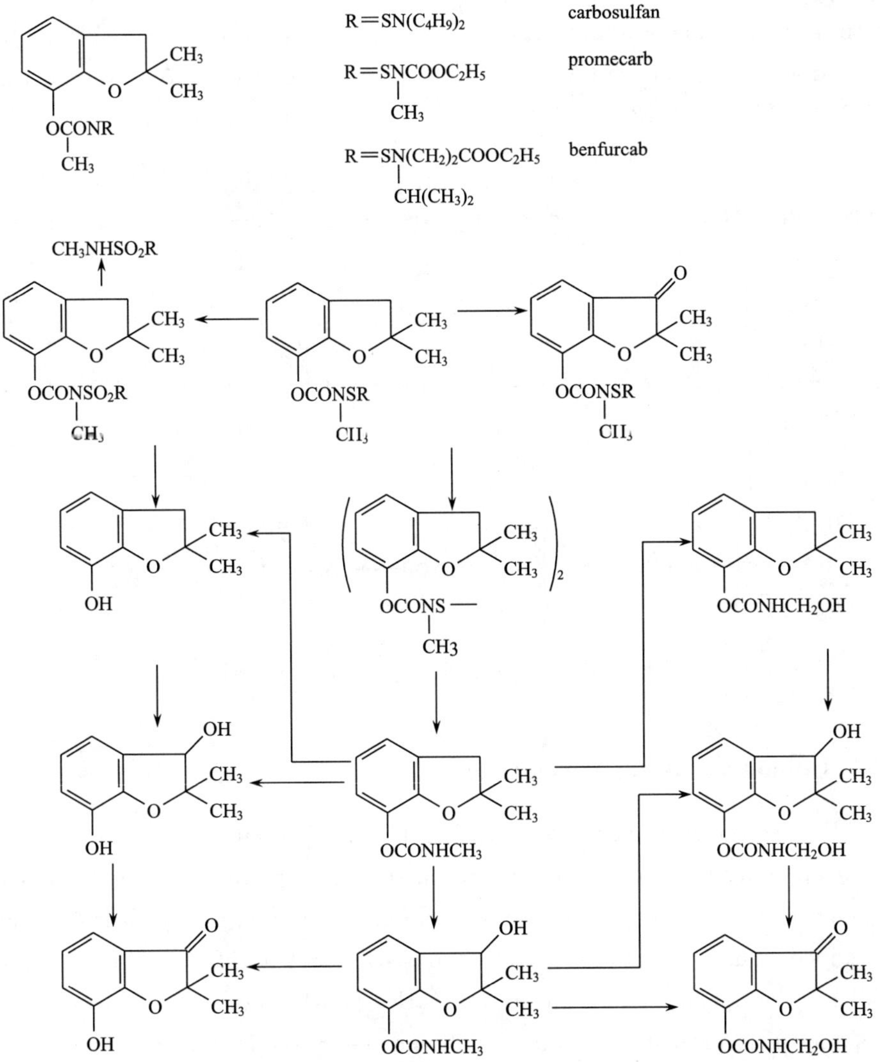

Figure 10-11   Metabolism pathway of carbosulfan in rats and drosophil

Sulfone products from aldicarb also named as oxygen aldicarb could be used as soil insecticide or nematicide individually. Aldicarb and its metabolites such as sulfoxide and sulfone are stable in soil. They may accumulate in the soil and leach into the groundwater due to frenquent application. Hydrolytic metabolism of aldicarb is very slow but the process will be promoted in chlorinated water. 80% of the pesticides will be broken down during 20 h.

Accumulation of aldicarb and its metabolites in radish, beet, potato, melon and other crops were studied.

It is confirmed that aldicarb or its toxic metabolites is found in the soil 400 days later from its application in the case of 15 kg/ha administration amount. Aldicar starts to transform into sulfoxide or sulfone on the 10th day from its application. 99% of sulfoxide will break down to sulfone in 120 days. Under anaerobic conditions, aldicarb will break down in 24 h (Figure 10-12).

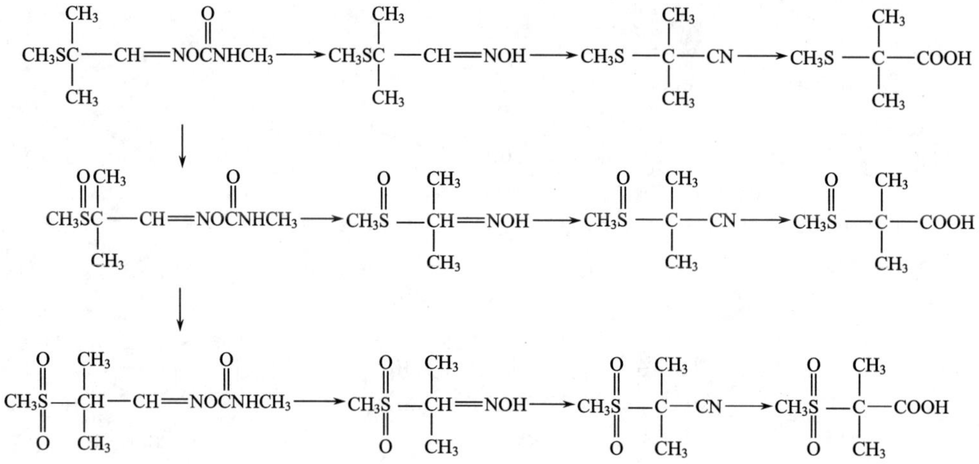

Figure 10-12　Metabolism pathway of aldicarb

10.3.2　Immunoassay for carbamates

Yu[28] carried out immunoassay on carbaryl. Two haptens of carbaryl were synthesised. The coupling method of hapten and carrier protein was optimized. A high titer and high specificity polyclonal antibody was obtained. A competitive ELISA for carbaryl was developed. The applications of this proposed ELISA in different food matrix including apples, pears, Chinese cabbages, rape, rice and barley were employed. The carbaryl residues were extracted with methanol and the extractions were diluted with assay buffer (0.5% fish skin gelatin-phosphate buffered saline) 20 times (apples, pears, rice and barley samples) or 50 times (Chinese cabbages and grape samples) for immunoassay. The LOD of carbaryl in different food matrix is less than 1 mg/kg. High performance liquid chromatography method for detection of carbaryl residues in samples was uesd to verify the accuracy of direct competitive ELISA. The consistency of the results of two methods was over 90%.

Based on the antibody above, Zhang[29] prepared test strips in dipstick and the flow-through format to detect carbaryl residues in food. Two tracers including horseradish peroxidase and colloidal gold were used. The dipstick assay was based on the competitive principle and an anti-carbaryl polyclonal antibody immobilized on positively charged nylon membrane was used. A strip of the nembrane coated with antibody was immersed into the solution mixture of antigen and HRP labelled antigen for 10 min. The assay was completed within 15 min and the limit of detection was 10 μg/L. Validity of the dipstick assay was verified by HPLC and it was proved to be an efficient method for carbaryl analysis. In colloidal gold immunoassay, flow-through and lateral-through strategies were tested. The LOD of flow-through format was 50 μg/L while the LOD of lateral-flow test strip was 100 μg/L.

Yang[30] synthesized two haptens of carbofuran (Figure 10-13) including 6-{[(2,3-Dihydro-2,2-Dimethyl-7-benzofuranyloxy)carbonyl]- amino} hexanoic acid (BFNH) and {(2,3-Dihydro-2,2-Dimethyl-7- benzofuranyloxy) carbonyl]- amino} butanoic acid (BFNB). The two haptens were coupled with carrier protein bovine serum albumin (BSA) and ovalbumin (OVA) to prepare immunogen and coating antigen. Polyclonal antibody was achieved to immunize rabbits with immunogens. The titer of purified antibody ranged from 1 ∶ 500,000 to 1 ∶ 4,000,000 and cross-reactivities with furan phenol or other carbomates were low. Purified antibodies were labelled with horseradish peroxidase to develop direct competitive ELISA for carbofuran and the linear range of assay was 0. 01-10 mg/L. Indirect competitive ELISA was also developed with linear range 0. 005- 10 mg/L. Both formats were used to test water, soil, vegetables and poisoning samples. Intraassay coefficient of variation between batches lower than 8. 0% and the recoveries were higher than 89. 62% in all matrixes but vomitus. Hu[31] established radioimmunoassay (RIA) for carbofuran residues in water samples based on this antibody. The results showed that the detection limit of RIA was 0. 175 ng/mL and a wider linear detection rang from 0. 27 ng/mL to 4,000 ng/mL. The proposed RIA was applified to

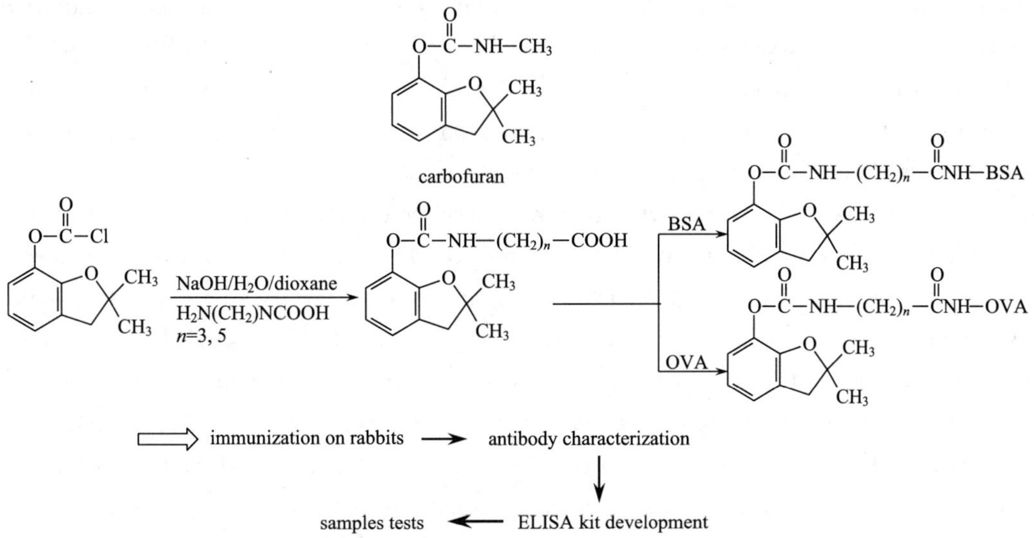

Figure 10-13   The process of ELISA kit development for carbofuran

detect carbofuran residues in water samples and high performance liquid chromatography was used to verify the results. The linear correlation coefficient of two detection method reached 0. 98.

Mao[32] developed gold immuno-chromatographic assay for carbofuran residue based on the antibody above. Average size of colloidal gold 17. 5 nm was synthesized with sodium citrate reduction method. Flow- through format was used (Figure 10-14). The conditions for conjugation between the antibody and gold nanoparticles were optimized (pH 7. 5-8. 0, antibody concentration 12 μg/mL). The coating antigen and goat anti-rabbit IgG were sprayed onto the nitrocellulose membrane for test line and control line with XY Dispenserwork station. This dipstick strip can give detection results in 10 min and the sensitivity of test strip was 0. 5 μg/mL. This strip can be used to analyze vegetable samples but not siutable for tea analysis due to serious pigment interferences.

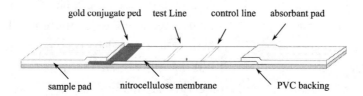

Figure 10-14　Assembly of gold immunochromatographic test strip

Liu[33] synthesized carbofuran hapten which reserved the structure of carbamate. The hapten was covalently conjugated to carrier proteins by modified active ester method. The antiserum with high affinity to carbofuran was obtained from New Zealand white rabbits. The polyclonal antibodies were purified by DEAE-cellulose column. The horseradish peroxidase (HRP) labelled antibody was prepared by means of modified NaIO$_4$ method and mixed anhydride method. The indirect competitive ELISA was developed with the linear detection range 10^{-4}-10 μg/mL and the detection limit was lower than 0. 01 ng/mL. There are no interferences from other commonly used carbamate insecticides in determination of carbofuran.

The hapten of the N-methyl carbamate insecticide metolcarb, 3-{[1-(3-(Methyl) phenyloxy) carbonyl] amino} propanoic acid (HOM), was synthesized by Zhang [34]. The hapten conjugated with the carrier proteins bovine serum albumin (BSA) and ovalbumin (OVA) respectively by the active ester method. The New Zealand rabbits were immunized by the conjugate of HOM-BSA and the titres of anti-metolcarb serum (1. 28×10^6) were determined by a noncompetitive indirect enzyme linked immunosorbent assay (ELISA) procedure. The cross reaction indicate that the antiserum could specially recognize the insecticide metolcarb with lower cross-reactivity to some related carbamates (less than 0. 1%). After optimization of the ELISA conditions such as ionic strengths, organic solvents, pH values, blocking agents and so on, A competitive indirect ELISA procedure for the determination of metolcarb was established. Based on statistical analysis, the linear range of the procedure was from 1-10, 000 $\mu g/L$, IC_{50} was 40. 74 $\mu g/L$ and the limits of detection were 0. 08-0. 1 $\mu g /L$. The intraassay relative standard deviatim (RSD) reached 2. 9 %, and inter-assay RSD reached 4. 6 %. The recoveries obtained by standard metolcar addition to the different samples such as rice, water and soil were 80 %, 93. 4 % and 107 % respectively.

10.4 Pyrethroid insecticides

10. 4. 1 Overview

10.4.1.1 Classification, structure and property

(1) Classification and structure

In the 15th century, it was found that flowers of pyrethrum could kill pests[36]. Later in the mid-19th century, people began to plant pyrethrum and the main producer was Kenya. After 40 years of studies, 6 active ingredients in pyrethrum flowers had been confirmed (Table 10-2 and Figure 10-15).

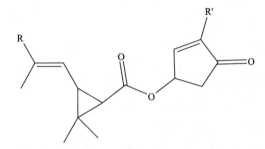

Figure 10-15   Structure of pyrethrum

Table 10-2   Active ingredients of pyrethrum flower

	Composition	R	R'	Content/%
1	Pyrethrin I	—CH₃	—CH₂—CH=CH—CH=CH₂	38. 7
2	Pyrethrin II	—COOCH₃	—CH₂—CH=CH—CH=CH₂	30. 7
3	Cinerin I	—CH₃	—CH₂—CH=CH—CH₃	10. 1
4	Cinerin II	—COOCH₃	—CH₂—CH=CH—CH₃	14. 5
5	Jasmone pyrethrin I	—CH₃	—CH₂—CH=CH—CH₂—CH₃	6. 1
6	Jasmone pyrethrin II	—COOCH₃	—CH₂—CH=CH—CH₂—CH₃	5. 4

Pyrethroids belong to bionic pesticides and its biological activity is on account of its spatial conformation. All pyrethroids contain two or three chiral centers, making them a family of pesticides with the highest number of enantiomers. Various geometric isomers, also known as *cis*- and *trans*-isomers were present in pyrethroids. An optical isomer can be named by the spatial configuration of its atoms. The D/L system does this by relating the molecule to glyceraldehyde. An enantiomer can also be named by the direction (+) and (d) in which it rotates the plane of polarized light. In general, biological activity of *trans*- isomers was higher than that of the *cis*- isomers. However, the *trans* isomer manifests higher bioactivity in halogenated pyrethroids. For pyrethroids, D-isomer shows higher bioactivity. Synthetic pyrethroids are a class of widely used insecticides, and their use may further increase in the near future as some organophosphate compounds are phased out for certain uses. The earliest pyrethoids are related to pyrethrin I and II by changing the alcohol group of the ester of chrysanthemic acid. This relatively modest change can lead to substantially altered activities. The Rothamsted team developed a second generation in the 1970s, notably permethrin, cypermethrin and deltamethrin, with improved light stability that made them much more suitable for use in agriculture. Now, pyrethroids account for up to 17% of global insecticide sales. Some typical pyrethroids are shown in Table 10-3.

(2) Properties

Pyrethroid belongs to broad spectrum insecticide with contact killing acion and also manifests strong stomach poisoning, ovicidal, antifeedant and repellent effects but no suction action or suffumigate application. Pyrethroids have a suitable potency and persistance in the environment, resulting in low pollution.

Pyrethroids are widely used to control a large spectrum of pests and parasites in animal husbandry including *lepidoptera*, *homoptera*, *hemiptera*, *diptera*, *coleoptera* and so on. Due to the high lipophilic properties of pyrethroids, the pestcide could penetrate into the body of insects. The dose of use is small compared with other insectides. Currently, widely used varieties include cypermethrin, *cis*-and *trans*-cypermethrin, deltamethrin, fenpropathrin, fenvalerate, lambda permethrin, bifenthrin and permethrin.

① Fenvalerate

The chemical structure of fenvalerate is shown as the following, the moleculas weight is 491. 9.

Table 10-3　Main varieties of pyrethroids

Nature Pyrethrin (6 compsitions)

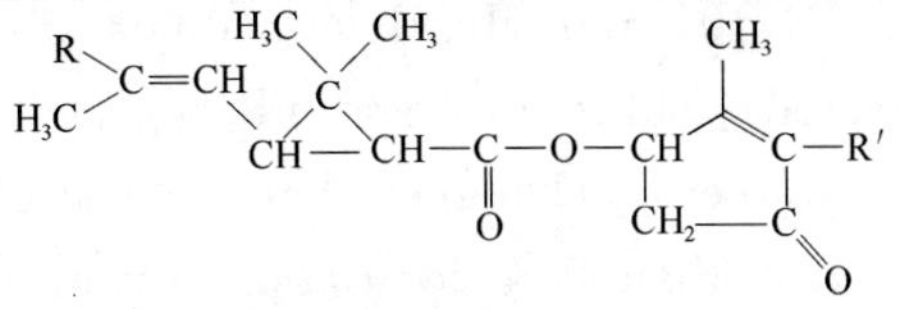

R: —CH₃, —COOCH₃
R': —CH₂CH=CHCH=CH₂
—CH₂CH=CHCH₃
—CH₂CH=CHCH₂CH₃

Light instability pyrethroid

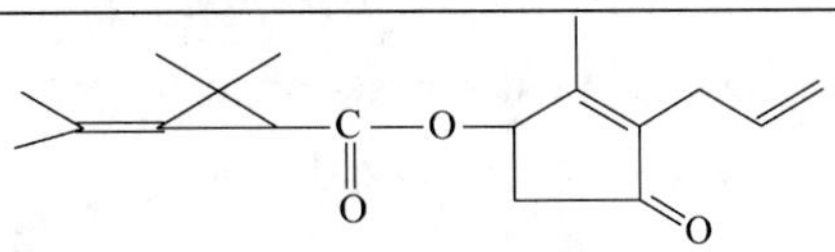

Allethrin

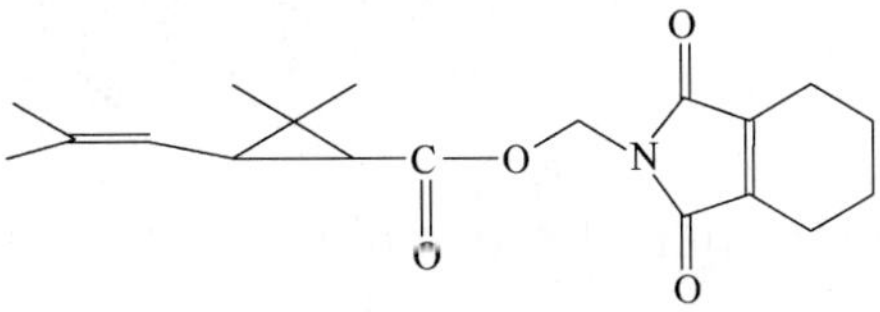

Tetramethrin

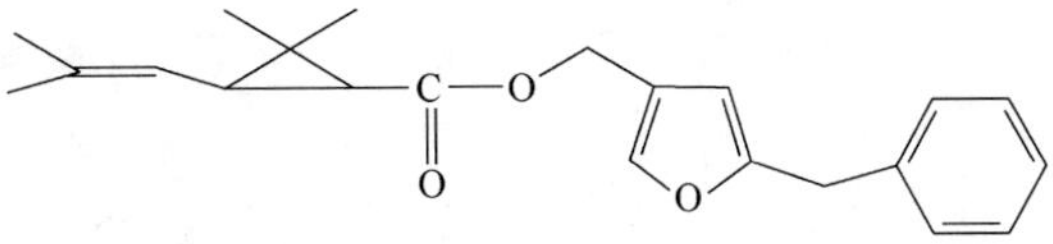

Resmethrin

Light stability pyrethroid

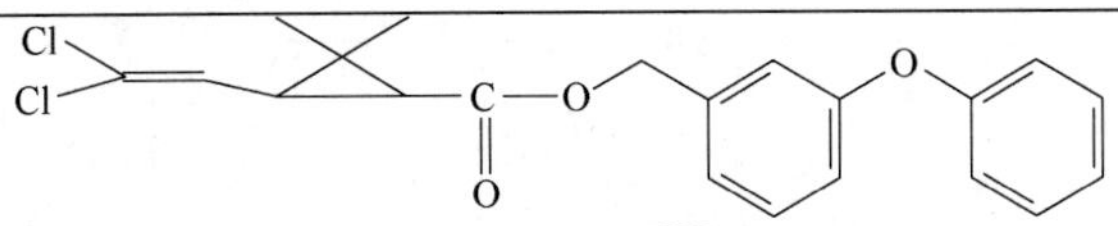

Permethrin

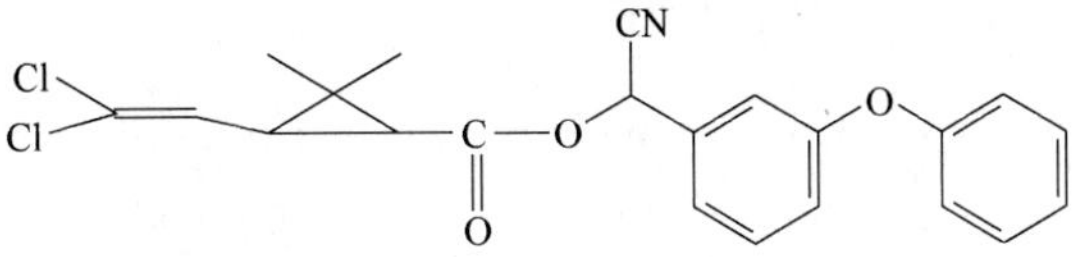

Cypermethrin

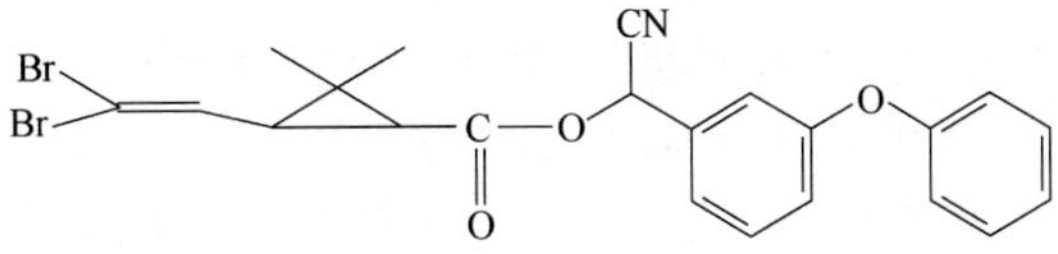

Deltamethrin

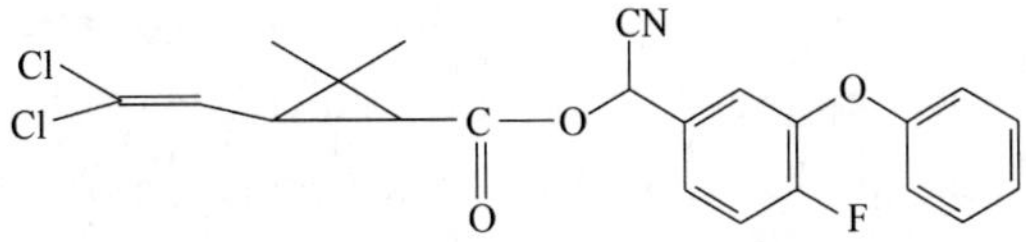

Cyfluthrin

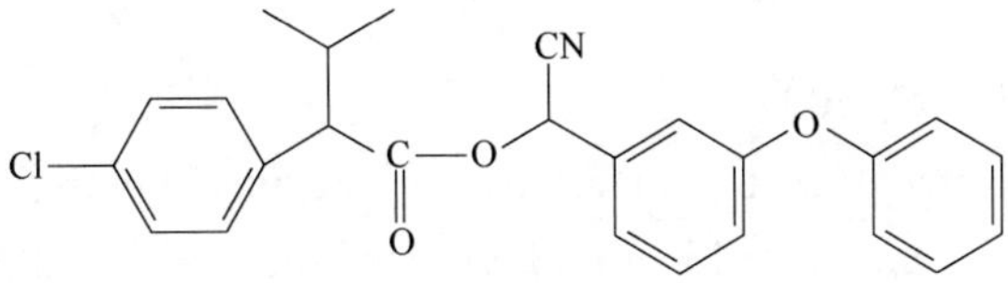

Fenvalerate

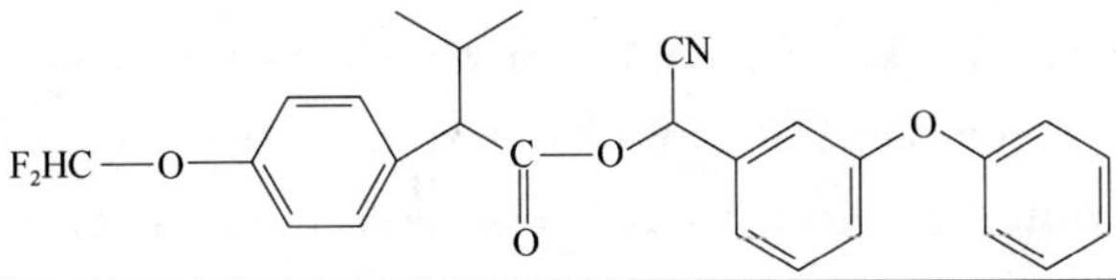

Flucythrinate

It is a colourless or yellowish oily liquid and its density is 1. 17 g/cm³. The vapor pressure at 25 ℃ was 2. 8 × 10⁻⁷ mm Hg. It was almost insoluble in water but soluble in methanol, ethanol, acetone, chloroform and other organic solvents. It is highly stable in acidic media but unstable if pH higher than 8. There are no significant changes when the temperature is lower than 75 ℃ for 100 h. However, gradual decomposition of fenvalerate will occure if the temperature is above 150 ℃.

Fenvalerate is particularly effective against *Lepidoptera larvae*, and also has a high insecticidal activity on *orthoptera*, *hemiptera* and *diptera* pests. Fenvalerate shows good prevention effects on pest's resistance to organophosphorus or carbamate insecticide. It is widely used for controlling the pests on cotton, fruit, vegetables, soybeans, potatoes, wheat, corn, tea and other crops and is also used for pests in homes and husbandry.

② Permethrin

The chemical structure of permethrin is as belows. The molecule weight is 391. 3. It is a colourless or pale yellow crystal with melting point about 35 ℃ and the boiling point 198-200 ℃.

Permethrin smells slightly aroma. It is almost insoluble in water but soluble in acetone, ethanol, benzene, xylene and most other organic solvents. This compound is stable in solvents and storage but rapidly hydrolysis in alkaline medium. Permethrin breaks down slowly in the air due to oxygen and sunlight.

Permethrin is particularly effective on killing the eggs of *Lepidoptera larvae*. It is used to control pests on cotton, rice, wheat, corn, soybeans, vegetables, fruit, tea, tobacco, forest and other crops. It is good for quick knock-down action on insects. In addition, it is also used for the control of cockroaches, mosquitoes, flies and parasites in the enviroment.

③ Valerathrin(S-5439)

The chemical structure of permethrin is as given below. The molecular weight is 394. 9. The purified compound is a yellow oily liquid with boiling point 248-250℃. It is insoluble in water but soluble in common organic solvents. Valerathrin shows stomach and contact toxicity on insects. It could be used to control insects on rice, wheat, corn, vegetables, tea, fruit trees, cotton and is also used in pests control indoors.

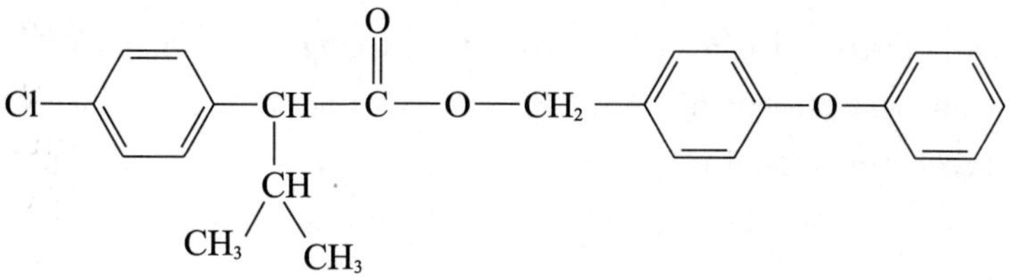

④ Cypermethrin

Cypermethrin is a synthetic pyrethroid insecticide used to kill insects on cotton and lettuce, and to kill cockroaches, fleas, and termites in houses and other buildings. Cypermethrin is a semivolatile nonpolar compound (vapor pressure 3.4×10^{-6} mm Hg at 25℃) with two chlorine atoms and its moleculas weight is 419.3.

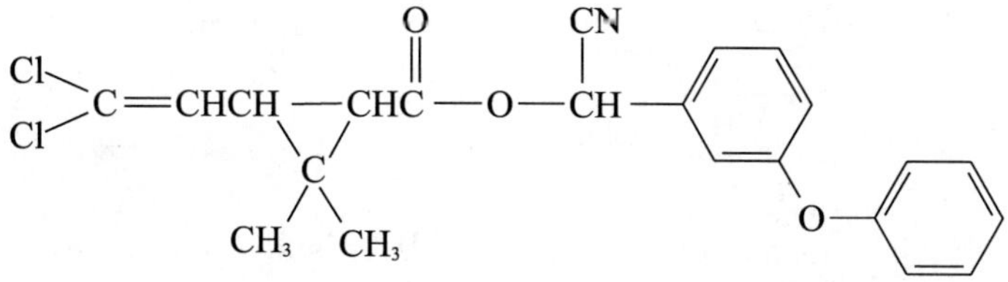

Industrial product of cypermethrin is a yellow or brown viscous liquid or solid in racemic formulation (*cis*-isomer : *tran*-isomers=4 : 6). It is soluble in acetone, hexane, cyclohexane, ethanol, xylene, chloroform and other organic solvents and the solubility in water is about 0.2 mg/L. It is stable in acidic or neutral solution, but breaks down in alkali medium. Cypermethrin manifests good thermal stability and can be used together with various pesticides.

Cypermethrin represents strong contact poisoning effects, but has no absorption or fumigation effects on insects. It is widely used on rice, cotton, vegetables, fruit, tea, soybean, cucumber and other crops.

⑤ Deltamethrin

Product of technical grade is off-white solid powder with vapour pressure 1.5×10^{-8} mm Hg and meluculas weight 505.2. Deltamethrin is insoluble in water but highly soluble in most organic solvents such as acetone, benzene, xylene, ethyl acetate, dioxane and so on.

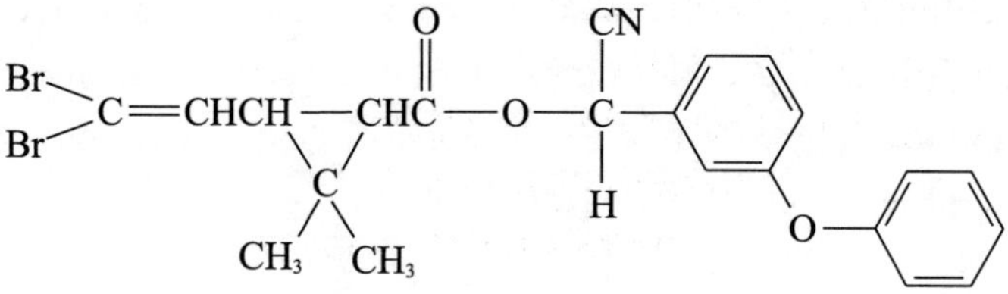

This pesticide is relatively stable in light, acidic or neutral solution but unstable in basic enviroment. Deltamethrin posseses contact toxicity, stomach posioning and antifeedant effect on pests featuring rapid knock down action. It is specially effictive on *lepidoptera larvae* and is used to control pests on rice, cotton, vegetables, tea, grape, citrus, apples, soybeans, wheat, corn, tobacco, storage of grain.

⑥ Flucythrinate

Flucythrinate is a synthetic pyrethroid used to control insect pests in apples, cabbage, field corn, head lettuce and pears, and to control *Heliothis* spp. in cotton. It is used primarily on cotton. It is available in emulsifiable concentrate and wettable powder formulations.

$$F_2CHO \text{—} \bigcirc \text{—CH—} \overset{\overset{O}{\|}}{C} \text{—O—} \overset{\overset{CN}{|}}{CH} \text{—} \bigcirc \text{—O—} \bigcirc$$
$$\underset{\underset{CH_3 \quad CH_3}{\diagup \ \diagdown}}{\overset{|}{CH}}$$

Flucythrinate is a dark amber, viscous liquid. It is stable in neutral and acidic conditions and it is stable in light with vapour pressure 8.7×10^{-9} mm Hg. It is soluble in most organic solvents such as acetone, xylene and isopropanol but almost insoluble in water. Flucythrinate belongs to highly effective broad spectrum pyrethroid insecticide with contact and stomach poisoning.

⑦ Beta-cyfluthrin

$$\underset{Cl}{\overset{Cl}{\diagdown}} C \text{=CHCH—} \overset{\overset{O}{\|}}{CHC} \text{—O—} \overset{\overset{CN}{|}}{CH} \text{—} \bigcirc \text{—F}$$

Industrial product of beta-cyfluthrin is a crystalline, viscous liquid. It is insoluble in water, slightly soluble in ethanol but soluble in ether, acetone, toluene, methylene chloride and other organic solvents. beta-cyfluthrin is stable in acidic medium but unstable in alkaline media. Its density is 0.997 g/cm^3 and can be used in mixtures of most insecticides. Beta-cyfluthrin possesses a strong contact and stomach toxicity but no suction effects. It slightly permeats through surface of insects. This insecticide has a long residual period and specially affects on *lepidopteran* pests on cotton, vegetables, fruit trees, soybeans, tobacco and other crops.

10.4.1.2   Mechanism of toxicity

Pyrethrins and pyrethroids act on tiny channels through which sodium is pumped to cause excitation of neurons. They prevent the sodium channels from closing, resulting in continual nerve impulse transmission, tremors, and eventually, death. In insects pyrethrins prolong the opening of the sodium channel, and produce instant paralysis. Nervous system stimulation proceeds from excitation to convulsions and tetanic paralysis. Muscular fibrillation and death may occur as with DDT. The typical effects of pyrehtroids are: Hyperexcitation; Loss of coordination, tremors, convulsions, tetanic spasms; Knock-down effect and dehydration; Death.

Pyrethrum insecticides show a poisoning effect dependency of temperature. With the decrease of temperature, the toxicity will increase. Metabolism of pyrethroids in insects mainly includes oxidation and esterlysis, thus, mixed function oxidase plays an important role in breakdown of pyrethroids. The compounds which inhibit mixed function oxiase or esterase are often used as synergists of pyrethroids. Some insecticides possess fumigation effects, including allethrin and prallethrin which are used for mosquito coils or electric mosquito propylene. Some pyrethroids such as tetramethrin and permethrin kill insects by contact posioning. Insects are prone to resistance to pyrethroids in agricultural use, which should be considered in applications.

10.4.1.3   Metabolism and Residues

Pyrethroids are ester compounds. Pyrethroid insecticides, such as fenvalerate, are thought to be rapidly metabolizing and quickly eliminated from animal tissues. The fenvalerate concentration in the blood reaches the peak in an hour for oraley treated rats or dogs and the blood concentration of fenvalerate will reduce to 3% of its peak value 24 hours later. The biological half-life period is 2 h for fenvalerate. In mammals, pyrethroids are rapidly metabolized by cleavage of the central ester linkage, leading to different metabolites, which can react with some molecules and ramain long time in the body. In insects, birds or mammals, the metabolites of pyrethroids include alcohol metabolite, benzene, carboxylic acids and other conjugates. Pyrethroid esters are rapidly detoxified by hydrolases and oxidases after oral, inhalative or dermal intake. Both acid and alcohol moieties of the pyrethroids are metabolized into their carboxylic acids, partially conjugated and finally eliminated with the urine. Metabolism of permethrin in mammals is shown in Figure 10-16.

Synthetic pyrethroids are extremely effective against insects, but are relatively safe to mammals and birds[6]. One potential problem of pyrethroids is their extreme toxicity to aquatic organisms, where often <1 $\mu g/L$ will produce toxic effects. However, animal and limited human studies also suggested possible associations of pyrethroid expo-

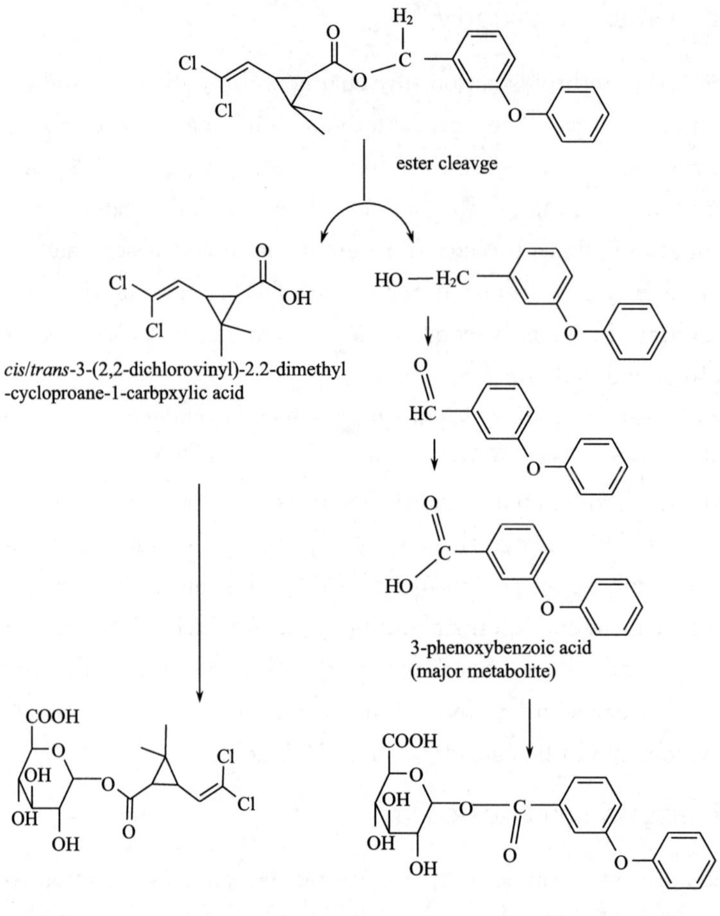

Figure 10-16　Major metabolism pathways of permethrin in mammals

sure with semen quality. Pyrethrins appear to disrupt the normal functioning of sex hormones.

10. 4. 2　Immunoassay for pyrethroids

Liu[35] coupled permethrin with carrier protein to prepare antigen. Indirect competitive ELISA (icELISA) was established to detect permethrin residues in environment samples. The LOD of this icELISA was 0. 1 μg/mL.

Two haptens were synthesized by Zheng[36] based on deltamethrin acid (DA). Conjugates of hapten -BSA were immunized mice to produce polyclonal antibody. Serum titer was detected by the indirect method. The antibody recognizes haptens well with IC_{50} 54. 5 ng/mL and 26. 9 ng/mL respectively and can be used to detect degradation products of deltamethrin. The cross-reactivities of polyclonal antibody with deltamethrin, tetramethrin, cypermethrin, deltamethrin acid are less than 0. 1%.

Li[37] established indirect competitive ELISA for determination of bifenthrin and fenveralate. The metabolite of bifenthrin (2′-methy-13-pheny 1 benzyl alcohol) was modified into [2-methyl -(1,1-biphenyl)-3-methoxy]-carbonyl propionic acid(LBc) and [2-methyl-(1,1-biphenyl)-3-methoxy]- carboxylic acid), which were used as haptens (Figure 10-17).

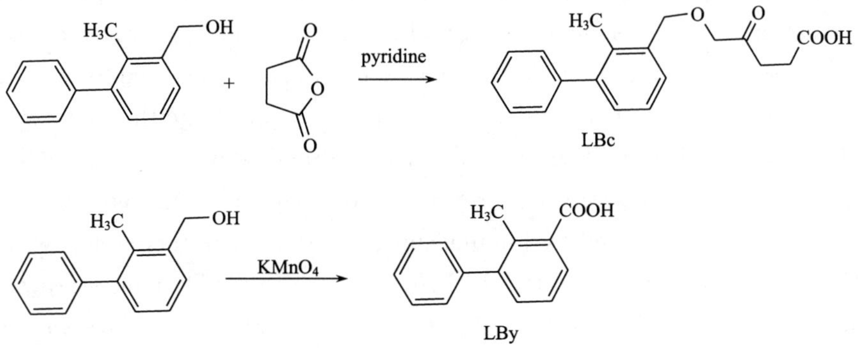

Figure 10-17   Synthesis of hapten for bifenthrin

Then, the haptens (LBc and LBy) were conjugated to BSA respectively using the carbodiimide method for immunogen (LBc-BSA and Lby-BSA) and conjugates of LBc-OVA and Lby-OVA were used for coating antigen with active ester method. Anti-serum of bifenthrin from rabbits was collected. Heterologous and homologous formats were tested in the experiments. The best sensitivity was achieved when antibody from LBc-BSA and Lby-OVA for coating antigen were used in ELISA system. The IC_{50} was 2. 16 mg/L and minimum detection limit (LDL) was 0. 016 mg/L. No cross-reactions with cyhalothrin, deltamethrin, cypermethrin, fenvalerate, fenpropathrin and 3-phenoxybenzoic acid were found. Furthermore, Li designed two haptens, the carboxylated propyl amino derivative of the acid moiety (HCA1) that is the hydrolyzed product of cyhalothrin, and the carboxylated ethyl carbonyl derivative of the alcohol moiety (HCB2) (Figure 10-18). HCA1 and HCA2 were conjugated with the carrier proteins to prepare immunogens and coating antigen. New Zealand white rabbits were immunized to produce polyclonal antibodies. The results showed that polyclonal antibody raised from HCA1-BSA recognized cyhalothrin and cyhalothric acid better wih IC_{50} 33. 12 mg/L and

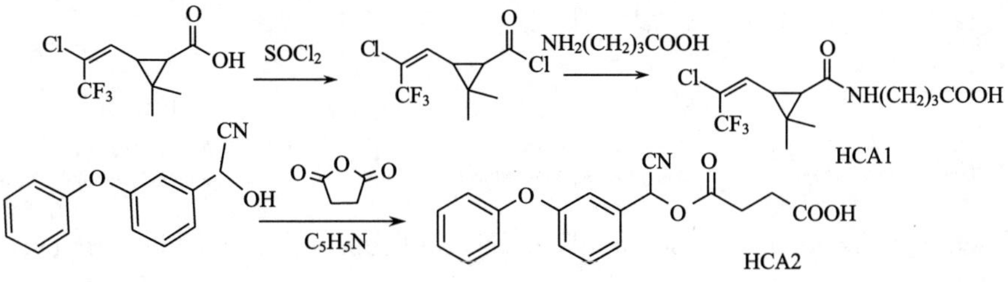

Figure 10-18   Hapten synsthesis for cyhalothrin

0. 95 mg/L respectively.

Han[38] prepared polyclonal antibody from cypermethrin and fluvalinate. The direct competitive ELISA was developed to detect cypermethrin based on labelled haptens with horseradish peroxidase (HRP). The linear range was 0. 01-100 μg/mL with IC_{50} 0. 46 μg/mL. The spiked cypermethrin cabbage samples at 0. 5 mg/kg and 5 mg/kg levels were analyzed with this direct ELISA and the recovery data was between 80%-120%. Similarly, the direct ELISA for fluvalinate detection was established with linear detection range 0. 01-10 μg/mL and IC_{50} 0. 1 μg/mL. You[39] synthesized two haptens for fenvalerate, Fen-Hp1 and Fen-Hp2. Active ester method was used to couple haptens with BSA and OVA respectively for preparation of complete antigens. Specific antiserum for fenvalerate was gotten by immunizing rabbits. Later, the hapten was labelled with HRP to set up direct competitive ELISA. Effects of the pH, ionic strength and organic solvents were optimized and the IC_{50} was 0. 16 μg/mL. Vegetables were analyzed with this direct ELISA. The recovery was 83. 76%-109. 5% at 0. 5 mg/kg spiked levels and the recovery was 93. 64% -109. 7% at 5 mg/kg spiked level.

Zhang[40] prepared The hapten of deltamethrin, [cyano-(3-amino-phenoxyphenyl) - methyl]-3- (2, 2-dibromoethenyl)-2, 2-dimethyl-cyclopropane-1-carboxylate. The structure of hapten was identified by1H NMR. Hapten was conjugated to bovine serum albumin (BSA) and ovalbumin (OVA) by diazotization method to prepare artificial antigens of deltamethrin. The coupling ratios between hapten and the carrier protein were 11 : 1 (hapten-BSA) and 8 : 1(hapten-OVA) respectively. Rabbits were immunized with hapten-BSA and the antiserum was examined by indirect ELISA. The cross reactivity (CR) of antiserum was tested. The CR was less than 1% for other pyrethroids but 19. 5% for 3-phenoxybenzoic acid. The optimized indirect ELISA for deltamethrin was extablished with IC_{50} value of 0. 16 mg/L. Besides, immunoaffinity column (IAC) was packed with the purified polycolonal antiboy to develop pretreatment tools for deltamethrin analysis. The column capacity of the prepared IAC of deltamethrin was higher than 8 μg.

Hao[41, 42] preparated common antigens for the pyrethroids with cyano group.

3-Phenoxybenzyl dehyde   alpha-cyano-3-phenoxybenzyl alcohol   5-(3-benzylphenyl)pentanoic acid

3-Phenoxybenzaldehyde, methanol, KCN, and succinic anhydride were used for the production of hapten. The hapten was conjugated with carrier protein by the active ester method and mixed anhydride method to get complete anigens. The coupling ratio of immunogen and coating antigen were 17. 9 ∶ 1 (molar ratio for hapten and BSA) and 13. 2 ∶ 1(molar ratio for hapten and OVA), respectively. Results showed that the obtained antiserum can recognize cyphenothrin, fenpropathrin, deltamethrin, cypermethrin, ucythrinate, and esfenvalerate. The cross reactivities of cyphenothrin, fenpropathrin, deltamethrin, cypermethrin, ucythrinate, and esfenvalerate were 100%, 81. 49%, 64. 68%, 42. 72%, 22. 96%, and 16. 25%, respectively. The low detection limit (LDL) of this assay was 0. 10 μg/L for cyphenothrin. This method was successfully used for the detection of alpha-cyano pyrethroids in Tai lake water.

10.5 Herbicides pesticides

10. 5. 1 Overview

10.5.1.1 Classification of herbicides

It is said that there are more than 5,800 kinds of weeds, which significantly do harm to the agricultural production and weed control has always been an important issue in agro chemical practice. Chemical agents, that is herbicides, are used widely in the world in protecting crops from undue competition from weeds.

The main chemical classes of herbicides include bipiridilium compounds, triazine derivatives containing three heterocyclic nitrogen atoms in the ring structure (atrazine, prometryn, propazin, etc.), chlorophenoxy acid derivatives (2,4-D;2,4,5-T), substituted chloro-acetanilides (alachlor, propachlor), derivatives of 2, 6-dinitroaniline (benfluralin, trifluralin), substituted phenylcarbamates (carbetamide, chlorbufam), urea derivatives (chlorbromuron, chlorotoluron), substituted sulphonylureas (amidosulfuron, trifusulfuron), etc. The intensive application of herbicides has resulted in the contamination of the atmosphere, ground and waste waters, agricultural products (wheat, corn, fruits, vegetables, beans etc.) and, consequently, in the direct or indirect pollution of food and food products and biological systems.

10.5.1.2 Selectivity of herbicide

The ability of a herbicide to kill certain plants without injuring others is called selectivity. Herbicides that kill or suppress the growth of most plant species are called non selective. Non selective herbicide use is limited to situations where control of all plant species is desired, or by directing the herbicide on the target weed and away from desirable plants. Herbicide selectivity is relative and depends on several factors, including environment, herbicide application rate, application timing, and application technique. Even the most tolerant plant species are susceptible to a herbicide if the rate applied is high enough. Herbicide selectivity may be based on herbicide placement, or differential spray retention, absorption, translocation, metabolism, or site exclusion of the herbicide in the plants.

(1) The biological cause of herbicide selectivity

Biochemical reactions of herbicides are different among plants, which is applified to kill the weeds and protect the crops.

1) Differences of activation of herbicides in plants. Some herbicides thermselvs have no or little toxicity to plants; however, they can be tranformed into a toxic substance after series of activation in the plants. The higher activation of herbicides will kill the plants while the plants in which lower activation of herbicides take place will survive. For example, 4-(2, 4-dichlorophenoxy) butyric acid (2, 4-DB) itself does no harm to plants, but it will be transformed into strong toxicic compound 2,4-D under catalyzation from β-oxidation bacteria. The content and activity of β-oxidation enzymes differ a lot among palnts. Some weeds such as nettle, gray dish and thistle have rich β-oxidation enzyme, which will produce more toxic substances, resulting into killing of plants. Many leguminous plants have less β-oxidation enzyme, which survive themselves in 2, 4-DB application. Another example is chlorazine which belonged to triazine herbicides. Chlorazine itself does not have activity to kill the grass. The strong killing grass activity will be obtained when chlorazine is transformed into trietazine and finally simazine under the calalyzation from N-dealkylation enzyme. Similarly, the content of N-dealkylation enzyme and its mode of action are much varied in different plants.

2) The varied desactivation of herbicides in plants. Herbicides, which belong to this class, are toxic to plants, but they will lose the toxic activity after its degradation in plants. Thus, the tolerance to herbicides depends on the catabolism capacity of plants. For example, rice and barnyard belong to the same family with similar apperance and growth habits. It is difficult to distinguish them, but propanil can selectively kill barnyard grass. The key of the selectivity lies on amide hydrolase enzymes contained in the ripe, which can break down propanil quickly into 2,4-dichloroaniline or propionic acid. Compared with ripe, the content of amide hydrolase enzyme is very limited, which results in death from propanil.

3) The difference of growth habits of plants. Grass plants tend to be more difficult to wet than broadleaf plants because grasses often have narrow, waxy leaves with upright orientation. The grass leaf presents a small target and there is a good chance the spray droplet will roll off the leaf upon contact. Many broadleaf plants are easy to wet because they present a large target with some pubescence and horizontal orientation.

(2) Non-biological reasons for the selectivity of herbicides

Some herbicides, lacking of biological selectivity specificly kill grass based on spray retention or other physical factors. Herbicide may do harm to seedlings, but its residual period is very short. We can choose suitable herbicide application time to emilate grass.

For example, sodium pentachlorophenate degrade rapidly in sunlight and its efficacy will disappear during 3-7 days. If a suitable spray time for application of sodium pentachlorophenate is chosen, the weeds will be killed while no harm is brought to seedlings. Besides, some herbicides selectively kill weeds based on the different depth distribution of plant roots in the soil, or leaching of herbicides themselves in soil. In general, roots of cultivated plants distribute deeply in the soil while roots of most weeds are located in shallow layer of soil. Some herbicides like nitrofen are sprayed on the soil and they can leach below the surface of soil in 1-2 cm depth. Thus, weeds will die after they get contact with herbicides while the roots of seedlings located in deeper layer of soil can survive. Similar principle is for herbicide application in garden. Most wooden plants have strong and deep roots in soil, thus, only very limited herbicides were absorbed, which does little harm to fruit tree. Some herbicides such as paraquat can kill weeds by strong inhibition of photosynthesis. However, it will lose activity when it enters into soil. So, paraquat often is used to control weeds in orchards, mulberry fields or rubber plants.

10.5.1.3 Mode of action

Herbicides now in use can be classified two types: one is contact killing type and the other is that the compounds move move to the site of action in the weed, and accumulate sufficient levels at the site of action to kill the plant[43].

(1) Herbicide site of action

Different herbicides gain entry into plants in different ways and different plants are susceptible to herbicide entry at different locations. Herbicides can enter plants through leaves, roots, shoots and stems.

(2) Foliar applied herbicides

Post emergent herbicides generally are absorbed through the leaves. Leaves are covered with a waxy cuticle layer that prevents water from escaping and, therefore, also prevents aqueous solutions from entering the leaf. The cuticle tends to be thinner at trichome bases and that is a preferential site of entry. Also, different plants have cuticles made of different types of waxes, some of which are more penetrable than others. Surfactants also help penetration by allowing greater contact between the herbicide solution and the leaf. However, surfactants may also alter the selectivity of a herbicide if the selectivity depends on differences in leaf absorption.

(3) Root inhibitors

Pre-emergent herbicides or soil active herbicides usually are absorbed by plant roots before they are active. However, some work through vapour action near the soil surface. The primary root region where herbicides are absorbed is 5 to 50 mm behind the root tip. In this area the xylem is functional, but the casparian strip that stops movement of materials into the roots is not completely lignified and therefore is not a significant barrier. Herbicides may be taken up passively by roots while others require the plant to expend energy to take them up. For the most part herbicides enter plant roots passively.

(4) Shoot inhibitors

The shoot inhibitors are soil applied for control of seedling grasses, some broad leaves and suppression of some perennials from tubers and rhizomes. Injury appears as malformed (twisted), dark green shoots and leaves on injured young plants. Grass crops with some tolerance to these compounds can be protected from injury with other chemicals [safeners (protectants)]. Crops include corn, large seeded legumes, small seeded legumes, beets, spinach, tomatoes, potatoes, and ornamental plants.

(5) The transportation of herbicides in plants

Once a herbicide enters the plant it may move to a location other than where it was absorbed to be effective. Two general pathways exist for compounds to move within plants. These are the xylem (apoplast) and the phloem (symplast). Movement in the xylem is passive and follows the same pathway as water. They enter the xylem and move along with the water to leaves and fruit. The driving force is water evaporating from leaves through transpiration. Movement in the phloem is generally from areas of high sugar concentrations (mature leaves) to areas where sugar is being utilized (developing fruit, growing points, young leaves). If herbicides that are to move through the phloem are applied to too high of concentrations, they kill or disrupt phloem movement and cannot be transported throughout the plant for a "complete kill".

(6) Herbicide Mode-Of-Action Summary

1) photosynthetic inhibitors. These herbicides translocate only apoplastically. Movement is upward with the transpiration stream (water moving through the plant from the soil and evaporating into the atmosphere at the leaf surfaces). Symptoms develop from bottom to top on plant shoots (older leaves show most injury; newer leaves least injury). Chlorosis first appears between leaf veins and along the margins which is later followed by necrosis of the tissue. Any potential control of established perennials must come from continued soil uptake and not movement downward through the plant from the shoots. Foliar activity alone can provide only shoot kill. Herbicides in these chemical groups have excellent soil activity. Most have foliar activity as well. These herbicides are used preplant incorporated, preemergence, and to a limited extent early postemergence, for selective control of weeds in annual and established perennial crops. Crops include corn, soybeans, potatoes, celery, parsnips, carrots, cotton, alfalfa, asparagus, mint, and woody species. They are also used for brush in pastures, rangeland, and non-cropland and for general vegetation control. Soil persistence varies from weeks to months depending on compound and dose and soil pH. Soil mobility varies from low

to high depending on the compound and soil characteristics.

2) Auxin growth regulators. Auxin growth regulator herbicides are used for control of annual, simple perennial, and creeping perennial broad leaves in grass crops (corn, small grains, sorghum, turf, pastures, sodded roadsides and rangeland) and in non-crop situations. All are organic acids which take on a negative charge after ionization of acids and salts. Esters are hydrolyzed to acids or salts in both plants and soils. Injury to off-target vegetation is a major problem associated with these herbicides.

3) Respiration inhibitors. Inhibitors of respiration (the arsenical herbicides DSMA and MSMA) interfere with the production of ATP, the major source of energy in plants. Also, the arsenicals may act by interfering with enzyme activity and by disrupting cell membranes. These herbicides act mainly on contact, with very little movement occurring in plants. Paraquat belongs to this group.

10.5.1.4   Common herbicides

① 2,4-D butyl ester

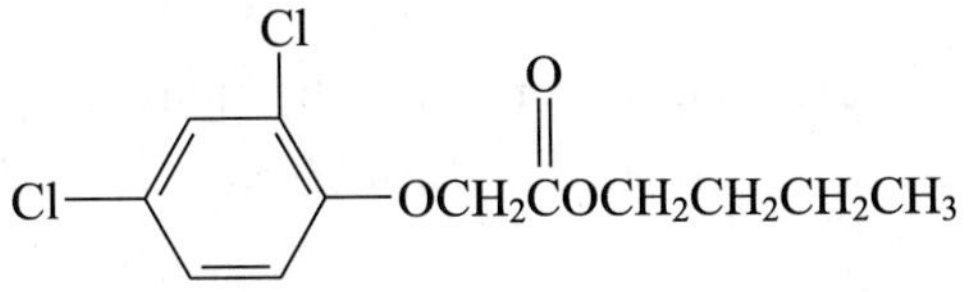

$$Cl$$

Cl—⟨benzene ring with Cl at top⟩—OCH₂COCH₂CH₂CH₂CH₃

$C_{12}H_{14}O_3C_{12}$

2, 4-D butyl ester is a colourless liquid and belongs to phenoxy acid herbicides. 2, 4-D butyl ester is stable but incompatible with strong oxidizing agents. It is used to control of weeds for protection of wheat (barley), maize, millet, oats, rice, also can be used on sorghum, sugar cane, grass pasture from *Chenopodium*, *Polygonum*, *Acalypha*, *Antidesma scandens* Lour. , *Capsella bursa-pastoris.* , *Equisetum*, *Ixeris polycephala* Cass. , *Cirsium setosum* (Willd.) Bieb. , *Xanthium*, *Convolvulaceae*, *Sagittaria trifolia* L. , *Monochoria*, *Bolboschoenus strobilinus* (Roxb.) V. I. Krecz. , *Monochoria vaginalis* (Burm. f.) C. Presl ex Kunth , *Achnatherum inebrians* (Hance) Keng ex Tzvelev and other inebrians.

② Gallant

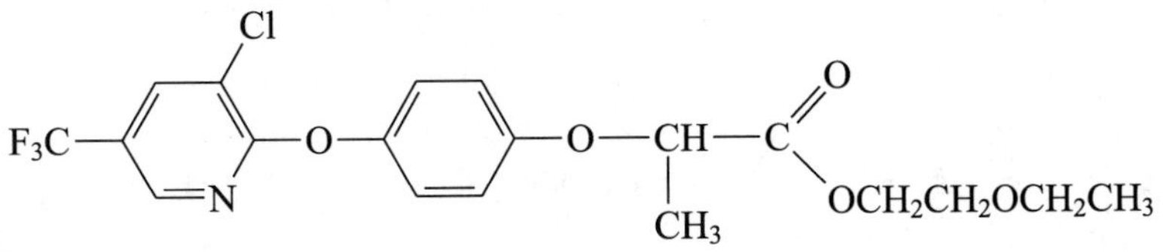

$$Cl$$

F₃C—⟨pyridine ring with Cl and N⟩—O—⟨benzene ring⟩—O—CH—C⟨=O / OCH₂CH₂OCH₂CH₃⟩ with CH₃ below CH

$C_{19}H_{19}F_3ClNO_5$

Gallant is mainly absorbed by the foliage of weeds, and transported to the entire plant. Also, it can be absorbed by plant root in use of postemergen period. It mainly suppresses meristem of stems and roots of grass, resulting in death of weeds. It is used to control annual and perennial weeds, such as barnyard grass, crab grass, *Digitaria*, *Eriochloa*, *Setaria*, *Avena fatua* L. , *Eleusine indica* (L.) Gaertn. , *Cynodon*, *Leptochloa*, *Poa*, *Agropyron* and other weeds in cotton, soybean, peanut, sugar beet and other crop field.

③ Fluazifop-butyl

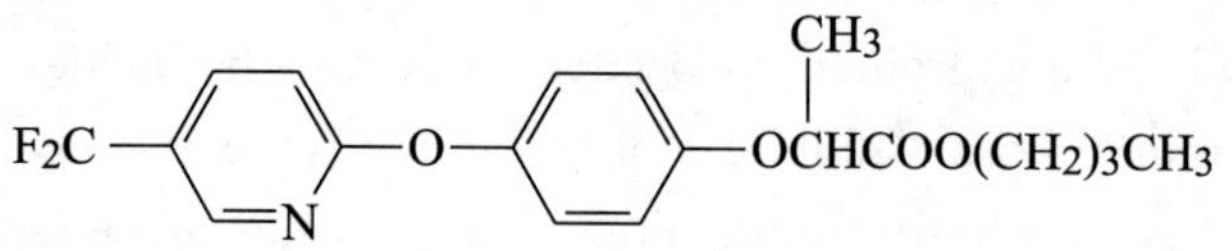

$$F_2C - \underset{=N}{\underset{}{\bigcirc}} - O - \bigcirc - \overset{CH_3}{\underset{}{O\,C\,H\,C\,O\,O(CH_2)_3CH_3}} \qquad C_{19}H_{20}F_2NO_4$$

Fluazifop-p-butyl kills annual and perennial grasses, but does little or no harm to broad-leaved plants (dicots). It kills by inhibiting lipid synthesis (lipids are necessary components of cell membranes), particularly at the sites of active growth. Fluazifop-p-butyl is a post-emergence phenoxy herbicide. It is absorbed rapidly through leaf surfaces and quickly hydrolyzes to fluazifop acid. The acid is transported primarily in the phloem and accumulates in the meristems where it disrupts the synthesis of lipids in susceptible species.

Both annual and perennial grasses can be controlled by fluazifop-p-butyl, including *Digitaria*, *Eriochloa*, *Setaria*, barnyard grass, *Brachiaria*, *Cynodon*, *Leptochloa*, *Paspalum* and so on.

④ Diclofop-methyl

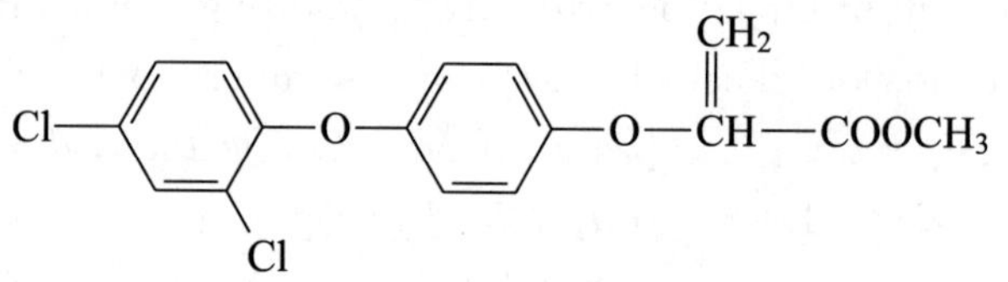

$$Cl - \bigcirc_{\underset{Cl}{}} - O - \bigcirc - O - \overset{CH_2}{\underset{\parallel}{CH}} - COOCH_3 \qquad C_{16}H_{14}Cl_{12}O_4$$

The active ingredient diclofop-methyl is found in a variety of commercial herbicides. It is a selective post-emergence herbicide for control of wild oats and annual grassy weeds in crop planting including wheat, barley, rye, soybean, beet ,potato, peanut, sunflower, celery ,rape, pea and so on.

⑤ Quizalofop-ethyl

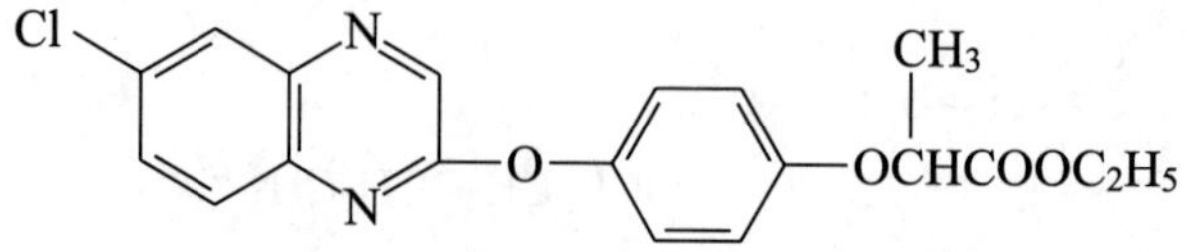

$$Cl \underset{N}{\overset{N}{\bigcirc}} - O - \bigcirc - \overset{CH_3}{\underset{}{O\,C\,H\,C\,O\,O\,C_2H_5}} \qquad C_{19}H_{17}N_2O_4$$

Quizalofop-P-ethyl is a selective, postemergence phenoxy herbicide. It is used to control annual and perennial grass weeds in crop management including potato, soybean, beet, peanut ,vegetable, cotton and flax. The compound is absorbed from the

leaf surface and is moved throughout the plant. It accumulates in the active growing regions of stems and roots.

⑥ Dicamba

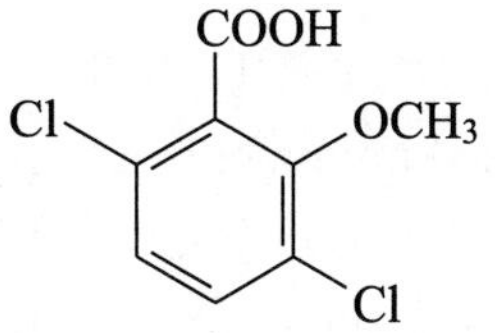

$$C_8H_6Cl_2O_3$$

Dicamba acts like a naturally occurring plant hormone and causes uncontrolled
growth in plants. At sufficiently high levels of exposure, the abnormal growth is so severe that the plant dies. There are various methods in application including ground or
aerial broadcast, soil (band) treatment, basal bark treatment, stump (cut surface)
treatment, frill treatment, tree injection, and spot treatment. It is mainly used for control *Artemisia scoparia* Waldst. et Kit. , *Amethystea caerulea* L. , *Artemisia scoparia* ,
Artemisia annua , *Bidens pilosa* L. , *Cephalanoplos* , *Commelina* , Xanthium sibiricum ,
Equisetum arvense , *Elsholtzia* , *Amaranthus* , *Acalypha australis* , *Polygonum* , and
Convolvulaceae. It also can be used to control weeds in lawns and pastures.

⑦ Chloramben

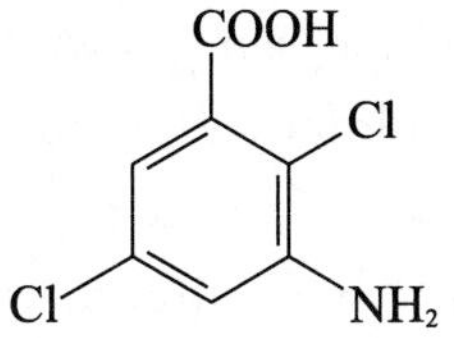

$$C_7H_5Cl_2NO_2$$

Chloramben is a selective, preemergence benzoic acid herbicide that is primarily
soil-applied to control annual grass and broadleaved weed seedlings. It has been widely
applied in crop production including soybean, *Phaseolus* , *maize* , *peanut* , *sunflower* ,
Brassica oleracea var. L. var. capitata L. , pepper and tomato.

⑧ Alachlor

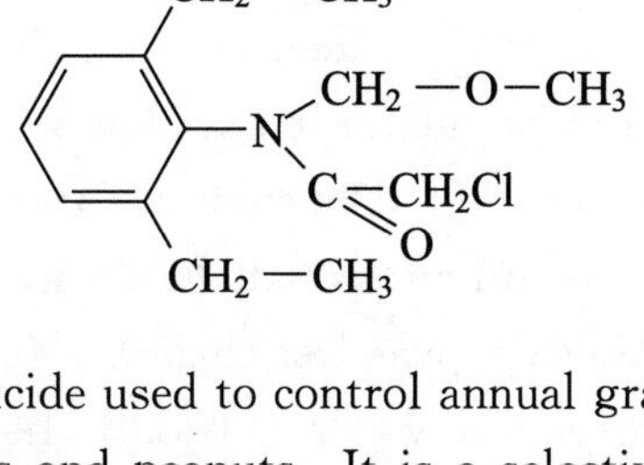

$$C_{14}H_{20}ClNO_2$$

Alachlor is an aniline herbicide used to control annual grasses and certain broadleaf
weeds in field corn, soya beans and peanuts. It is a selective systemic herbicide, absorbed by germinating shoots and by roots. The compound works by interfering with

the ability of a plant to produce (synthesize) protein and by interfering with root elongation.

10. 5. 2   Immunoassay for herbicides

Shan[44] designed three haptens (hapten 1, 2, 3) according to the structural features of triazine herbicides and synthesized immunogens to obtain three polyclonal antibodies. The scheme of hapten synthesis is shown in Figure 10-19. The antiserum from hapten1-BSA recognized atrazine very well and the cross-reaction rates for simazine, prometryn, simetryn were 10%, 7. 1% and 1. 4% respectively. Antibody from hapten 2-BSA showed good specificity for simazine with cross-reaction rates for atrazine, prometryn and simetryn 44. 6%, 0. 31%, and 0. 25% respectively. The third antibody from hapten 3-BSA recognized prometryn well with cross-reaction rates for atrazine, simazine, simetryn and hexazinone 50. 1%, 8. 9%, 25. 1% and 0. 08% respectively. No cross reactions with non-triazine herbicides such as trifluralin, dicamba, alachlor or fomesafen were found. The detection limits of established ELISA for detection of atrazine, simazine and prometryn were 1 ng/mL, 0. 5 ng/mL and 2 ng/mL.

Yu[45] used sulfosulfuron and N-[2-Mcthoxycarbonyl phenylsulfonyl] urea to prepare two haptens and two immunogens were gotten by coupling haptens with carrier protein. New Zealand white rabbits were used to prepare antiserum. The direct competitive ELISA was developed by labelling haptens with horse-redish peroxidase (HRP). The effects of pH value, phosphate buffer concentration and organic solvem on antigen-antibody affinity were evaluated. Under optimized conditions, the antibody from sulfosulfuron showed good specificity with sulfosulfuron with IC_{50} value 8. 3 ng/mL. Little cross-reactivity with other sulfonylurea herbicides was found (<0. 04%). The antibody from common structure N-[2-Mcthoxycarbonyl phenylsulfonyl] urea showed specific affinity with metsulfuron-methyl, ethametsulfuron and tribenuron with IC_{50} valueof 0. 08 μg/mL, 0. 12 μg/mL and 0. 14 μg/mL respectively.

An indirect competitive (ELISA) was developed for isoproturon by Sun[46]. Two haptens of isoproturon in different spacer arm were synthesized (Figure 10-20). One is [1-(3-carboxypropyl)-3- (4-iso-propylphenyl)-1-methylurea] and the other is [1-(5-carboxypentyl)-3-(4-isopropylphenyl)-1 -methylurea]. The polyclonal antibody against isoproturon was preparated by immunization of hapten-BSA with IC_{50} value of 70 ng/mL. The limit of detection was 5 ng/mL. The cross reactivities between antibody and chlorbromuron, monolinuron, tebuthiuron, thidiazuron, neburon, and fluomethuron were tested and all cross-reactivity was less than 0. 1%. The developed ELISA method was applied to detect isoproturon in water and soil. The recovery rate in the water samples was 90%-115% and the coefficient of variation was 1. 75%-7. 02%. The recovery

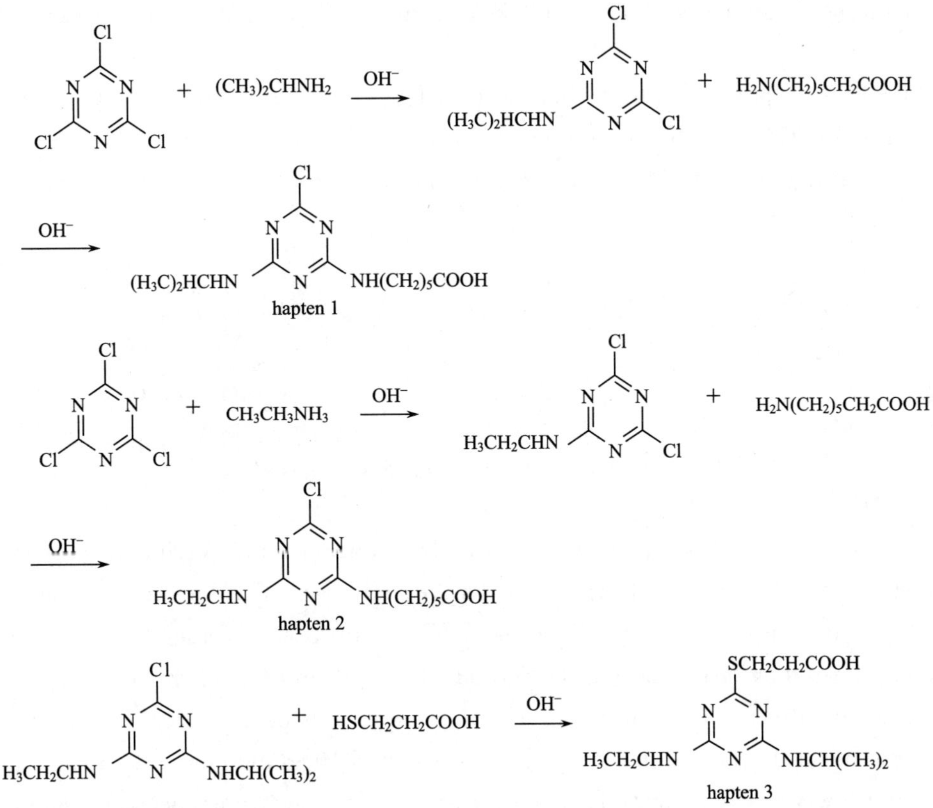

Figure 10-19　The scheme of hapten synthesis for triazine herbicides

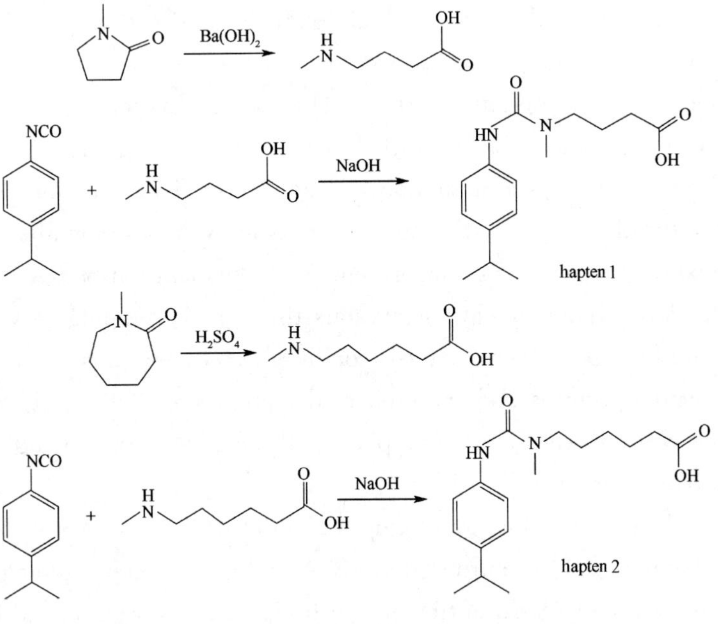

Figure 10-20　Systhesis of hapten for isoproturon

rate in the soil samples was 117%-160% and the coefficient of variation was 3. 10%-6. 65%.

Liu[47] developerd a hapten(CSH)derived from 2-chloro-phenylsulfonylamide (CP-SA, Figure 10-21), which was conjugated to a carrier protein to prepare synthetic antigens. The antiserum with high affinity to chlorsulfuron was obtained from New Zealand white rabbits.

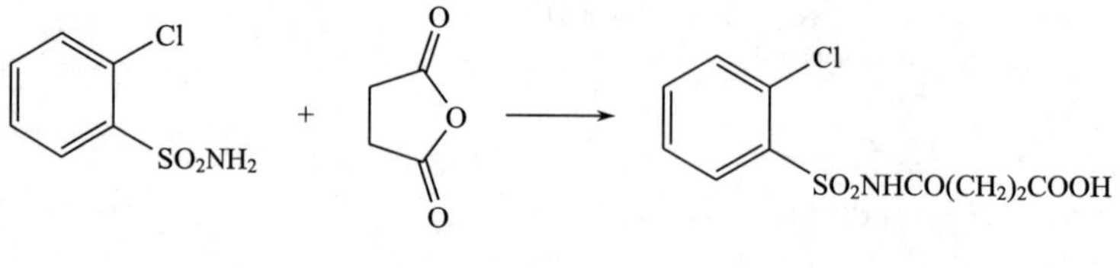

Figure 10-21　Systhesis of hapten for chlorsulfuron

The polyclonal antibodies were separated from antiserum by salting out method and purified by reversed phase adsorption of DEAE-cellulose. The horseradish peroxidase (HRP) labelled antibody was prepared by $NaIO_4$ modificationand the HRP labelled hapten was prepared by mixed anhydride method. Thus, an indirect competitive ELISA and a direct competitive ELISA for chlorsulfuron were established. Under optimized conditions, the linear detection range for chlorsulfuron was 100-1,000 ng/mL with the detection limits lower than 0. 1 ng/mL. No interferences for the two assays were found from some analogues of chlorsulfuron, such as metsulfuron-methyl and tribenuron. Based on this antibody, Shao[48] prepared immunoaffinity column (IAC) to capture chlorsulfuron residues in soil samples. Under optimal conditions, dynamic capacity of IAC column was up to 3. 5 ng/mL gel. Besides, fluorescence polarization immunoassay (PFIA) was developed in homogeneous competitive format. The hapten was labelled with fluorescence FITC. Good sensitivity was achieved with IC_{50} value of 0. 08 $\mu g/mL$.

A piezoelectric (Pz) crystal immunobiosensor for the determination of metsulfuron-methyl was developed by Shi[49]. Piezoelectric crystals were pre-modified with γ-aminopropyl triethoxysilane (APTES) and protein A. Polyclonal antibodies were coated on gold electrodes of 10 MHz piezoelectric crystals through the binding between protein A and Fc fragment of antibody. The analysis conditions including ions, surfactants and organic solvents were optimized and the limit of detection was 0. 03 $\mu g/L$. Some sulfonylurea herbieides with methoxyearbonyl phenylsulfonamide group could be also recognized by this biosensor.

Wang[50] used ethyl succinic anhydride and intermediates of chlorimuron-ethyl to synthesize the hapten of chlorimuron-ethyl, BSA and OVA were used as carrier protein for complete antigen. Polyclonal antibodies with high titer were obtained by immunization of rabbits. An indirect ELISA was developed with IC_{50} value of 36. 6 ng/mL. Cross-reactivity rates with chlorsulfuron (0. 27%), 2-aminopyrimidine (0. 2%), metsulfuron-methyl (0. 32%), ethametsulfuron (32%), halosulfuron methyl (4. 8%) were detected, respectivily.

10.6 Other pesticides

10. 6. 1 Triazole pesticides

Triazole compounds have shown great potential in pest control. Triazophos, adopted since the early of 1970's is still in use as a highly effective pesticide. Triadimefon as a fungicide, firstly used in 1974, further demonstrated the potential development of triazole pesticides. Until 1980s, more studies focused on plant growth regulators containing triazole group. Now, the triazole pesticides have gained a renewed interest for new pesticide development[51].

(1) Triazole fungicides

This chemical family of fungicides, introduced in the 1980s, consists of numerous members -difenoconazole, fenbuconazole, myclobutanil, propiconazole, tebuconazole, tetraconazole, triadimefon, and triticonazole (Table 10-4). They are important tools against diseases of turfgrasses, vegetables, citrus, field crops and ornamental plants. Triazole fungicides are applied as foliar sprays and seed treatments, but are diverse in use, as they may be applied as protective or curative treatments. If applied as a curative treatment, triazole applications must be made early in the fungal infection process. Once the fungus begins to produce spores on an infected plant, the triazoles are not effective. The fluorine atoms introduced into the active center of triazole compounds increase the bioactivity and until now, fluorinated triazoles including flutriafol, trifluoromethyl oxacillin and flusilazole have been marketed.

(2) Insecticide and miticide of triazole

Triazophos synthesized by Farbuwerke Hoechst Company in 1970, is a broad spectrum insecticide, nematicide and miticide. The introduction of triazole group to the structure of organophosphorus insecticides can delay the generation of resistance and expand the insecticidal spectrum. In 1973, Isazofos was devolepped by Ciba-Geigy and demonstrated contact killing, stomach poisoning and suction effects. Isazofos has been used widely in cotton, rice, sugar beet, and vegetables fields.

Table 10-4　Triazole fungicides

Name/Time/Producor	Structure	Targets
Triadimefon/1974/Bayer	Cl—C6H4—O—CH(—N-triazole)—C(=O)—C(CH3)2—CH3	Powdery mildews, rusts and other fungal pests on cereals, fruits vegetables, turf, shrubs and trees.
Triadimenol/1978/Bayer	Cl—C6H4—O—CH(—N-triazole)—CH(OH)—C(CH3)2—CH3	Seed-born diseases
Bitertanol/1979/Bayer	biphenyl—O—CH(—N-triazole)—CH(OH)—C(CH3)2—CH3	Apple scab、powdery mildews、rust and leaf spot
Diniconazole /1984/Sumitomo Chemical Company	2,4-Cl2—C6H3—CH=CH—C(OH)=...—C(CH3)3 (N-triazole)	Broad spectrum fungicide
Tebuconazole/1986/Bayer	Cl—C6H4—CH2CH2—C(OH)(CH2CH3)—C(CH3)2—... (N-triazole)	Control of smuts, bunt, seed rots and seedling blights on barley, oats and wheat as a seed treatment and for the control of Fusarium Head Blight on wheat as a post-emergent treatment.
Cyproconazole/ 1986/ Sandoz	Cl—C6H4—C(OH)(CH2CH3)—CH(CH3)—cyclopropyl (N-triazole)	Water-based fungicide used to protect above-ground wood
Flutriafol/1983/ICI Agrochemicals	F—C6H4—C(OH)(C6H5)—CH2—N-triazole	Powdery mildew, rust, Alternaria, *Helminthosporium* sp.
Fluotrimazole/ 1973/Bayer	(C6H5)2C(—C6H4-CF3)—N-triazole	Rust
Quinazolinone/1984/DuPont	F—C6H4—Si(CH3)(—C6H4-F)—CH2—N-triazole	Phytopathogenic diseases of crop plants and against infestation on propagation stock of plants or on other vegetable material, especially phytopathogenic fungi.
Tetraconazole/1988 /Agrimont. S. P. A	2,4-Cl2—C6H3—CH(CH2OCF2CF2H)—CH2—N-triazole	Powdery mildew,rust,scab

triazophos　　　　　　　　　　　　isazofos

(3) Triazole herbicides

Compounds containing triazole group can be used as herbicides, which was reported in early 1970s. Isazofos is used to control annual grass weeds, which was marketed by Boots companies. Other triazole herbicides include flupoxam, epronaz and triazofennanide and so on.

(4) Other application of triazole compounds

Triazoles can also be used as plant growth regulators. They are widely used in the production of rice and vegetables to improve yield.

10. 6. 2　Benzimidazole pesticides

Benzimidazoles are a large chemical family characterized by a broad spectrum of activity against roundworms (nematodes), an ovicidal effect, and a wide safety margin. The most effective of the group are those with the longest half-life, such as oxfendazole, fenbendazole, albendazole, and their prodrugs, because they are not rapidly metabolized to inactive products. Effective concentrations are maintained for an extended period in the plasma and gut, which increases efficacy against immature and arrested larvae and adult nematodes, including lungworms.

10. 6. 3　Immunoassay

Wang[52] prepared the conjugates of human serum albumin and triadimefon, which was used as immunogen to immunize rabbit. The polyclonal antibodies against triadimefon were obtained and an ELISA was established. The ELISA was successfully applied in the analysis of triadimefon residues in the cucumber, pear and other foods with the minimum detection limit of 40 ng/g.

Phytohormone is a natural substance which can regulate the life course and the physiological activities of plant. Phytohormone research is very significant to agriculture, horticulture and other related fields, and calls for a rapid, precise and convenient measurement procedure for phytohormone assay in plants. Indolencetic acid is an auxin plant hormone. Li[53] developed an electrochemical method to quantitatively detect indolencetic acid. A gold electrode is modified with a uniform, stable and orderly mercapto monolayer, which was covalently bonded with antibody against indolencetic acid. Good linear response between current and content of indolencetic acid was found in cyclic voltammetry format ranging from 5.68×10^{-7} to 2.83×10^{-5} mol/L.

Li[54] used bovine serum albumin (BSA) as a carrier protein to synthesize complete antigens of β-indole acetic acid (IAA), isopentenyl adenine (iPA), gibberellic acid (GA). The corresponding antibodies were obtained through immunization of rabbits. Various immobilization methods of biological molecules were optimized to develop enzyme sensor, immunosensor and piezoelectric immunosensor for determination of endogenous hormone substances in plants. For IAA detection, three sensors were developed. A quartz crystal microbalance (QCM) was prepared based on single amplification and competitive immunoreaction. The detection range was 0. 5-5 μg/mL IAA. With the orderly, stable mercapto self-assembled monolayer formed on the golden surface of the microelectrode, an amperomrtric immunosensor was prepared for detecting IAA in the mungbean sprout. The detection range was 5.68×10^{-7}-2.83×10^{-5} mol/L. Using sol-gel technology in conjunction with an enzyme-linked competitive immunoreaction, an amperometric immunosensor was developed to determine IAA. Sodium alginate was used to modify carbon electrode, which increased the negative charge of the electrode surface. Antigen labelled with HRP was used for identification. Self- assembly on the electrode surface was carried out through the electrostatic adsorption. The detection range was 5-500 μg/mL based on this self-assembly electrode. An amperometric immunosensor was constructed by electropolymerization and immobilization of antibody to determine iPA in plant samples in the range of 5-300 μg/mL. An electrochemical immunosensor for GA based on competitive immunoreaction with copper cation labelled antigen was developed using square wave stripping voltametry technique. The level of GA in plant sample was determined with detection range 1-150 μg/mL.

Wang[55] designed haptens to prepare antibody against benomyl (Figure 10-22). The first hapten, 1-oxycarbonylethyl propionate-2-benzoimidazol-2-ylaminoformate, was synthesized by activating the imine group on benzimidazole ring and insert activated hydroxyl group. The affinitive sepecific groups such as benzimidazole ring and aminoformate were maintained (Figure10-23, Figure 10-24). The second hapten 2-succinimide benzimidazole was prepared by hydrolysis of carbendazim under alkaline conditions and acylization between the resulting products 2-amino-benzimidazole-2-ylaminoformate with succcinic acid. Both haptens were coupled with carrier protein by mixed anhydride method and active ester method to prepare coating antigen and immunogen Antiserum obtained after immunization wtih good affinity with benomyl and carbendazim. The cross-reactivity with thiabendazole reached 8% and is less than 2% with thiophanate and thiophanate-methyl. Under optimized conditions, an indirect competitive ELISA for benomyl was established with the minimum detection limit 0. 05 ng/mL. The IC_{50} values of carbendazim (the metabolite of benomyl) and benomyl was 4. 0 ng/g. The soil and water samples were analysed and the results showed good consistency with HPLC analysis.

To develop immunoassay for heterocyclic insecticide, Zheng[56] prepared hapten for

Figure 10-22   Chemical structure of benomyl and its analogues

Figure 10-23   Synthesis route of hapten 1

Figure 10-24   Synthesis route of hapten 2

imidacloprid and fipronil respectively. Under alkaline conditions, imidacloprid reacted with β-mercaptopropionic acid to get hapten for imidacloprid, which was coupled with carrier protein through activated ester method to get complete antigens. Similarly, cyano group of fipronil was transferred to carboxyl, which was conjugated with carrier protein by glutaraldehyde method (Figure 10-25). Rabbits were immunized to obtain polyclonal antibody. High titer anti-imidacloprid antibody (1 : 2.56×10⁴-1 : 5.12×10⁴) was obtained while low titer anibody (1 : 400) for fipronil was found. All antisera were purified and freeze-dried before storage at −20℃. Indirect and direct competitive ELISA was established for abalysis of imidacloprid and fipronil respectively. The IC_{50} value of indirect competitive ELISA for imidacloprid was 21.8 μg/L. The IC_{50} value of ELISA for fipronil was 15.5 μg/L and the cross-reactivity rates for four of itd metabolites were more than 80%. Under optimized conditions, the two ELISAs were sucessfully used to analyze soil and water samples

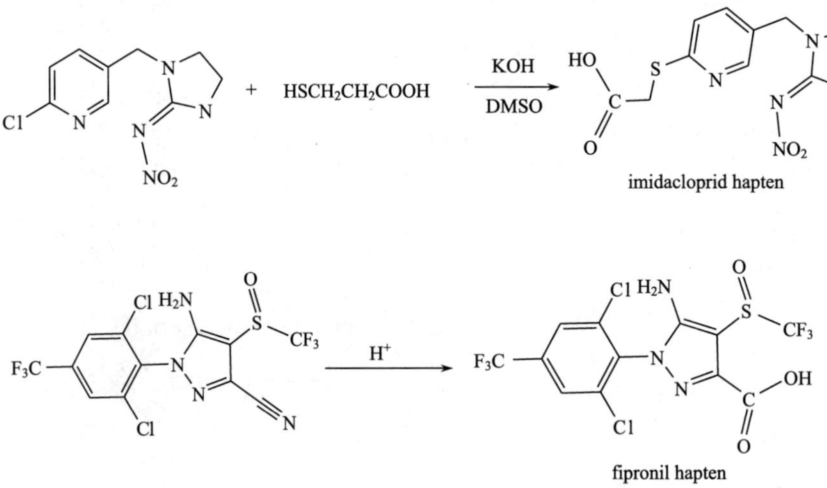

Figure 10-25　Hapten sysntesis for imidacloprid and fipronil

Luo[57] modified toosendanin by reaction with succinate and glutarate, respectively to prepare haptens (toosendanin-succinil and toosendanin-glutaryl), which was conjugated with carrier protein with the mixed-anhydride method and the active ester method. Polyclonal antibodies were produced through immunizing rabbits. The indirect competitive ELISA was developed with IC_{50} value of 1.52 μg/mL.

10.7　Immunoassay based on class specific antibody

New high-throughput immunoassay methods for rapid point-of-care diagnostic applications or food analysis represent an unmet need and current focus of numerous innovative methods. A general and broad selective immunoassay was based on the class selective antibody, which was prepared with the shared antigenic determinants.

The pyrethroids are one of the most heavily used insecticide classes in the world. It is important to develop sensitive and rapid analytical techniques for environmental monitoring and assessment of human exposure to these compounds. Because major pyrethroids contain a phenoxybenzyl group and phenoxybenzoic acid (PBA) is a common metabolite form or intermediate, PBA might be used as a biomarker of human exposure to pyrethroids. Luo[58] insert 6 carbon spacer arm (6-aminocaproic acid) at the carboxyl group of PBA and then conjugated with carrier protein for complete antigen. Polyclonal antibody was prepared by immunization of rabbits. The raised broad-spectrum antibody specificly recognized permethrin (IC$_{50}$ 4. 07 μg/mL), fenpropathrin (IC$_{50}$ 4. 23 μg/mL), cypermethrin (IC$_{50}$ 4. 87 μg/mL), cyhalothrin (IC$_{50}$ 4. 98 μg/mL), deltamethrin (IC$_{50}$ 8. 98 μg/mL) well. Li[59] also prepared antibody based on PBA hapten. The antibody showed specific affinity to cyhalothrin, fenvalerate, cyhalothrin and permethrin.

Wang[60] prepared four haptens (Figure 10-26), which were conjugated with the same carrier protein through covalent bonds. Various carrier proteins including BSA, KLH, HSA and OVA were used. The specific broad-spectrum antibody was achieved by immunization of rabbits, which showed acceptable affinity and specificity to the four classes of related pesticides. The direct competitive ELISA showed that IC$_{50}$ values were 0. 14, 3. 41, 8. 39 and 3. 20 ng/mL for triazophos, carbofuran, chlorpyrifos and methyl parathion respectively.

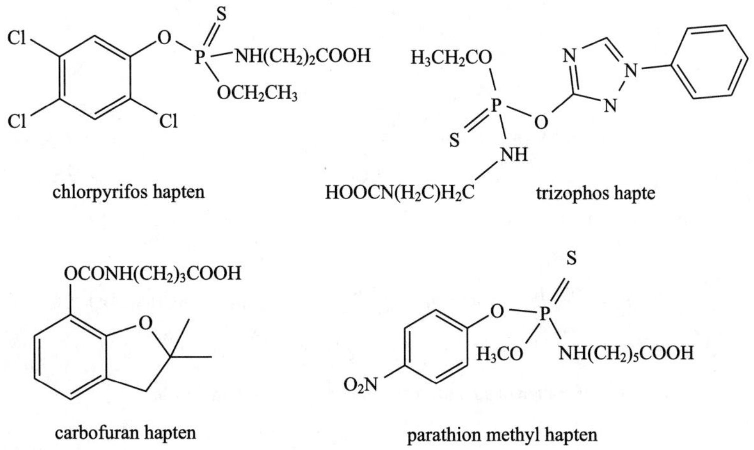

Figure 10-26   Four haptens used for preparation of broad- spectrum antibody

From the chemical structure of organophosphate pesticides, they can be divided into ① aromatic ring phosphorothioates, including methyl parathion, fenitrothion, pirimiphos-methyl, etc. ② straight-chain phosphorothioates, including dimethoate, malathion, etc. ③ vinyl phosphates, including monocrotophos, phosphorus amine, mevinphos, etc. Han[61] designed three comman antigens for each class of organophosphates to develop class-specific antibody (Figure 10-27). The results showed that high titer antisera from rabbits were obtained from haptens for aromatic ring phosphorothioates (MP) and straight-chain phosphorothioates (SP). The antibody for MP recognized methyl parathion best with IC$_{50}$ value of 0. 13 μg/mL and the cross-reactivity with fenitrothion and fenthion were found with IC$_{50}$ values of 0. 50 μg/mL and 0. 35 μg/mL. The antibody raised by SP antigen showed specific affinty to acephate, dimethoate, methyl parathion and malathion, with IC$_{50}$ values of 2. 66 μg/mL, 3. 31 μg/mL, 8. 26 μg/mL and 14. 9 μg/mL, respectively. Based on these antibodies, Han developed immunoassays in indirect competitive format to detect targeted organophosphate pesticide residues in cucumber samples and the results were in agreement with GC detections.

Based on the antibodies against carbofuran, triazophos and chlorsulfuron, Wei[62]

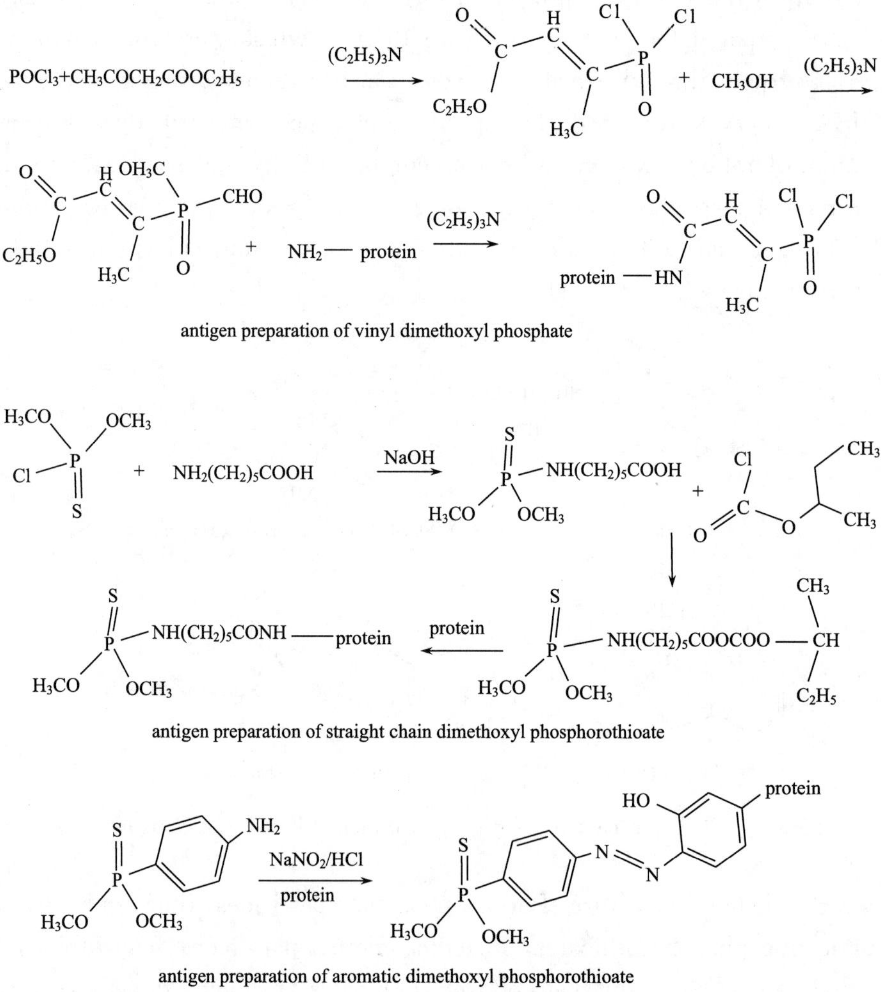

antigen preparation of vinyl dimethoxyl phosphate

antigen preparation of straight chain dimethoxyl phosphorothioate

antigen preparation of aromatic dimethoxyl phosphorothioate

Figure 10-27 Common antigen preparation for organophosphate pesticides

prepared immunoaffinity chromatography column (IAC) to purify these pesticides simutaneously from matrix. The purified anti-chlorsulfuron antibody, anti-carbofuran antibody and anti-triazophos antibody were conjugated to 1,1'-carbonyldiimidazole(CDI) activated sepharose CL-4B (particle size 45-165 μm, 4% agarose). The resulted immunosorbents of three antibodies were mixed together to fill columns. Thus, the IAC can be used to enrichment of carbofuran, triazophos and chlorsulfuron in samples. The dynamic column capacity of chlorsulfuron, carbofuran and triaophos were 1.81 μg/mL, 2. 29 μg/mL and 1.89 μg/mL gel. The purication of these pesticides from soil and water samples showed its potential use in the future.

References

[1]　Xiong JL. Introduction of phosphorus chemicals. Chemical Industry Press, Beijing: China 1994, 417

[2]　Bai NF. Treatment of organophosphorus pesticide poisoning. Beijing: People's Education Press, 1996, 51

[3]　Zeng FZ. Modern treatment of acute poisoning. Beijing: Science and Technology Press, 1993, 26

[4]　Hui XJ. Environmental toxicology. Beijing: Chemical Industry Press, 2003, 160

[5]　Lin YS, Gong RZ, Zhu ZL. Pesticides and environment protection. Beijing: Chemical Industry Press: 2000, 54

[6]　Tai SR. Pesticide application. Chongqing:Chongqing Science and Technology Press, 1985: 123

[7]　Rui YK and others. Establishment of rapid ELISA method for parathion. Chinese Journal of Food Hygiene, 2005, 17(4): 315-317

[8]　Lv XD. Study on a rapid immunology examination method for parathion residue in Rice [Dissertation]. Nanjing: Nanjing Agricultural University, 2004

[9]　Liu FQ, Xue ZG, Wang JS. The development of an ELISA for quantitation of methamidophos and its application. Journal of Agricultural Biotechnology, 1998, 6(2): 140-146

[10]　Zhang DL. Immunoassay of methamidophos [Dissertation]. Beijing:Chinese Academy of Agriculture,2005

[11]　Han LJ,Jia MH, Qian CF and others. Study on ELISA of parathion-methyl journal of agro-environmental science, 2005, 24 (1): 187-190

[12]　Li XX. Develop enzyme-linked immunosorbentassay kit of dimethoate [Dissertation]. Shenyang:Shenyang Agricultural University, 2006

[13]　Yin LM. Determination of organophosphorus pesticides multi-residues by immunoassay methods [Dissertation]. Wuxi:Jiangnan University, 2009

[14]　Yin XG, Yin LM, Xu CL. Advances of immunoassay on organophosphorus pesticide. Food Science, 2008. 29 (10): 684 -688

[15]　Chen ZL. Study on enzyme-linked immunosorbent assay and ELISA kit for triazophos [Dissertation]. Hangzhou:Zhejiang University, 2006

[16]　Wang Y. Preparation and application of chlorpyrifos monoclonal antibody [Dissertation]. Nanjing:Southeast University, 2006

[17]　Wang H. Study on the direct elisa of detection for organophosphorus pesticide residue [Dissertation]. Wuhan: Central China Agricultural University, 2007

[18]　Lu C. Development of piezoelectric immunosensor for methyl parathion [Dissertation]. Beijing:Chinese Academy of Agriculture, 2008

[19]　He Y. Studies on fast Detection of organophosphorus pesticides by an enzyme-based biosensor [Dissertation]. Hangzhou:Zhejiang University, 2003

[20]　Shi XM. Food safety and hygiene. Beijing:Agriculture Press, 2003: 217

[21] Huang ZM, Ma YC. Forensic pathology. Beijing: People's Public Security University Press, 2002: 750

[22] Zhang ZB. Fan defang and qian chuanfan environmental toxicology of insecticide. Beijing: Agriculture Press, 1989: 83

[23] Dong YH, Liu RY, Xu DY and others. Detection of DDT in seawater and shellfish using indirect competitive ELISA fisheries science, 2007, 26 (4): 229-233

[24] Zhang J. Study on the rapid detection method of endosulfan residues in agricultural products [Dissertation]. Tianjin: Tianjin University of Technology, 2006

[25] Shao XL. Study on enzyme-linked immunosorbent assay for the residues of pentachloronitrobenzene [Dissertation]. Beijing: China Agricultural University, 2004

[26] Yang MY, Mao W, Bei WB and others. Surface acoustic wave immunosensor for detection of DDT. Chinese Journal of Sensors and Actuators 2007, 20 (1): 1-4

[27] Chinese Medicine Encyclopedia of Nutrition and Food Hygiene. Shanghai: Shanghai Science and Technology Press, 1988: 72

[28] Yu CD. Enzyme immunassay for the determination of carbaryl residues in agricultural products [Dissertation]. Tianjin: Tianjin University of Technology, 2006

[29] Zhang C. Development of membrane strip tests for the rapid detection of cabaryl [Dissertation]. Tianjin: Tianjin University of Technology, 2006

[30] Yang T. Study on preparation of ELISA kit for carbofuran and detoxic function of anti-carbofuran antibody [Dissertation]. Hangzhou: Zhejiang University, 2002

[31] Hu XQ, Zhu GN, Xu BJ. Radioimmunoassay of carbofuran residues. ACTA Agriculturae Nucleatae Sinica, 2002, 16 (5): 320-324

[32] Mao LJ. Study on application of gold immuno-chromatographic assay for rapid test of carbofuran residue [Dissertation]. Hangzhou: Zhejiang University, 2005

[33] Liu SZ, Feng DH, Chen MJ. Study on a highly specific enzyme linked immunosorbent assay for cabofuran. Journal of Analytical Science 2000, 16 (5): 373-378

[34] Zhang Q. The research on immunoassay technology for chemical pestcides such as metolcarb [Dissertation]. Nanjing: Nanjing Agricultural University, 2004

[35] Liu TF, Liu YZ, Sun C. Indirect ELISA assay of permethrin pesticides. Environmental Science, 2006, 27 (2): 347-350

[36] Zheng WH. Immunoassay of deltamethrin and its metabolites [Dissertation]. Urumqi: Xinjiang Agricultural University, 2004

[37] Li Bo. The research on enzyme-linked immunosorbent assay for bifenthrin and cyhalothrin [Dissertation]. Nanjing: Nanjing Agricultural University, 2007

[38] Han AH. Studies on immunochemistry for analysis of cypermethrin and flumethrin [Dissertation]. Yangzhou: Yangzhou University, 2007

[39] You HQ. Immunoassay of fenvalerate. [Dissertation]. Yangzhou: Yangzhou University, 2005

[40] Zhang XZ. Immunoassay for detection of deltamethrin pesticide residues [Dissertation]. Beijing: Chinese Academy of Agriculture, 2008

[41] Hao XL. Determination of α-cyano pyrethroids multi-residues by immunoassay methods [Dissertation]. Wuxi: Jiangnan University, 2009

[42] Yin XG, Hao XL, Xu CL. Development of research on residue determinations of cyano-pyrethriods pesticides based on immunoassay method. Food Science. 2008, 29 (9): 664-667

[43] Zhang DJ, Cheng MR. Application guidance of herbicides. Beijing: Rural Readings Press, 1987: 98

[44] Shan GM, Qian CF. Study on enzyme linked immunosobent assay (ELISA) for residue determination of triazine herbicides. Journal of China Agricultural University 1996, 1 (3): 52 -58

[45] Yu W. Studies on immunochemistry for analysis of sulfosulfuron [Dissertation]. Yangzhou: Yangzhou University, 2007

[46] Sun F. Development of an enzyme-linked immunoassay for isoproturon [Dissertation]. Nanjing: Nanjing University of Technology, 2006

[47] Liu SZ, Feng DH, Shao XJ. Studies on enzyme-linked immunosorbent assay of chlorsulfuron. Journal of Analytical Science 2000, 16 (6): 461-465

[48] Shao XJ. Studies on immunochemistry for residue analysis of chlorsulfuron [Dissertation]. Yangzhou: Yangzhou University, 2002

[49] Shi GQ. Metsulfuron piezoelectric immunosensor and its binding residues in soil [Dissertation]. Beijing: Chinese Academy of Agricultural Science, 2000

[50] Wang QH. Study on enzyme-linked immunosorbent assay for residue of chlorimuron-methyl [Dissertation]. Harbin: Northeast Agricultural University, 2002

[51] Zheng FN. Application of pesticide. Beijing: Chinese Agricultural Press, 2000: 149

[52] Wang Y. Enzyme-linked immunosorbent assay of triadimefon in food. Chinese Analytical Chemistry, 1993, 21(9): 1055-1057

[53] Li CX. Amperometric immunosensor based on mercapto carboxylic acid self assembled monolayer for phytohormone indoleacetic acid assay. ACTA, Cmeica, Sinica 2003, 61 (5): 790-794

[54] Li J. The electrochemical sensing technology on plant hormones and other substances [Dissertation]. Changsha: Hunan University, 2003

[55] Wang GX. Enzyme-linked immunoassay technique on carbendazim. [Dissertation]. Beijing: Chinese Academy of Agricultural, 2006

[56] Zheng ZT. Insecticide imidacloprid and fipronil in the enzyme-linked immunosorbent assay (ELISA) technique [Dissertation]. Hangzhou: Zhejiang University, 2004

[57] Luo L. Study on indirect competitive enzyme-linked immunosorbent assay (IC-ELISA) for Toosendanin [Dissertation]. Yangling: Northwest Agricultural & Forest University, 2008

[58] Luo AL. Enzyme immunoassay method on pyrethroid pesticide residues. China Agricultural Science, 2005, 38(2): 308 -312

[59] Li ZR. Antibody against a family of pyrethroid pestcides praparation and immunoassay technology for cyhalothrin [Dissertation]. Nanjing: Nanjing Agricultural University, 2004

[60] Wang ST. Immuno-specificity of multi-determinat artificial antigen and gold immuno-chromatography assay for multi-residue of pesticide [Dissertation]. Hangzhou: Zhejiang University, 2008

[61] Han LJ. Immunoassay for organophosphorus pesticides [Dissertation]. Beijing: China Agricultural University, 2003

[62] Wei LH. Immunoaffinity chromatography of pesticide residues [Dissertation]. Yangzhou: Yangzhou University, 2005

Chapter 11   Immunoassay for Veterinary Drugs

11.1   Aminoglycoside antibiotics

11.1.1   Introduction

11.1.1.1   Categorization, structure and properties

Aminoglycoside antibiotics (AGs) are a class of glucoside which consist of amino sugars and amino alcohols, and with similar structure and physico-chemical properties. As their structure contains multiple amino and hydroxyl, these drugs are alkaline, and with high polarity and water solubility. AGs have strong antibacterial effects and are commonly used antibiotics[1].

The commonly used aminoglycoside contains one 1,3 di-aminocyclitol or 1,4 di-amino-cyclitol and up to 2 amino sugars (monosaccharides or disaccharides). According to the structure of aminocyclitol, AGs can be divided into two categories: streptamines and 2-deoxy mold amines. 2-deoxy-streptamines are further divided into neomycin group (substituent at C-4 and C-5 sites), gentamicin group and kanamycin group (substituent at C-4 and C-6 sites) according to the number and location of substituting group in ring. Streptamines mainly consist of streptomycin and dihydrostreptomycin (streptomycin group). According to their source, AGs can also be divided into two classes: the first class is produced by Streptomyces, including streptomycin family, kanamycin and neomycin family; the second is hazimicin which mainly includes gentamicin group.

The structures of commonly used aminoglycoside antibiotics are listed in Table 11-1. Their sulphate are generally used in clinical pratice. The naming and chemical structure of streptomycin sulphate can be seen from Figure 11-1.

There is non-conjugated double bond in the structure of AGs, so these drugs only have weak end absorption (less than 230 nm). But their structure contains more active or polar groups and can be detected or separated by derivatization, such as amino (first amino or secondary amino), guanidine, hydroxyl, aldehyde and so on. AGs have properties of hy-droxylamine and amino acid, for example they can generate ninhydrin reaction. Moreover, AGs consist of multiple chiral carbons and have optical activity.

Table 11-1　Chemcial constitution of aminoglycoside antibiotics

Categorization	Drugs	R_1	R_2	R_3	R_4	R_5
Phytomycin	Phytomycin	—CHO	—	—	—	—
	Dioxygen-phytomycin	—CH₂OH	—	—	—	—
Amkin	Amkin A	—OH	—OH	—OH	—NH₂	—H
	Amkin B	—NH₂	—OH	—OH	—NH₂	—H
	Amkin C	—NH₂	—OH	—OH	—OH	—H
	Dibekacin	—NH₂	—H	—H	—NH₂	—H
	Tenebrimycin	—NH₂	—H	—OH	—NH₂	—H
	Butyro-amkin	—OH	—OH	—OH	—NH₂	H
Cidomycin	Cidomycin C₁	—CH₃	—CH₃	—	—	—
	Cidomycin C₂	—H	—H	—	—	—
	Cidomycin C₃	—CH₃	—H	—	—	—
	Sagamicin	—H	—CH₃	—	—	—
	Florimycin	—	—	—NH₂	—	—
	Ethylsisomicin	—	—	—NH CH₂ CH₃	—	—
Fradiomycin	Fradiomycin B	—NH₂	—	—	—	—
	Aminosidin	—OH	—	—	—	—

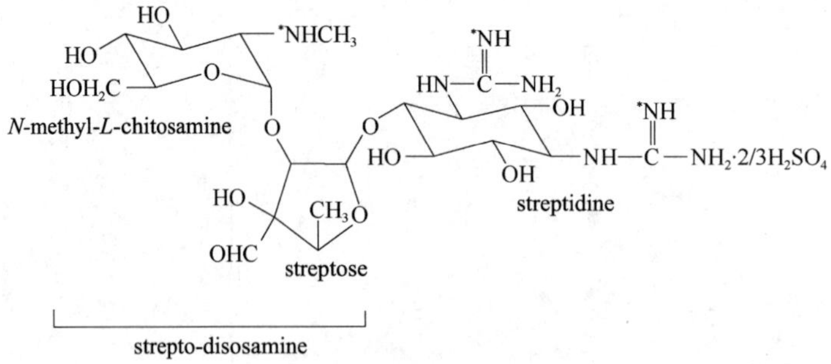

Figure 11-1   Streptomycin sulfate: O-2-decxy-2-methylamid-α-L-pyranose(1→2)-O-5-deoxy-3-C-
Formyl-α-L-Lysol furyl-(1→4)-N,N'-dis(amid-imino)-methyl-D-streptamine sulphate
* alkalinity center

AGs are of alkaline natuse and easily react with inorganic acids (such as sulfuric acid or hydrochloric acid) or organic acid to generate salts. The sulphate of AGs is white or crystalline powder, and insoluble in most organic solvents (slightly soluble in methanol and other high polar solvents, almost insoluble in hydrophobic solvent). In aqueous solution, amine groups of AGs usually exists with positive charges. According to the number of amine groups in structure, pKa values of amine groups range from 7 to 8. For example, gentamicin has 5 amine groups with similar alkaline (basic centers), and its pKa is about 8; kanamycin contains 4 amine groups, the pKa is about 7; neomycin has 6 amine groups and its pKa about 8.8. Streptomycin has three basic centers in its structure, where the pKa of 2 alkaline guanidines on streptidine is 11.5 and pKa of amine group on glucosamine is 7.7.

The basic aminoglycoside structure of AGs is quite stable in acid, alkali or under heating conditions, but hydrolysis will take place under the conditions of strong acid or alkali. The hydrolysis products and their properties can be used for the identification of AGs. For example, streptomycin can be firstly hydrolyzed to streptidine (by Sakaguchi reaction) and strepto-disaccharide. The further hydrolysis products of N-methyl-L-glucosamine will appear under acidic conditions (by Elson-Morgan reaction) while under alkaline conditions, streptidine rearrangement reaction will happen to produce maltol (the solution will be purple in case of iron reagent).

11.1.1.2 Pharmacology and toxicology

AGs mainly act on the ribosome of bacterial (30 subunit), which causes tRNA mistakes in the translation of mRNA. Thus, that abnormal protein was synthesized to obstruct the protein synthesis, inhibiting bacterial growth. AGs are bactericide for rest period and are mainly used to control Gram-negative bacteria such as *E. coli*, *Salmonella*, *Klebsiella aerogenes* and so on. Streptomycin is effective to *Mycobacterium tuberculosis*. The combination of AGs, such as streptomycin and penicillin, may be associated with synergies. However, bacterial is easy to produce resistance to AGs[2].

Owing to the good water solubility, AGs is hardly absorbed by oral intake, and is usually used as injected agents with half-life of 2-3 h in plasma. AGs are easily accumulated in animal kidneys after intramuscular injection (especially the cortex). Gentamicin concentration in human kidney tissue is 161 times of that in muscle tissue. Thus, the withdrawal period of gentamicin and neomycin in the muscle tissue are typically less than 5 d, but in the kidney tissue their withdrawal period are 60-90 d. AGs isn't recommended to people with renal insufficiency.

Ototoxicity and renal toxicity are the common side effects of AGs. AGs can selectively damage the 8th cranial nerve, resulting in vestibular and cochlear nerve injury. The former is more common in streptomycin, kanamycin and gentamicin, and the latter is more common in kanamycin and amikacin. Renal toxicity mainly reflects on proximal renal tubules damage, resulting in proteinuria, hematuria, renal dysfunction and so on. Kanamycin, florimycin and Gentamicin have higher nephrotoxicity. Infants are sensitive to AGs, which can damage hearing through the placenta. Due to the side effects and drug resistance, streptomycinp has been banned for use in China. The maximum residue limits of some AGs in meat was shown in Table 11-2.

Table 11-2　MRLs of aminoglycoside antibiotics in animal food(China)

Antibiotics	MRL/(mg/kg)			
	Cattle	Sheep	Pork	Poultry
Dihydrostreptomycin	No detection		No detection	
Cidomycin			0. 4(fat)	
			0. 4(kidney)	
			0. 3(liver)	
			0. 1(muscle)	
Fradiomycin	0. 25(edible tissue)	1. 25(fat)	1. 00(fat)	0. 50(fat)
	1. 00 (fat)	1. 25(kidney)	1. 00(kidney)	1. 00(kidney)
	0. 75(kidney)	1. 25(liver)	0. 75(liver)	0. 75(liver)
	0. 50(liver)	0. 25(muscle)	0. 25(muscle)	0. 25(muscle)
	0. 25(muscle)			
	0. 15(milk)			
Phytomycin	No detection (edible tissue)	No detection (edible tissue)	No detection (edible tissue)	
Actinospectacin	No detection (milk)	No detection (milk)		
			0. 1 (edible tissue)	

11. 1. 2   Immunoassay

Zu[3] synthesized immunogen and coated antigen of amikacin, and successfully validate by UV and animal immunity method. The polyclonal antibody of amikacin was also prepared and identified. Both of the two rabbits in the experiment generated anti-amikacin antibody. By optimizing conditions of indirect competitive ELISA, the IC_{50} value was 0. 38 $\mu g/mL$ with the minimum detection limit 40 ng/mL. It was also found that there were no cross reactions among amikacin, its analogues. The proposed ELISA was used to detect amikacin residues in milk and kidney tissue of pig. The recoveries of amikacin in milk samples were 75%-81% and 72%-79% in kidney samples.

In order to develop a fast and effective method for streptomycin detection in milk, Tang[4] prepared monoclonal antibody and developed ELISA for streptomycin det based on synthesized artificial antigens. The prepared ascites titer was 1:1, 280,000 with IC_{50} value of 93. 22 ng/mL. Weak cross-reactivity with other antimicrobial agents (less than 0. 1%) was found. Milk samples were analyzed with this ELISA with detection limit of about 5 ng/mL.

After oxime treatment of oxygen-carboxymethyl hydroxylamine, Fan[5] introduced carboxyl groups in the chemical strucuture of streptomycin(SM). The hapten was coupled with carrier protein with carbodiimide, glutaraldehyde and 1,4-butanediol ethyl oxide ether as linker respectively. Mice were immunized to prepare monoclonal antibody. A hybridoma can stably produce anti-SM antibody used to prepare of ascites in vivo. The chromosome number in hybridoma was 96-102, (mean value of 98. 6). The monoclonal antibody was characterized as isotype of IgG_{2a}/κ with affinity constant (Ka) of 7. 5 $\times$ $10^{10} L/mol$. The monoclonal antibody recognized streptomycin well with IC_{50} value of 8. 99 μg /L and showed cross-reactivity of 93. 6% for dihydrostreptomycin (DHSM). Also, New Zealand white rabbits were immunized to produce polyclonal antibody. The IC_{50} value of 5. 81 $\mu g/L$ was detected for antisera with indirect ELISA. The proposed ELISA was used for SM residual analysis withmonoclonal antibody and sensitivity of SM was 0. 52 $\mu g/L$ in milk urine analysis. Fan used colloidal gold particles (25 $\pm 1.$ 0 nm) to label antibody, which was usedas a tracer to develop an immunochromatography strip for detection of DHSM and SM. It was visually examined that the detection limits of SM and DHSM was 4. 14 and 4. 0 $\mu g/L$ respectively and no cross-reaction with other aminoglycoside or antibiotics. Urine of pig and milk samples were analyzed with ELISA, test strip and LC-MS and good correlations among these three methods were found.

Zhang[6] synthesized complete antigen of streptomycin by condensation reaction between aldehyde group of streptomycin and amine group from carrier protein. Monoclonal antibody was prepared with isotype of IgG2b. The ascites was prepared and IC_{50} value was determinated as 33. 03 ng/mL. an indirect competitive ELISA was developed to analyze chicken and milk samples with limit of detection being 1. 0 μg/kg.

Liu[7] coupled neomycin with bovine serum albumin (BSA) and hemocyanin (KLH) respectively with carbodiimide method. New Zealand rabbits were immunized to prepare antiserum. The indirect ELISA was developed based on the antisera with IC_{50} value of 174. 58 μg/mL for neomycin. The cross-reaction rates with gentamincin and kanamycin was 2. 04% and 0. 02%, respectively. No cross reactivity was found for ampicillin, erythromycin and tetracyclin.

Primary amine group of gentamicin (GM) was coupled with carboxyl of carrier protein through a linker. The anti-GM monoclonal antibody was successfully prepared by Xu[8]. Then, an indirect ELISA method for determination of GM was established with the detection limit of 3. 85 ng/mL and quantification limits being 6. 6 ng/mL in milk analysis.

11.2   Macrolide antibiotics

11. 2. 1   Introduction

Macrolide antibiotics (MALs) is a large and important group of antibiotics. The macrolides are a group of drugs (typically antibiotics) whose activity stems from the presence of a macrolide ring, a large macrocyclic lactone ring to which one or more deoxy sugars, usually cladinose and desosamine, may be attached. The lactone rings are usually 14-, 15-, or 16-membered(Figure 11-2). Macrolides belong to the polyketide class of natural products. In 1957, Woodward firstly used "macrolides" as the name of these compounds. Most macrolide antibiotics are stem from *Streptomyces*, and only a minority is produced by *Micromonospora*. Erythromycin was the first MALs to be approved in clinical application. At present more than 100 kinds of MALs have been identified[9].

MALs have been used in veterinary since the late fifties of 20th century, when erythromycin was only as a substitute for penicillin. With the emergence of more MALs, MALs has been widely used in chemotherapy for bacteria and mycoplasma infection of animals. Especially a small dose of MALs can well promote growth of animals.

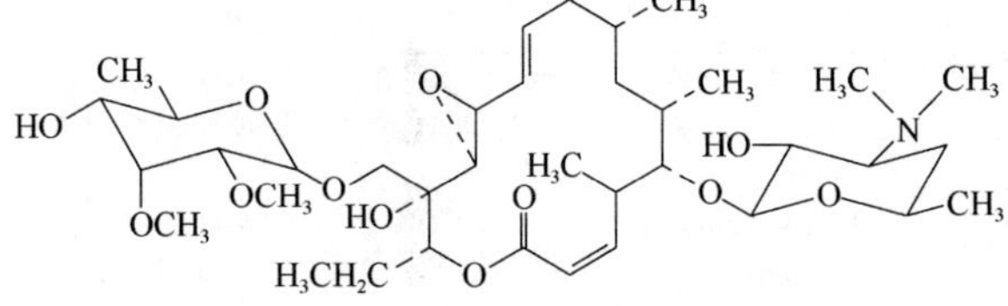

microsamicin 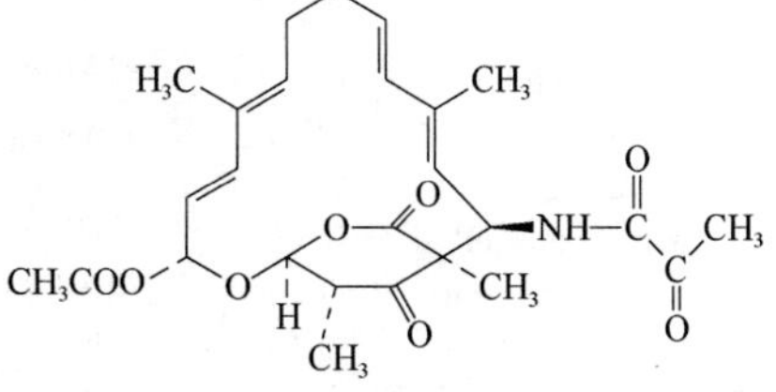   sedcamycin

Figure 11-2   Chemcial constitution of macrolide antibiotic(Continued)

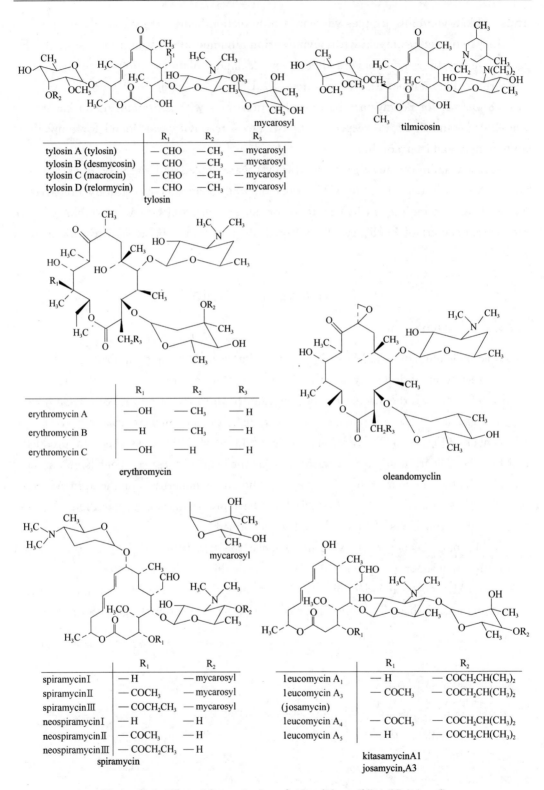

	R_1	R_2	R_3
tylosin A (tylosin)	— CHO	— CH$_3$	— mycarosyl
tylosin B (desmycosin)	— CHO	— CH$_3$	— mycarosyl
tylosin C (macrocin)	— CHO	— CH$_3$	— mycarosyl
tylosin D (relormycin)	— CHO	— CH$_3$	— mycarosyl

tylosin

	R_1	R_2	R_3
erythromycin A	—OH	— CH$_3$	— H
erythromycin B	—H	— CH$_3$	— H
erythromycin C	—OH	—H	— H

erythromycin

oleandomyclin

	R_1	R_2
spiramycin I	—H	— mycarosyl
spiramycin II	— COCH$_3$	— mycarosyl
spiramycin III	— COCH$_2$CH$_3$	— mycarosyl
neospiramycin I	—H	—H
neospiramycin II	— COCH$_3$	—H
neospiramycin III	— COCH$_2$CH$_3$	—H

spiramycin

	R_1	R_2
leucomycin A$_1$	— H	— COCH$_2$CH(CH$_3$)$_2$
leucomycin A$_3$ (josamycin)	— COCH$_3$	— COCH$_2$CH(CH$_3$)$_2$
leucomycin A$_4$	— COCH$_3$	— COCH$_2$CH(CH$_3$)$_2$
leucomycin A$_5$	— H	— COCH$_2$CH(CH$_3$)$_2$

kitasamycinA1
josamycin,A3

Figure 11-2　Chemcial constitution of macrolide antibiotic(Continued)

Thus, MALs were also used as important additives in husbandry and some have become animal-specific antibiotics such as tylosin, tilmicosin and kitasamycin.

11.2.1.1 Classification and physico-chemical properties

MALs contain a highly substitutive fourteen or sixteen intra-ring glycoside, which is their structural feature. Lactone ring connects with one or two sugar chain (dimethyl amino sugar and neutral sugar) through glycoside bond. In addition to lactone structure, the structure of glycosides contain alkyl, hydroxy, alkoxy, epoxy, keto or aldehyde groups, and a majority also contain conjugated diene or α, β-unsaturated ketones. Therefore, according to the number of atoms constituting a lactone skeleton, or UV absorption characteristics, MALs can be classified as listed below (Table 11-3).

Table 11-3 Classification of macrolide antibiotic

Drug(principal constituent)	Molecular formula/ Molecular mass	Melting point/℃	$[\alpha]_D$	pK_a	λ_{max}/nm
Erythromycin (ERM)	$C_{37}H_{67}O_{13}N$/733. 4613	135-140	−73. 5(methanol)	8. 8	280 (weak)
Oleandomyclin (OLD)	$C_{35}H_{61}O_{12}N$/687. 4194	177	−50. 0(methanol)	8. 6	287 (weak)
Spiramycin I (SPM)	$C_{45}H_{78}O_{15}N_2$/886. 5403	134-137	−96. 0(methanol)	—	232 (strong)
Tylosin (TYL)	$C_{46}H_{77}O_{17}N$/915. 5192	128-132	−46. 0(methanol)	—	282 (strong)
Tilmicosin (TIL)	$C_{46}H_{80}O_{13}N_2$/869. 5661	—	+12. 8(methanol)	7. 4 8. 5	283 (strong)
Kitasamycin (KIT)	$C_{44}H_{68}O_{14}N$/785. 4652	—	−66. 0(chloroform)	6. 7	232 (strong)
Josamycin (JOS)	$C_{42}H_{69}O_{15}N$/827. 4668	120-121	−55. 0(chloroform)	7. 1	232 (strong)
Microsamicin (MIS)	$C_{37}H_{61}O_{13}N$/727. 4143	102-106	−31. 0	—	218 (moderate)
Sedecamycin (SED)	$C_{25}H_{37}O_9N$/505. 2312	—	—	—	226 (strong)

11.2.1.2 Pharmacology, metabolism and residue

MALs have outstanding antibacterial activity against Gram-positive bacteria and mycoplasma, which are also effective to spirochetes, rickettsia, and mycoplasma. Especially, ERM has the strongest bactericidal effect on Gram-positive bacteria, and TYL present the strongest effect on mycoplasma. MALs are clinically used for the treatment of respiratory, gastrointestinal and urogenital infections caused by susceptible bacteria, such as pneumonia, bacterial enteritis, puerperal infection and mammitis. The intra-muscular injection dose is usually 2-50 mg/kg bw[10].

The antibacterial mechanism of MALs is to interfere with the synthesis of bacterial protein. MALs bind with the ribosomes of 50s subunits of sensitive bacteria to inhibit the catalytic function of transpeptidase and hinder the formation and extension of the peptide chain. In the moderate concentration, MALs mainly show inhibitory effect of bacteria, and manifest bactericidal action in higher dose. It is difficult to permeate the outer membrane of Gram-negative bacteria for these drugs. Therefore, MALs are inva-lid to Gram-negative bacteria. The bacteria generate resistance to MALs mainly through the modification of ribosomal binding site. Some MAL-resistant strains can produce enzyme to catalyze MALs lactone. MALs resistance genes are located in the bacterial plasmid or chromosome. There is cross-resistance among MALs.

MALs have low toxicity. For example, the LD_{50} for intravenous mice was 150-650 mg/kg, and LD_{50} was 1,500-8,000 mg/kg by oral intake. MALs are widely distribu-ted in the body due to their weak alkalinity and liposolubility. Their distribution is char-acterized by high ratio of tissue/plasma (5-10 : 1) and higher accumulation in the organ-izations with low pH value. The accumulation order is usually liver> lung >kidney> plasma, and very limited accumulation in muscle and fat tissue. The main problem caused by MALs residues in food are allergy reaction and spread of strains carrying resistance factors.

MALs belong to a minority of antibiotics which have a high accumulation concen-tration in the lung tissue. Most of MALs and their metabolites can be rapidly excreted by the bile duct while a small part of that are excreted by kidneys. Therefore, the bile usually keeps high concentrations (dozens of times higher than tissue concentrations). Small MALs in the gastrointestinal tract may be reabsorbed. Route of administration would affect the distribution of residues. For example, oral intake of TYL would result in highest levels in liver tissue and muscle injection resulted in highest levels in the kid-ney. In addition, the injection site often maintains high levels of MAL residues.

In oral administrations of a small amount of MALs, gastric fluid can acidolyse 1 or 2 sugar chains of MALs. MALs intaken mainly go breakdown in the liver. It has been found that TYL metabolites in the body of pigs and cattle include tylosin B (acid hydrol-ysis product), tylosin C (demethylation product), tylosin D (reduction product) and di-hydrotylosin B, but the level of these metabolites is lower than the parent drugs.

The metabolic process of SPM is complex, the mainly known products include the new Spiral Ray (the acidolysis product of SPM by gastric acid) and cysteine condensa-tion product, which have been indicated that these metabolites belong to the main me-tabolites of residual SPM. The main metabolic pathway of ERM is N-demethylation re-action in deoxy-amino sugar chains and the degraded products basically lost antibacterial activity. The main metabolic pathway of KIT and JOS is hydroxylation. In the residue monitoring of MALs, liver generally acts as target organ and the prototype of drug are marker residues, such as TYI, TIL, ERM and OLD.

11. 2. 2   Immunoassay of MALs

Ivermectin (IVM), a member of avermectin pesticides family, is a semisynthetic macrocyclic lactone compound, with broad-spectrum anti-parasitic and arthropod activity. It is a modified product of natural insecticide avermectin.

Cui[11] designed hapten of IVM (Figure 11-3). Using succinic anhydride method, chemical transformation was performed at C5 hydroxyl group of Ivermectin. The resluting IVM-5-succinyl half ester was coupled with carrier protein with carbodiimide method. With these complete antigens, Cui prepared monoclonal antibody against IVM and developed an ELISA with IC$_{50}$ value of 7. 95 ng/mL. The cross-reactivity rate with avermectin B1 was 51%, and less than 0. 2% for streptozotocin oxime compounds. Vegetable samples were successfully analyzed with this proposed ELISA.

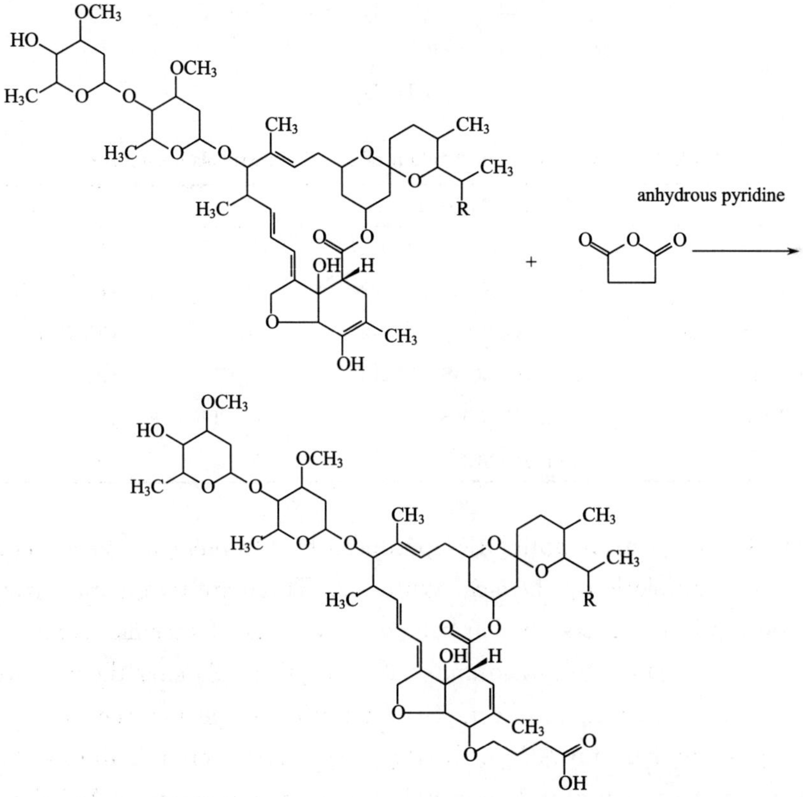

Figure 11-3   Hapten synthesis route for ivermeetin
ivermectin H2B1a R=C2H5(≥95%); ivermectin H2B1b R=CH3(≤2%)

Lu[12] synthesized artificial antigens for ivermectin (IVM) with succinic anhydride method and carbodiimide method. Hybridoma cell line that can steadily secrete specific monoclonal antibody anti-IVM was obtained. The titer of prepared ascites was 1 : 32,000, and cross-reactivity with other antimicrobial agents was less than 0.1%. An ELISA was established with IC_{50} value of 200 ng/mL for IVM. Pork liver and muscle were analyzed with this ELISA with good recovery.

11.3 Chloramphenicols drugs

11.3.1 Introduction

11.3.1.1 Structure and property

Chloramphenicols (CAPs) include chloramphenicol and its derivatives. CAPs are also known as amide alcohols and the typical structures were shown in Table 11-4.

Table 11-4 Chemcial constitution of chloramphenicols antibiotics

Drugs	Formula/Molecule weight	R_1	R_2
Chloramphenicol(CAP)	$C_{11}H_{12}O_5N_2Cl_2$/322.0124	—NO_2	—OH
Chloramphenicol succinate	$C_{15}H_{16}O_8N_2Cl_2$/422.0284	—NO_2	—$OCOCH_2CH_2COOH$
Chloramphenicol palmitate	$C_{22}H_{42}O_6N_2Cl_2$/560.242	—NO_2	—$OCOC_{15}H_{31}$
Thiamphenicol(TAP)	$C_{12}H_{15}O_5NCl_2S$/355.0048	—SO_2CH_3	—OH
Florfenicol(FF)	$C_{12}H_{14}O_4NCl_2SF$/357.0005	—SO_2CH_3	—F
Cetofenicol	$C_{13}H_{15}O_4NCl_2$/319.0378	—$COCH_3$	—OH

CAP is a broad-spectrum antibiotic, produced by *Streptomyces Venezuela*. CAP is the first artificial antibiotic by chemical synthesis. There are two chiral carbon atoms, resultingn four optical isomers, which compose two sets of racemic isomers: D-(—)-threo, L-(+)-threo, D-(+)-erythro and L-(—)-erythro. Among them, only D-(—)-threo possesses an antibacterial effect. Syntho-mycin is the racemic isomer of CAP, whose efficacy is only half of CAP[13]. CAP is very stable even in boiling water for 5 h. It can maintain its bioactivity for 25 h in pH2-9 at room temperature. Dried crystals can be stored for more than two years. In the conditions of acid, alkali (more sensitive) and heating, hydrolysis of CAP occur. The artifial CAP is white or slightly yellow needle like crystal, with melting point 149-153 ℃ and doesn't show optical rotation.

The paratope nitro in CAP molecular is substituted with mesyl to make thiamphenicol, whose antibacterial spectrum and applications are basically the same with CAP, but with lower toxicity. Thiamphenicol is white crystalline powder, with melting point 164. 5-166. 5 ℃ (decomposed), $[\alpha]D^{25}+12. 9°$ (C = 1, absolute alcohol). Its solubility is low in water, and water solubility at room temperature is about 0. 5%-1%. Thiamphenicol is soluble in methanol, slightly soluble in acetone, ethanol, but insoluble in benzene, chloroform and ether. Thiamphenicol is Stable in light and heating and easily absorbs moisture. Florfenicol is a derivative of thiamphenicol, and has been applied in aquaculture industry at home and abroad.

11.3.1.2   Pharmacology and toxicology

CAP is a broad-spectrum antibiotic, and can inhibit bacterial protein synthesis. It has a strong antibiosis on gram-negative bacteria such as *E. coli*, *Salmonella*, *Typhoid bacillus*, *Paratyphoid bacillus*, *Shigella*, *Aerobacter aerogenes*, *Klebsiella pneumoniae*, *Pneumococcus*, *Streptococcus*, *Staphylococci*, *Diphtheria bacilli* and *Anthrax bacilli*. It also has a certain effect on a variety of rickettsia, protozoa and some viruses. CAP has been a good choice for the treatment of typhoid, paratyphoid, and salmonellosis since 1948. It is also a good drug for treatment of mastitis. CAP is well absorbed orally and it shows systemic distribution after being absorbed, and can permeate through the blood-brain and placental barrier. But the absorption form injection of CAP is slow. Most of intaken CAP is devitalized after binding with glucuronic acid in liver acid while a small part of them is degraded into aromatic amines. About 10% of CAP is excreted by kidney in prototype and they can also be excreted through milk secretion. Liver or kidney dysfunction would extend elimination period of CAP, resulting in cumulative intoxication[14].

The most serious adverse effect associated with chloramphenicol treatment is bone marrow toxicity, which may occur in two distinct forms: bone marrow suppression, which is a direct toxic effect of the drug and is usually reversible, and aplastic anemia, which is idiosyncratic (rare, unpredictable, and unrelated to dose) and in general fatal. CAP has been widely used in clinical veterinary and is still used in a considerable range. Many countries have strictly prohibited the use of CAP in food animals (especially the hens and cows), and MRL of CAP is lower than 0. 01 mg/kg.

11. 3. 2   Immunoassay

Zhao[15] activated the carboxyl group of chloramphenicol succinate and coupled it with carrier protein to get immunogens. a polyclonal antibody for chloramphenicol was prepared and a direct competitive ELISA was established for chloramphenicol with the IC_{50} value of 12. 9 ng/mL. The cross-reactivity between antibody and amber chloramphenicol was 174. 54% and no cross reactivity with other antibiotics was found (less than

0.01%). The established ELISA method was successfully used in analysis chloramphenicol of milk. Wang[16] applied colloidal gold immunochromatographic assay (GICA) to successfully develop a rapid test strip for CAP residues. The detection limit of this CAP strip was 0.5 μg/L. Cross-reaction with chloramphenicol sodium succinate was determined as 160%. Based on polyclonal antibody against CAP, Lin[17] developed a simple, solid-phase chemiluminescence immunoassay (CLIA) for the measurement of CAP in foodstuffs. CAP was labelled with horseradish peroxidase (HRP), which competitively bound to immobilized antibody. The luminol system was used as chemiluminescent substrate for HRP. The sensitivity of this CLIA is less than 0.05 ng/mL.

Two haptens (CAP-HS$_1$ and CAP-HS$_2$ in Figure 11-4) of chloramphenicol were synthesized by Huang[18] using CAP and succinic anhydride. Due to the different site of OH group in CAP, the mixture of two haptens was produced. After purification, they were coupled with BSA as immunogen respectively. Two anti-chloramphenicol polyclonal antibodies were obtained. The antibody was labelled with horseradish peroxidase (HRP) with modified sodium iodide method. The antibody rainsed from CAP-HS$_1$ showed higher affinity to CAP with IC$_{50}$ value 6.0 ng/mL. The cross-reacticity with CAP analogues including thiamphenicol and florfenicol was 0.03% and 0.17%, respectively. An ELISA kit was developed and assay buffer was optimized. The kits were used to analyze shrimp, pork, chicken and egg samples with good reproducibility.

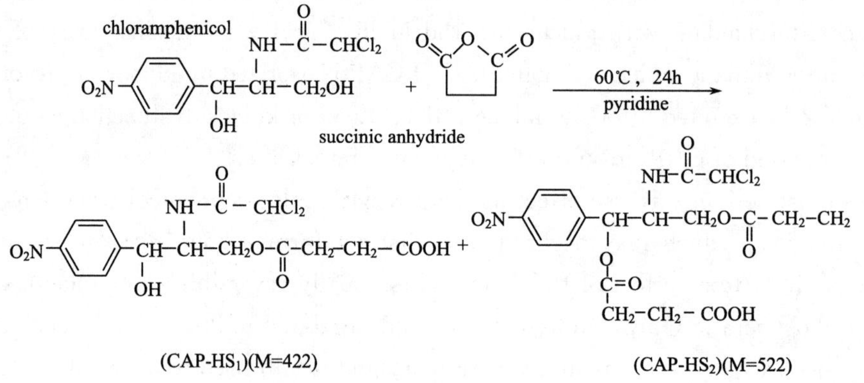

Figure 11-4   Hapten sysnthesis for chloramphenicol

Hao[19-22] used succinic anhydride to introduce a carboxyl into thiamphenicol (TAP),which was used to conjugate with carrier protein with mixed anhydride method. New Zealand white rabbits were immunized to produce antisera. The obtained antibody showed good specificity with TAP with 0.4% cross-reactivity to CAP. Based on the antibody, an indirect competitive ELISA was established with linear detection range of TAP 0.1-34.7 ng/mL.

11.4 β-lactam antibiotics

11. 4. 1 Introduction

11.4.1.1 Structure and classification

β-lactam antibiotics represented by penicillins and cephalosporins are the oldest antimicrobial agents, and also the largest and most important class of antibiotics. β-lactam antibiotics is characterized in inhibiting activity of bacterial peptidoglycan transpeptidase and preventing the formation of cell walls, which show a strong bactericidal activity. Because there is no cell wall strucuture in mammalian cells, the side effects of these antibiotics is small. Due to the problems in application of these antibiotics, including drug resistance, allergies and poor stability, more new generation of β-lactam antibiotics have been developed with powerful performance and small side effects, such as broad-spectrum and resistance to enzymes, acid[23].

β-lactam antibiotics contains β-lactam nucleus, which is rare in nature (Figure 11-5). According to the differences in parent nuclear structure, these antibiotics can be divided into penicillins (PENs), cephalosporins (CEPs), cephamycins (methoxy cephalosporins, oxacephems), carbapenems and monobactams. Among them, the PENs and CEPs family developed rapidly, resulting in various species. Their structure is as follows.

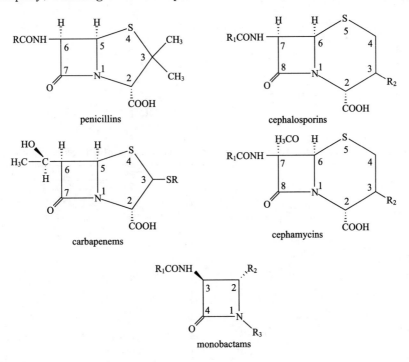

Figure 11-5 Structure of β-lactam antibiotics

Firstly, the structure of β-lactam antibiotics consists of nucleus and side chain. The nucleus contains a four-membered β-lactam cycle, which is fused with the second five-or six-membered ring through N atom and 3 carbon atoms nearby. The fused ring of PENs is hydrogenated thiazole ring, whose parent nucleus was 6-aminopenicillanic acid. The fused ring of PEPs is hydrogenated thiazole ring, with nucleus of 7-amino cephalosporinic acid. Secondly, site of C-6 or C-7 contain an amide side chain (RCONH-). Thirdly, the two fused rings in nucleus are not coplanar. For PENs, two side chains fold along axis N1-C5 while CEPs fold along axis N1-C6. Founthly, there are 3 chiral carbon atoms in nucleus of PENs, resulting in eight isomers. Only the absolute configuration 2S, 5R, 6R have activity. For CEPs, there are two chiral centers in its parent nucleus and it was proved that absolute configuration of 6R and 7R have activity. Their bioactivity is affected by chiral carbon atom in substituent group (RCONH-). The common natural and semi-synthetic PENs and their chemical structure are listed in Table 11-5.

Natural PENs stem from *Penicillium*, contain five ingredients (penicillin F, G, X, K and dihydrotestosterone F). Benzylpenicillin (penicillin G) has the strongest effects on bacteria and the highest yield. PENs are mainly used in control of Gram-positive bacteria such as *Staphylococcus*, *Streptococcus*, *Meningococcus*, *Diphtheria bacillus*, *Bacillus anthracis*, *Clostridium tetani*. Semi-synthetic penicillin is formed based on the transformation of 6-aminopenicillanic acid (6-APA) including oxacillin, ampicillin and amoxicillin.

Table 11-5   The chemical structure of some natural and semi-synthetic cephalosporin drugs

Drugs	R	Drugs	R
Benzylpenicillin	⬡—CH₂—	Oxacillin	(structure)
Penicillium V phenoxymethy l penicillium	⬡—O—CH₂—	Cloxacillin	(structure)
Phenoxyethyl penicillium	⬡—O—CH— C₂H₅	Dicloxacillin	(structure)

Continued

Drugs	R	Drugs	R
Azidocillin		Flucloxacillin	
Belfacillin		Oxacillin	
Nafcillin		Carbenicillin	
Ampicillin		Sulbenicillin	
Amoxicillin		Metampicillin	
Tazobactam		Etacillin	
Ticarcillin			
Piperacillin			

CEPs are broad-spectrum antibiotics (Table 11-6). The semi-synthetic CEPs was used in clinical diagnostic applications due to their strong bactericidal activity, stability and low sensitization. The update of CEPs is very fast and it has developed to the fourth generation with the typical drug cefalexin (also known as Pioneer 4)

Table 11-6　The chemical structure of some natural and semi-synthetic cephalosporins drugs

Drugs	R_1	R_2
Cefalotin		$-O-COCH_3$
Cefazolin		
Cefapirin		$-O-COCH_3$
Ceftiofur		
Cefalexin		$-H$
Cefadroxil		$-H$
Cefradine		$-H$
Cefuroxime		$-OCONH_2$
Cefotaxime		$-OCOCH_3$
Cefmenoxime		
Cefepime		

11.4.1.2　Physico-chemical properties

　　The free carboxylic of PENs and CEPs have very strong acidity (pKa of 2. 5-2. 8), which are easy to form salts with alkali. Potassium or sodium salts are used widely in clinical therapeusis[24]. Nucleus part of the (6-APA) has no UV absorption properties, so, the UV absorption usually comes from the side chain on the benzene ring. The β-lactam ring of CEPs has conjugated system $O = C-N-C=C$, resulting in strong UV absorption at 260 nm. The tension of CEPs quaternary-six member fused ring system is smaller than the quaternary-five fused ring system in PENs, so CEPs are more stable than PENs.

β-lactam ring is the most unstable part of chemical structure of these drugs. Hydrolysis or molecular rearrangement in their structure results in bioactivity loss. Acids, alkalis, some heavy metal ions (oxidant) or bacterial penicillinase (a kind of β-lactamase) can accelerate the degradation. Among them, degradation of penicillin G has been fully investigated. Penicillin G is fairly stable in dry conditions, but in the aqueous solution β-lactam ring would be easily broken to produce a series of degradation products (Figure 11-6). Under acid catalyzation, molecular rearrangement in β-lactam ring haptens due to electron transfer. Penillic acid or penicillenic acid is formed in different acidity. When heated, penicillic acid can be further decomposed into D-penicillamine and penicilloaldehyde.

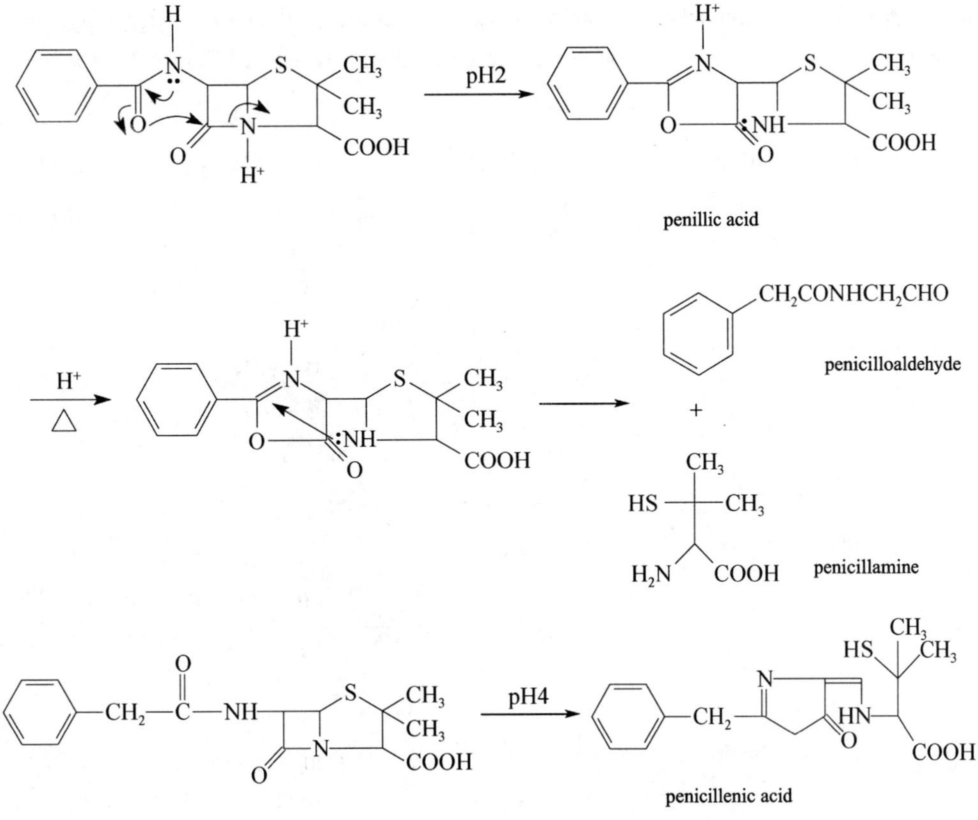

Figure 11-6 the breakdown of β-lactam ring(Continued)

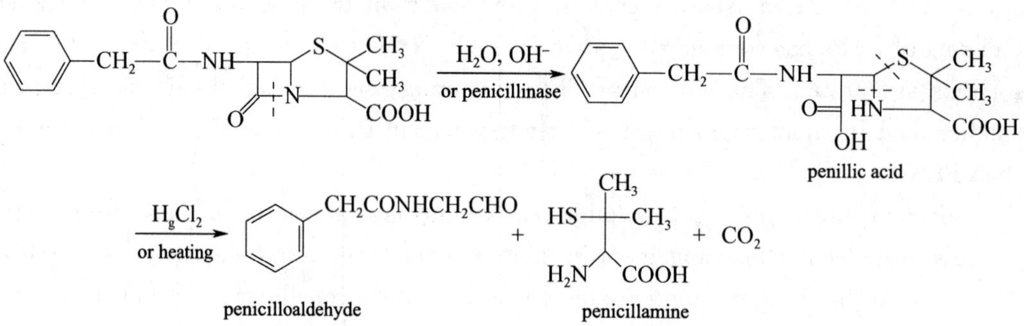

Figure 11-6   the breakdown of β-lactam ring(Continued)

β-lactam ring would breakdown in alkaline condition. It will breakdown into penicilloic acid (also known as penicillic acid) firstly and is further decomposed into D-penicillamine and penicilloaldehyde in existance of mercuric chloride or heating. Similar degradation will take place in the existance of alcohols or penicillinase.

Penicilloic acid combines with free amine groups from proteins, polypeptide or hydroxylamine, resulting in penicilloyl protein, penicilloyl polypeptide, α-penicilloyl hydroxamic acid and penicilloic acid polymer, which is called penicillin reaction (Figure 11-7).

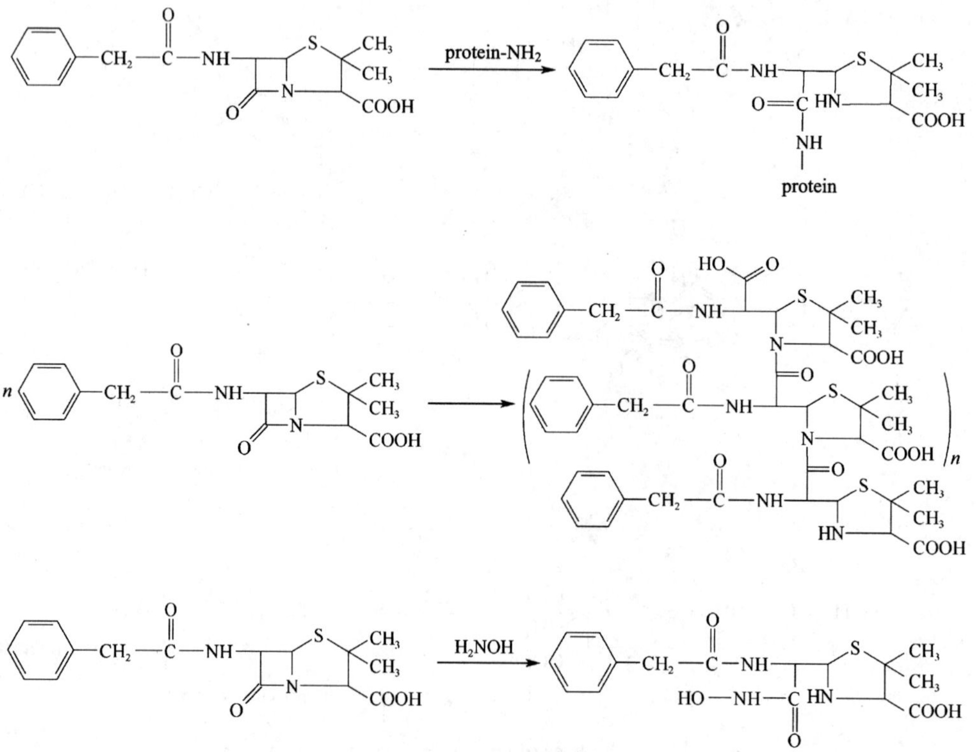

Figure 11-7   The penicillin reaction

11.4.1.3   Residues of β-lactam antibiotics

β-lactam antibiotics are widely used. Because of allergic reactions and bacterial resistance, many goverments limited the use of β-lactam antibiotics in animal production. In 1996, FDA of USA screenedβ-lactam residues in 4,480,530 milk samples and 6,148 samples (0.14%) were confirmed positive. The main varieties of β-lactam antibiotics in milk were benzyl penicillin, cephalosporins, aspirin, ampicillin, amoxycillin and cloxacillin. EU has set MRLs for some typicalβ-lactam antibiotics (Table 11-7).

Table 11-7   Maximum residue limit of β-lactam antibiotics in food established by EU(μg/kg)

β-lactam antibiotics	Meat	Milk	β-lactam antibiotics	Meat	Milk
Ampicillin	50	10	Dicloxacillin	300	—
Amoxicillin	50	10	Nafcillin	300	—
Benzylpenicillin	50	4	Cefapirin	100	20
Oxacillin	300	—	Ceftiofur	12,000	50
Cloxacillin	300	—			

11.4.2   Analytical method

He[25] prepared polyclonal antibody of ampicillin, and established immunoassay techniques for ampicillin. Two strategies of antigen development were used. One is coupled ampicillin with caiier protein, and the other is to applicate glutaraldehyde as a linker. Rabbits were immunized to produce antisera. The specific recognization of antibody to ampicillin was identified with agar diffusion analysis. An indirect competitive ELISA was established and the assay conditions were optimized. The limit of detection of this ELISA was 4.17 ng/mL. The cross reactivity beween antiserum and amoxicillin was up to 78% because the structure difference between amoxicillin and ampicillin is only one hydroxyl. The cross-reactivity rate between antiserum and penicillin potassium was 1.5%, and cross-reactivity rate with oxacillin is 0.4%. No cross-reactivity of streptomycin sulphate, kanamycin, and gentamicin sulphate was found (< 0.01%).

Lu[26] prepared immunogen by coupling ampicillin (Amp) in carbodiimide (EDC) method with hemocyanin (mcKLH) and cationic bovine serum albumin (cBSA) respectively. The conjugate of KLH was used to immunize Balb/C mice. Coupling ratio of hapten and carrier protein (Amp-KLH) was 30 : 1 and 10 : 1 for Amp-BSA. By hybridoma technology, the monoclonal antibody against Amp was prepared. The antibody was characterized as IgG_{2a}, and molecular weight was 154 kDa. The cross-reactivity rates of this monoclonal antibody between ampicillin-sodium and penicillin-sodium were 0.36% and 0.007%, respectively. The cross-reactivity rates were less than 0.001% with cephradine, ceftiofur and other commonly used antibiotics, such as kanamycin, streptomycin, sulfadiazine, and gentamicin. An indirect competitive ELISA method was used for Amp analysis in milk samples with the minimum detection limit of commercial sterilized milk 0. 4 ng/mL.

Zhang[27] used carbodiimide method (EDC) and penicillin reaction to couple penicillin G with bovine serum albumin (BSA) and ovalbumin (OVA) for the synthesis of immunogen. Rabbits were immunized to produce antiserum. Different concentration IgG which had been obtained coupled with different concentration SPA to prepare SPA diagnosticum. Then SPA coagglutination test was done to detect penicillin. The minimum detectability of SPA-COA in detecting penicillin was tentatively determined as 10 μg/L.

Yang[28] used ampicillin hapten, penicillin G to conjugate with carrier protein by glutaraldehyde method and penicillin reaction. It showed higher titer antibody was rainsed with conjugates from penicillin reaction. A direct competitive ELISA was established with detection limits of 1. 58 ng/mL for ampicillin, 0. 38 ng/mL for penicillin G, 1. 90 ng/mL for amoxicillin, 0. 13 ng/mL for cloxacillin and 0. 44 ng/mL for oxacillin.

Hao[29] immunized mice using ceftiofur (CEF) antigen using direct cross-linking method, glutaraldehyde method, carbodiimide method and maleoyl amide activation. The conjugates of ceftiofur and carrier protein (KLH-CEF and CEF-BSA) were used as immunogens. The antisera from mice was identified with indirect ELISA and the mouse secreting high affinity with CEF was used in cell fusion for monoclonal antibodies preparation. 5 hybridoma strains were selected which can steadily secret object antibody. The amplified and cultured cell strains were used to induce the growth of mice ascites. After identification and purification of monoclonal antibody, the competitive ELISA for CEF test showed that all of the five monoclonal antibodies obtained had good specificity with the IC_{50} value of CEF 36-105 ng/mL. The highest cross-reactivity rate among 7 analogues (penicillin, ampicillin, amoxicillin, cefazolin, cephalexin, cloxacillin andcefotaxime) was 20. 8%. Among them, monoclonal antibody from $3B_3$ showed that highest specificity to CEF with cross-reactivity to cefotaxime 6. 13% and no recognization with penicillin, ampicillin, amoxicillin, cefazolin, cephalexin and cloxacillin. A method of indirect competitive ELISA was established for detection of CEF residues with IC_{50} of 21. 6 ng/mL and detection limit of 10 ng/mL.

Xie[30-32] used cephalexin, 7-ACA, cefadroxil, cefotaxime, cephalothin as haptens to conjugate with BSA by various methods to prepare 5 immunogens. Also, Xie coupled ceftiofur and KLH with glutaraldehyde method to obtain the sixth immunogen. Each immunogen was immunized two rabbits. The antisera from cephalexin-BSA showed higher titer and the IC_{50} value of cephalexin and cefadroxil were 1. 5 ng/mL and 2. 6 ng/mL. The antisera from ceftiofur-KLH showed best specificity to ceftiofur (IC_{50} value of 10 ng/mL) with less than 10% cross-reaction rate to other cephalosporins. The polyclonal antibodies against cephems were conjugated to colloidal gold particles as the detection reagent for ICA(immunochromatographic assay) strips to test for cephems. This method achieved semi-quantitative detection of cephems in <5 min, with high sensitivity to cephalexin and cefadroxil (both 0. 5 ng/mL). At the same time, cefatiofur, cefapirin, cefazolin, cefalothin and cefotaxine were detected at <100 ng/mL in spiked processed-milk samples. This method was compared with an enzyme-linked immunosorbent assay by testing 40 milk samples, and the positive samples were validated by a high-performance liquid chromatographic method, with an agreement rate of 100% for both comparisons.

11.5　Nitrofuran drugs

11. 5. 1　Introduction

11.5.1.1　Structure and classification

Nitrofurans are a class of drugs typically used as antibiotics or antimicrobials. Furan compounds were found in 1780 by Scheele, but the derivatives of nitrofuran compounds was synthesized 150 years later, in 1930. In 1944, studies by Dodd and Stuman suggested that introduction of nitro on position 5 of carbon atoms and the introduction of other groups on position 2 of carbon atoms for many nitrofuran derivatives such as furan nucleus would produce significant antibacterial effects. And studies about the introduction of other groups on position 2 of carbon atoms in the side chain were carried out. So far, 1, 000 kinds of nitrofuran have been reported[33]. Some typical nitrofuran compounds are shown in Figure 11-8. The defining structural component is a furan ring with a nitro group.

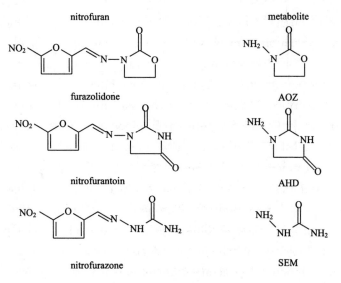

Figure 11-8　The chemical structure of nitrofuran drugs and their metabolites(Coutinued)

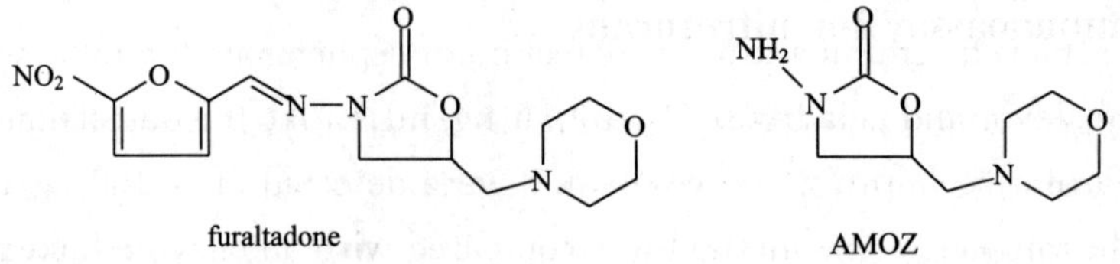

Figure 11-8　The chemical structure of nitrofuran drugs and their metabolites(Coutinued)

11.5.1.2　Physico-chemical properties

Nitrofuran drugs are usually yellow crystals, and most are insoluble in water or vegetable oil, slightly soluble in ethanol and chloroform. They are more easily soluble in dimethyl formamide. The solution of nitrofuran compounds is sensitive to light, so their solution or powder is stored avoiding sunlight and fluorescent light. In addition, the solution is unstable in an alkaline environment, and easily breaks down when pH is higher than 10. 0.

Nitrofurans are broad spectrum agents, characterized by effectiveness against Gram-positive and Gram-negative bacteria, including Salmonella species. This group of drugs is also effective against Giardia, amoebae, trichomonads and some species of Coccidia. Some nitrofurans are carcinogenic and future use is questionable; however, nitrofurans are used presently in both veterinary and human medicine Nitrofurazone exhibits an antibacterial spectrum similar to nitrofurantoin but is utilized primarily to treat mastitis, metritis and wounds in cattle. Presence of blood, pus or milk reduces the effectiveness of the drug.

Nitrofurazone is also used as a feed additive in food producing animals to control intestinal bacterial infections and coccidiosis. However, in the European Union use of nitrofurans in food production animals is prohibited. The nitrofuran parent compounds degrade rapidly in the treated animal and are stored in the tissue and products where they can remain long after treatment[34]. Due to this instability, effective monitoring of their illegal use has been difficult. The short in vivo half-life of the parent drugs (7 to 63 min) results in rapid depletion of nitrofurans in blood and tissue. However, the formed metabolites (AOZ, AMOZ, AHD and SEM) bind to tissue proteins in the body for many weeks after treatment.

Nitrofuran antibiotics, employed for the treatment of bacterial diseases in livestock production, were banned from use in the European Union (EU) in 1995 due to concerns about the carcinogenicity of their residues in edible tissue. This review provides an overview of nitrofuran toxicity, metabolism, and also specific aspects of legislation surrounding their prohibition. Special attention is devoted to semicarbazide-a nitrofuran metabolite and food contaminant.

11.5.2   Immunoassay for nitrofurans

3-amino-2-oxazolidinone (AOZ) is a stable metabolite of furazolidone. Xu[35] used aldehyde benzoic acid and AOZ to synthesize hapten (CPAOZ), which was coupled with carrier protein for preparation of immunogens. After immunization, polyclonal antibody was obtained. The antibody recognized CPAOZ well and the cross-reactivity for AOZ was 1%. The antibodies were used to develop direct competitive ELISA for CPAOZ with detection limit 0.2 μg/L.

Xu[36-38] synthesized 3-amino-2-oxazolidinone (AOZ), the metabolite of furazolidone, which reacted with 4-formylbenzoic acid to produce hapten CPAOZ. Similarly, CPSEM and CPAHD were derived from SEM (semicarbazide) and AHD (1-aminohydantoin) with 4-formylbenzoic acid. The structures of haptens are shown in Figure 11-9.

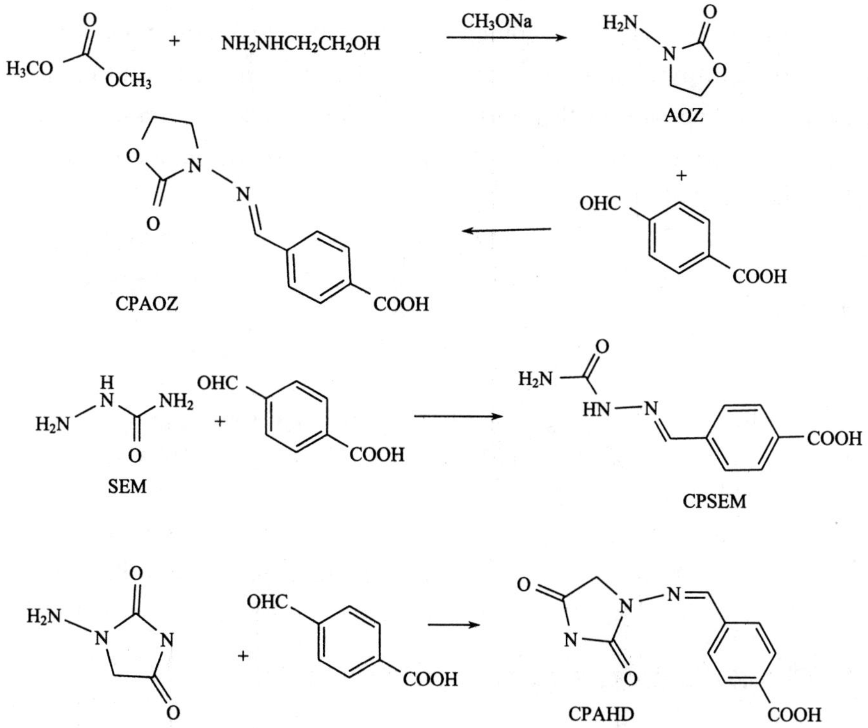

Figure 11-9   Synthesis of hapten CPAOZ, CPSEM and CPAHD

Then, immunogens were synthesized and immunized on rabbits for production of polyclonal antibodies. The IC_{50} value of AHD was 3.6 ng/mL. A immunochromatographic strip for SEM was assemblied with a visual detection limit of 40 ng/mL.

Luo[39] coupled furazolidone (FZ) with bovine serum albumin (BSA) and human serum albumin (HSA) with improved diazotization method. The conjugates were used as immunogen and injected into New Zealand white rabbits. Good titer of antisera was gotten using FZ-BSA conjugates as immunogens. An indirect competitive ELISA was established with FZ-HAS as coating antigen. The linear range of inhibition curve for FZ was 1-100 ng/mL and the detection limit low to 1 ng/mL. The cross-reactivity rate with nitrofurazone was 43%.

11.6　Quinolones

11.6.1　Introduction

Since their discovery in the early 1960s, the quinolone group of antibacterials has generated considerable clinical and scientific interest. Nalidixic acid, the first quinolone to be developed, was obtained as an impurity during the manufacture of quinine. Since that time, many derivatives have been synthesized and evaluated for their antibacterial potency. Two major groups of compounds have been developed from the basic molecule: quinolones and naphthyridones. Manipulations of the basic molecule (Figure 11-10), including replacing hydrogen with fluorine at position 6, substituting a diamine residue at position 7 and adding new residues at position 1 of the quinolone ring, have led to enhanced antibacterial efficacy.

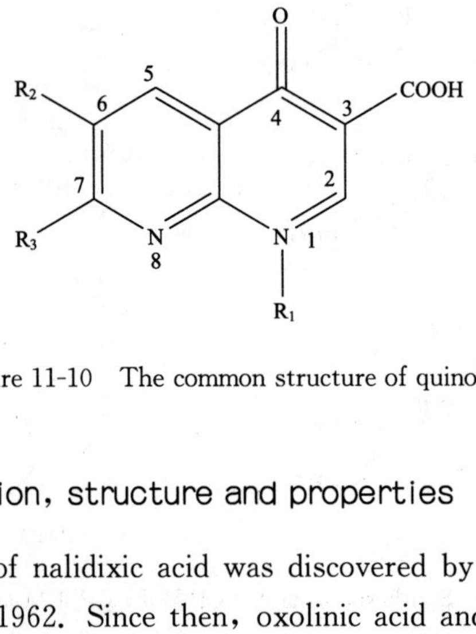

Figure 11-10　The common structure of quinolones

11.6.1.1　Categorization, structure and properties

The first quinolone of nalidixic acid was discovered by institute of Sterling-Winthrop of United States in1962. Since then, oxolinic acid and piromidic acid appeared, which were known as the first generation of QNs. Due to their narrow antimicrobial spectrum, short half-life, side effects and high bacterial resistance, they are not in use now. In 1970s, pipemidic acid and siragen acid were developed as the second-generation QNs. The third generation QNs appeared in 1980s. These QNs showed broader antimicrobial spectrum and more powerful bactericidal activity. Except pyrylium keto acids, the common feature of the third generation of QNs is fluorine atoms in C-6 site of quino-line ring (a naphthyridine ring individual), and piperazine or pyrrole was connected at C-7 site, which is also known as 6-fluoroquinolone or fluoride ketones. Now, fluoro-quinolones have gotten great progress and trovafloxacin is a typical representative of the fourth generation. At present, some have been approved for use in animal production such as norfloxacin (NOR), enrofloxacin (ENRO), sarafloxacin (SAR), danofloxacin (DAN), ciprofloxacin (CIP), diflucloxacillin (DIF), ofloxacin (OFL) and Marbofloxa-cin (MAR).

Parent QNs nucleus can be divided into two primary sturctures. One is two pyri-dine nuclei at 1-site and 8-site of nitrogen atoms, which is known as the nalidixic acid nucleus and the other only contain one nitrogen nucleus (at the 1-site), which is called quinolone nucleus. Both of the two stucutures contain 4-site keto and 3-site carboxylic acid side chain, which are necessary for the antibacterial activity of QNs[40]. Some typi-cal QNs are shown in Table 11-8.

Table 11-8   Chemical constitution of quinolones drugs

Drugs	Formula	Mmass	Constitution
Amifloxacin (AMI)	$C_{16}H_{19}FN_4O_3$	334.14	
Ciprofloxacin (CIP)	$C_{17}H_{18}FN_3O_3$	331.13	
Danofloxacin (DAN)	$C_{19}H_{20}FN_3O_3$	357.15	

Continued

Drugs	Formula	Mmass	Constitution
Difloxacin (DIF)	$C_{21}H_{19}F_2N_3O_3$	399.14	
Enoxacin (ENO)	$C_{15}H_{17}FN_4O_3$	320.13	
Enrofloxacin (ENR)	$C_{19}H_{22}FN_3O_3$	359.17	
Lomefloxacin (LOM)	$C_{17}H_{19}FN_3O_3$	351.14	
Marbofloxacin (MAR)	$C_{17}H_{19}FN_4O_4$	362.14	
Norfloxacin (NOR)	$C_{16}H_{18}FN_3O_3$	319.13	

Continued

Drugs	Formula	Mmass	Constitution
Ofloxacin (OFL)	$C_{18}H_{20}FN_3O_4$	361.14	
Pefloxacin (PEF)	$C_{17}H_{20}FN_3O_3$	333.15	
Sarafloxacin (SAR)	$C_{20}H_{17}F_2N_3O_3$	385.13	
Flumequine (FLU)	$C_{14}H_{12}FNO_3$	261.25	
Nalidixic acid (NAL)	$C_{12}H_{12}N_2O_3$	232.24	
Oxolinic acid (OA)	$C_{13}H_{11}NO_5$	261.23	

Continued

Drugs	Formula	Mmass	Constitution
Cinoxacin (CIN)	$C_{12}H_{10}N_2O_5$	262.22	
Sparfloxacin (SPA)	$C_{19}H_{22}F_2N_4O_3$	392.17	
Orbifloxacin (ORBAX)	$C_{19}H_{20}F_3N_3O_3$	395.38	
Pipemidic acid (PIP)	$C_{14}H_{17}N_5O_3$	303.32	
Temafloxacin (TEM)	$C_{21}H_{18}F_3N_3O_3$	417.13	
Tosufloxacin (TOS)	$C_{19}H_{15}F_3N_4O_3$	404.11	

11.6.1.2 Mechanism of toxicity

QNs are inhibitors of bacterial DNA synthesis, and the targeted enzyme is DNA-grase, also known as topoisomerase II (Topo II), which plays a key role in replication of DNA. QNs manifests powerful bactericidal action on gram-negative bacteria, mycoplasma and some gram-positive bacteria including *E. coli*, *Streptococcus pneumoniae*, *Klebsiella pneumoniae*, *Chlamydia pneumoniae*, *Staphylococcus aureus*, *Enterococcus*, *Bacillus bacteria degeneration*, the avian flu strain and *Pseudomonas aeruginosa*. These antibiotics are used in therapeusis of infection or sepsis at respiratory, intestinal, urinary or skinissue. QNs is concentration-dependent antibacterial, with a longer antimicrobial effect (up to 4-8 h)[41].

11.6.1.3 Metabolism and residues

The common biological conversion of QNs occurs in 7-piperazine substituent (I -phase reaction), including N-dealkylation, N-oxidation, hydroxylation reactions, acetylation and sulfonation reactions. Glucuronide reaction (II -phage reaction) occurs in 3-carboxylic acid group such as norfloxacin glucuronide. The antibacterial activity of metabolites is generally significantly reduced compared with prototype drug. But dealkylation products of ENR and PEL, CIP and NOR, have become excellent varieties of QNs with better bioactivity. QNs residues in organs are mainly in prototype drug. Most metabolites can be eliminated faster. Some QNs metabolites, such as demethylation and deethyl metabolites, are still capable of biological activity, which should be considered in QNs residue monitoring. When exposed to QNs, liver and kidney tissue accumulate higher concentration of QNs, followed by muscle and the skin attached to fat while fat and plasma have minimum residues[42].

With the increase of clinical applications, their adverse reactions have been reported, which have caused widespread concerns in the world and many govements set MRLs in foods (Table 11-9).

The main adverse effects from QNs include : ① Gastrointestinal reactions. Symptoms include nausea, vomiting, loss of appetite, abdominal pain and diarrhea. The incidence of adverse reaction caused by gatifloxacin, moxifloxacin and trovafloxacin was 6% to 10%. ② Central nervous system reactions. The main symptoms were headache, dizziness, and poor sleep. It can also result in psychiatric symptoms, confusion, hallucinations and epileptic seizures. The incidence of dizziness caused by gatifloxacin, gemifloxacin, levofloxacin and moxifloxacin was 1% to 5%. ③ Skin allergy reactions. The allergy from quinolones varied a lot and the main symptoms are erythema, urticaria, rash and itching. It was reported that the incidence of skin rashes caused by gatifloxacin and moxifloxacin was <1%, while it was up to 4.8% for gemifloxacin. ④ Liver toxicity.

Trovafloxacin has serious liver toxicity or even can cause death. Although the incidence is low, its application has been strictly limited. Sparfloxacin and grepafloxacin can cause increase of liver transaminase, with an incidence rate of 2%-16%. ⑤ Toxic effects of cartilage. Ciprofloxacin or norfloxacin was reported in a higher incidence of cartilage toxicity and tendon disorders (tendinitis, tendon rupture, etc.).

Table 11-9　MRLs of quinolones drugs

Drugs	Animal	Tissue	MRL/(ng/g)	
			EU	CAM
DAN	cattle, chicken	muscle	200	200
		fat	100	100[a]
		liver, kidney	400	400
		milk	100	200
		muscle	100	30
	pig	skin + fat	50	50
		skin + fat	—	1,000
		liver	200	200
		kidney	200	200
DIF	cattle, pig	muscle	400	400
		fat, skin+fat	100	100
		liver	1,400	1,400(800[b])
		kidney	800	800
	chicken, turkey	muscle	300	300
		skin+fat	400	400
		liver	1,900	1,900
		kidney	600	600
ENR+CIP	cattle, sheep	muscle, fat	100	100
		liver	300	300
		kidney	200	200
		milk	100	100
	pig, poultry, rabbit	muscle, skin + fat, fat	100	100
		Liver	200	200
		kidney	300	300
FLU	cattle, sheep, pig	muscle	200	500
		fat, skin+fat	300	1,000
		liver	500	500
		kidney	1,500	3,000
		milk	50	50
	chicken, turkey	muscle	400	500
		skin +fat	250	1,000
		liver	800	500
		kidney	1,000	3,000
	salmon	muscle + skin	600	500[d]

Continued

Drugs	Animal	Tissue	MRL/(ng/g)	
			EU	CAM
MAR	cattle,pig	muscle,live,kidney	150	—
		fat	50	
		milk	75	
OA	cattle, pig, chicken	muscle	100	100
		fat, skin + fat	50	50
		liver, kidney	150	150
		egg	100	
	fin fish	muscle + skin^c	300	300^d
SAR	chicken	muscle	—	10
		skin+fat	10	—
		fat	—	20
		liver	100	80
		kidney	—	80
	turkey	Muscle	—	10
		fat	—	20
		liver, kidney	—	80
	salmon	iver,kidney muscle	30	30^d

CMA: China and America; a: only for cow, sheep and goat; b: only for China; c: the total mass from two tissues; d: exclude America

11.6.2　Immunoassay for QNs

Liu[43] conjugated enrofloxacin (Enro) with carrier protein with carbodiimide method and activated ester method. The carrier protein was modified with ethanediamine to increase the free amino group of the molecule. After the identification of ultraviolet spectrum, the coupling ratio of hapten and carrier protein was up to 75 : 1. An indirect competitive ELISA was applied for the determination of antibody titers. The antibody titer was 1 : 250,000. Under the optimized conditions, detection limit of this proposed ELISA was 1ng/mL for Enro with linear range from 1 ng/mL to 1,000ng/mL. Lin[44] coupled enrofloxacin (Enro) to carrier protein with modified N-hydroxysuccinimide active ester method. The prepared immune antigens of Enro-BSA were used to immunize mice and spleen of immuunized mouse was selected for cell fusion. Two hybridoma which secreted anti-enrofloxacin monoclonal antibody, 3A8 and 5D7 were selected for preparation of ascites. The ascites were purifed with octanoic acid-ammonium sulphate precipitation method were applied for purification of ascites and sub-class identification showed that these anibody belong to IgG1 type. The antibody from 3A8 hydrima recognizes enrofloxacin and ciprofloxacin with cross-reaction to ciprofloxacin 102.3%. The optimized indirect competitive ELISA has good sensitivity with minimum detection limit of 1.56 ng/mL and 0.39 ng/mL. Li[45] also used N-hydroxysuccinimide active ester to synthesize two artificial antigens of Enro-BSA and Enro-OVA. Enro-BSA as antigens was to immune Balb/C mice and the positive mice spleen cells were fused with SP/0 cells to select hybridoma which can produce anti-Enro monoclonal antibodies. Two cell lines were selected and the resulted antibody was identified as IgG2b type. The two hybridoma cells were injected into mice for preparation of ascites. Indirect competitive ELISA was established for Enro residual detection with linear detection range of 0. 3-10,000ng/mL and limit of detection (LOD) 0. 3 ng/mL. The IC_{50} value of Enro was detected as 24. 5 ng/mL. The cross-reactivity rate of anti-Enro monoclonal antibody with ciprofloxacin, ofloxacin, gentamicin, potassium penicillin, sulfadiazine and tetracycline were less than 0. 51%.

Wang[46] connected norfloxacin (NFLX) with carrier protein BSA (bovine serum albumin) and HSA (human serum albumin) by carbodiimide (EDC) method, respectively. By UV spectrophotometry, the two antigens were identified with the coupling ratio of 60 : 1 (NFLX-BSA) and 60 : 1 (NFLX-HSA). The two artificial antigens were immunized mice to obtain antibodies against norfloxacin. The positive spleen cell was fused with myeloma cells to obtain cell lines stable secreting specific anti-NFLX monoclonal antibody. A hybridoma was obtained, named 3C2. Based on the monoclonal antibody from 3C2 hybridoma, an indirect competition ELISA was established to detect NFLX residue. Optimum detection range was in the range of 10 ug/mL-10 ng/mL, and detection limit was 10 ng/mL. Zhan[47] synthesized two NFLX artificial antigens NFLX-BSA and NFLX-OVA with carbodiimide method and mixed anhydride method. The coupling ratios of NFLX-BSA and NFLX-OVA were about 9. 7 : 1 and 8. 2 : 1, respectively. Mice were immunized with NFLX-BSA and a monoclonal antibody against norfloxacin was screened, named as 3E9. The hybridoma in the logarithmic phase was injected into the pretreated peritoneal cavity of mice to prepare ascites. The collected ascites was purified by octanoic acid-ammonium sulphate and stored at−20 ℃ for use. The titer and work concentration of monoclonal anti-NFLX were measured by indirect ELISA. The results showed that IC_{50} value of NFLX was 2. 51 μg/mL and the detection limit was 0. 12 μg/mL. There are no cross-reaction with several other fluoroquinolones, such as enrofloxacin sodium, ciprofloxacin hydrochloride, diflucloxacillin hydrochloride, sarafloxacin hydrochloride, pefloxacin mesylate and danofloxacin mesylate.

Liu[48] coupled lomefloxacin hapten with carrier protein bovine serum albumin and ovalbumin for the preparation of immunogen (LMF-BSA) and coated antigen LMF-OVA) with carbodiimide method and mixed anhydride method. The prepared LMF-BSA immunized New Zealand white rabbits. The prepared antiserum was detected by indirect competitive ELISA on titer and specificity, with the titer of lomefloxacin antiserum of above 1 : 640,000 and good specificity. By optimizing conditions, the ELISA method of lomefloxacin has linear detection range of 0. 05-5 μg/mL with detection limit 0. 02 μg/mL. Recovery in milk samples was from 85% to 120%.

Du[49] coupled pefloxacin (PEF) with bovine serum albumin (BSA) and ovalbumin (OVA) by carbodiimide method and mixed anhydride. The conjugation ratios of pefloxacin and carrier protein were 11 : 1 and 12 : 1, respectively. Five hybridomas (2D7, 2E10, 2G6, 2E5 and 1D7) which can stably secret monoclonal antibody against PEF were obtained. The monoclonal antibody produced from 2D7 hybridoma was selected for the establishment of indirect competitive ELISA due to its high affinity and specificity to PEF. A standard inhibition curve of ELISA was established with detection range 0.38-2,000 ng/mL and the minimum detection limit of 0.38 ng/mL. Chicken muscle and liver tissue were successfully analyzed with this ELISA.

Pan[50] conjugated ofloxacin (OFL) and difloxacin (DIF) with two carrier proteins for synthesis of four artificial antigens using carbodiimide method. The immunogens were used to immunize rabbits to prepare antisera of OFL and DIF. The indirect competitive ELISA based on anti-OFL antibody was established with IC_{50} value of 180.32 ng/mL. Similarly, the IC_{50} value of 119.2 ng/mL for DIF was determined based on indirect competitive ELISA. Yu[51] also coupled ofloxacin with bovine serum albumin by carbodiimide method, and immunized BalB/c mice with these immunogens. With hybridoma technology, seven cell strains (1E8, 2A3, 2F10, 2H3, 3D8, 3G7 and 4F4) stable secreting monoclonal antibody against OFL were obtained. The ascites from 4F4 hybridoma was prepared and an indirect competitive ELISA for OFL detection was established based on this antibody. Using the optimized experimental conditions, the minimum detection limit was 1.43 ng/mL of OFL, with linear detection range of 1.43-4,000 ng/mL. Spiked tests in chicken muscle and liver tissues showed that average recoveries of OFL were 87.0%-107.9%.

Li[52] prepared immunogens with N-hydroxysuccinimide method of enrofloxacin (ENRO), norfloxacin (NFLX) and danofloxacin (DFLX). The antisera for each of these QNs were prepared by immunization on New Zealand white rabbits. Then using indirect competitive ELISA, cross-reactions were performed with several other QNs (enrofloxacin, sarafloxacin, norfloxacin, danofloxacin, ciprofloxacin, lomefloxacin, ofloxacin, pefloxacin) respectively. The results showed that a class-specific antibody was obtained with the IC_{50} value of DFLX 20.02 ng/mL, 88.22 ng/mL of ENRO, 58.16 ng/mL of sarafloxacin, 18.01 ng/mL of ofloxacin, 35.26 ng/mL of NFLX and 69.34 ng/mL of lomefloxacin.

Lu[53] modified carrier protein with ethylenediamine first and then conjugated it with haptens including danofloxacin, pefloxacin and lomefloxacin with arbodiimide method, N-hydroxysuccinimide method and mixed anhydride method. New Zealand white rabbits were immunized to obtain polyclone antibody. The results showed that a specific anti-danofloxacin antibody was obtained with IC_{50} value of 2 ng/mL. Cross-reactivity of 20.6% and 21.7% for fleroxacin and pefloxacin respectively was found and there was no recognition with other fluoroquinolones (cross-reactivity rate less than

0.5%). The IC_{50} value was 6.7 ng/mL for antisera from pefloxacin-BSA and the coss-reactivity rates for fleroxacin, ofloxacin, enrofloxacin were 116%, 10% and 88%. A high specific antibody for lomefloxacin was achieved with IC_{50} value of 0.35 ng/mL and lower coss-reactivity rates for fleroxacin, norfloxacin (8.75% and 17.5%). Three indirect competitive ELISA were developed based on these antibodies and they are sucessufully used to analyze chicken liver and milk samples.

11.7　Sulfonanlides

11.7.1　Introduction

Sulfonanlides drugs (SAs) are the general name for a class of drugs which have sulfanilamide structure. They are chemotherapy drugs used for the prevention and treatment of bacterial infectious diseases. Since the discovery of good inhibitory effect on *Streptococcus* and *Staphylococcus* of prontosil in 1932, more Sas have been developed. They manisfest significant effect on *hemolytic streptococcus* and other bacterial infections. Some typical SAs are listed in Table 11-10.

Table 11-10　The information on some sulfonamides

Struture	Drug name	Chemical name
	Sulfadimidine (SM₂)	N-(4,6-dimethyl-2-pyrimidinyl)-4-aminobenzen e sulfonamide
	Sulfametoxydiazine (SMD)	N-(5-methoxyl-2- pyrimidinyl)-4- aminobenzene sulfonamide
	Sulfathiazole (ST)	N-2- thiazole -4- aminobenzene sulfonamide
	Sulfaguanidine (SG)	N-guanyl -4- aminobenzene sulfonamide
	phthalylsulfathiazol e(PST)	2-[[4-[(2- thiazole amino) sulfuryl] phenyl]- amido] carbonyl] benzoic acid
	sulfadimethoxine (SDM)	N-2,6-dimethoxyl -4-pyrimidinyl -4- aminobenzene sulfonamide
	Sulfisoxazole (SIZ)	N-(3,4- dimethyl -5-isoimidazolyl)-4- aminobenzene sulfonamide

11.7.1.1　Categorization, structure and property

Derivatives of sulfanilamide stem from parent sulfanilamide with different substituents in N_1 and N_4 (the nitrogen atom at sulfonamido group and para-amino), most clinical drugs most are substituents of N_1 with R_2 mostly heterocyclic formats(Figure 11-11)[54].

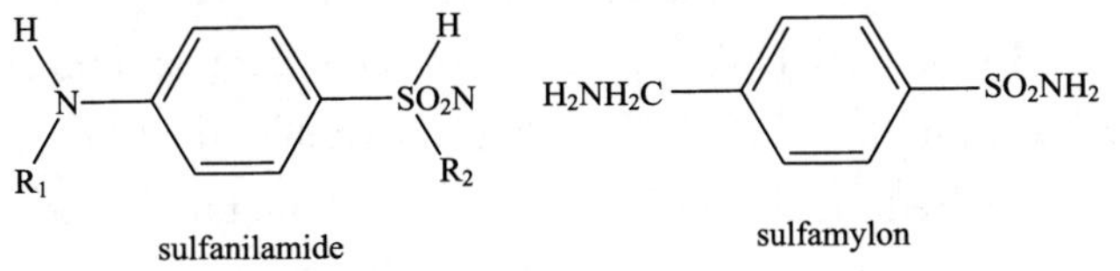

sulfanilamide　　　　　　　　sulfamylon

Figure 11-11　Parent strucuture of SAs

SAs are generally white or light yellow crystalline powder with odollrless or basically tasteless. The molecular weight of SAs is 170 to 300. SAs are slightly soluble in water, ethanol and acetone, but almost insoluble in chloroform or ether. Because of the primary amine and the sulfonyl containing in SAs, they appear amphotericity and soluble in acid or alkali solution. Generally, they are quite stable, and can be stored for several years if stored properly.

11.7.1.2　Mechanism of toxicity

SAs inhibit (bacteriostatic) gram-positive and gram-negative bacteria, Nocardia, Chlamydia trachomatis and some protozoa. Certain microbes require p-aminobenzoic acid (PABA) in order to synthesize dihydrofolic acid which is required to produce purines and ultimately nucleic acids. Sulfonamides, chemical analogues of PABA, are competitive inhibitors of dihydropteroate synthetase. Sulfonamides, therefore, are reversible inhibitors of folic acid synthesis and bacterostatic not bacteriocidal. It was determined that the length and size of sulfanilamide with N-substituent (Figure 11-12) were almost the same as that of para-amino benzoic acid ions[55].

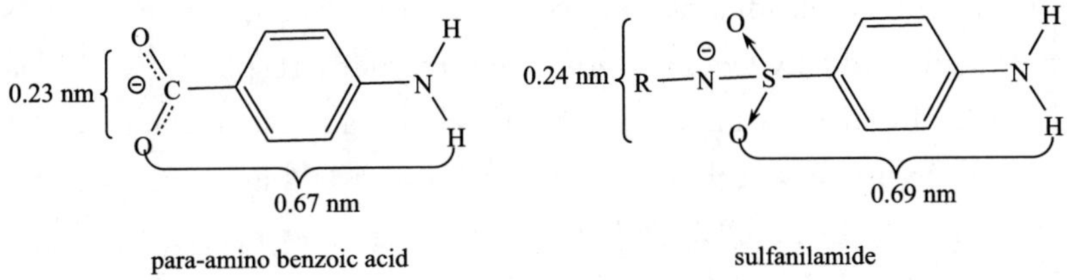

para-amino benzoic acid　　　　　　　sulfanilamide

Figure 11-12　Chemical structure of para-amino benzoic acid and sulfanilamide

11.7.1.3 Metabolism and residues

Soluble SAs can be quickly absorbed by oral intake and distributed through the body and the drug concentrations in plasma reach the highest level 2-3 h later. The concentration of SAs distributed in body fluids accounts for 50%-80% of concentrations in blood. The concentration of SAs in stomach, kidney, mucosa and liver, are high while the concentration of SAs is only half of blood in other organs and muscles. The concentration in tears or breast milk is 5%-15% of that in the blood. The distribution in bone or fat tissue is very limited. There are three kinds of metabolic pathways in vivo by acetylation, hydroxylation and combination. Due to the long lasting period and metabolic process for SAs in the body, SAs are likely to accumulate in the body after intake. It is harmfal to human health if more SAs are accumulated in the body. What is more, the long-standing of SAs in human body can cause resistance of many bacteria. Therefore, the United Nations Codex Alimentarius Commission (CAC) and many countries regulated that the total amount of SAs and other individual SAs of sulfamethazine in food and feed should not exceed 0. 1 mg/kg. SAs residues in food in Japan is not known. In China, the total SAs is less than 0. 3 mg/kg. The limit on sulfamethazine in milk is not to exceed 0. 025 mg/kg[56].

11. 7. 2 Immunoassay for SAs

Liu[57] linked sulfadiazine (SD) with human serum albumin (HSA) by diazotization. The prepared antigens SD-HAS was immunized on mice. The positive mouse spleen was fused with myeloma cell to obtain hybridoma secreting monoclonal antibody. Liu coupled SD with ovalbumin (OVA) to prepare coating antigen and prepared horseradish peroxidase (HRP) labelled SD (SD-HRP) with sodium iodide method. Based on the molecules above, a direct competitive ELISA method was established. It showed good specificity to SD, with IC_{50} value of 41. 7 ng/mL and detection limits of 5-20 μg/kg in pork, chicken, chicken liver, eggs and milk samples. Xu[58] prepared conjugates of sulfadiazine (SD)-bovine serum albumin (BSA) and SD-thyroglobulin (Tg) using carbodiimide method, glutaraldehyde and diazotization method, respectively. Polyclonal antibody (PcAb) was obtained by immunization of rabbits and monoclonal antibody (mAb) was prepared with hybridoma techiques. The PcAb showed high affinity to SD (constant of affinity up to 2. 9 $\times$ 10^{9}mol/L) with low cross-reaction to sulfamethazine (less than 0. 2%) and sulfaquinoxalinum (less than 0. 01%). Also mAb against sulfadiazine were obtained from four hybridoma (SD_2D_5, SD_3E_8, SD_3E_{12} and SD_4D_{11}). The cross-reaction rates with sulfamethazine was <0. 1% and less than 0. 01% for sulfaquinoxalinum. The mAb from SD_2D_5 was used to establish an indirect competitive ELISA with sensitivity of SD detection 0. 023 ng/mL.

Chen[59] coupled sulfadoxine-methoxy-pyrimidine (SMD) with the carrier protein bovine serum albumin (BSA) using glutaraldehyde method. The conjugate of SMD-BSA was used as immunogen. Similarly, coating antigen SMD-OVA (ovalbumin) was prepared. Antisera were prepared from immunized rabbits. The double agar diffusion test showed that this PcAb specificly recognized SMD. An indirect competitive ELISA was developed based on this PcAb with detection limits of 63 $\mu g/L$. The spiked milk samples of SMD were analyzed with recovery 94.7%.

Ying[60] coupled sulfamethazine (SM_2) with carrier protein by diazotization method for preparation of synthetic antigens SM_2-BSA and SM_2-HSA. New Zealand white rabbits were immunized by there conjugates. The antiserum was showed to be specific for SM_2. There is no cross-reaction with other SAs except sulfadiazine (11 %). The indirect competitive ELISA for SM_2 was established. The standard curve of indirect competitive ELISA indicats that the minimum detection limits is 1.89 $\mu g/L$ with a favourable linear detection range of 10-200 $\mu g/L$.

Liu[61] prepared the conjugates of sulfaquinoxaline (SQ) and carrier protein with carbodiimide method. Polyclonal antibody (PcAb) was prepared by immunization of rabbits and monoclonal antibody (mAb) was also obtained from hybridoma. An indirect competitive enzyme-linked immuno assay (ELISA) was developed using PcAb to detect the residues of sulfaquinoxaline (SQ) in food. The detection range of standard curve is 0.3-100 ng/mL. The sensitivity of the assay is 0.2 ng/mL. The recoveries tests of SQ in milk, honey, chicken and eggs is 93.1%-107.0%, 69.0%-104.0%, 40.0%-73.6%, and 50.0%-84.0%, respectively.

Yang[62] parepared antigen by conjugation of sulfa-5-methoxy-pyrimidine (SMD) and carrier protein. Six cell strains steadily secreting antibody of anti-SMD (B_6, D_{12}, E_2, H_5, G_8, G_{11} and H_5) were obtained using hybridoma techques. The ascites were prepared by injection of H_5 hybridoma into the abdominal cavity of mice. Under the optimized conditons, an ELISA was established to monitoring the residues of SMD in chicken. The storage of antibody in glycerol was been evaluated and the results showed that the antibody still kept high activity after six month storage.

Shen[63] prepared monoclonal antibodies for sulfamethoxazole (SMX) by hybridoma technology. The antibody subtypes was IgG_1 with molecular weight of 62,000 dalton. The affinity constant of antibody was 2.3 $\times$ 10^8 mol/L with SMX and no cross-reaction with other SAs. The antibodies showed good tolerance to three organic solvents (methanol, acetonitrile and acetone). The determination of free sulfamethoxazole was performed by ELISA. The established standard curve of ELISA was in linear range of 10pg/mL-10 ng/mL and the IC_{50} value of this ELISA was 0.82 ng/mL. Shen prepared complete antigen for sulfamethoxazole and immunized it to rabbits. Three formats of ELISA including direct format, indirect format and double-antibody sandwich were tested to develop test kit for sulfamethoxazole. Finally, double-antibody sandwich format was chosen with linear detection range of 10-100 ng/mL and sensitivity 20. 91 ng/mL. Based on the conjugates of sulfadimethoxine (SDM) and carrier protein, Shen prepared polyclonal antibodies against SDM, which showed specific to SDM. The antibody was linked with thermosensitive hydrogel. The antigen labelled with fluorescence competitively bind to antibody in hydrogel with free sulfadimethoxine. The detection range is 1-100 ng/mL wih IC_{50} value of 0. 64 μg/mL. Similarly, flourence detection by immunoassay was carried out for sulfonamide (SN). The detection range is 1 ng/mL-100 μg/mL with IC_{50} value of 60. 45 ng/mL.

11.8　Tetracyclines

11.8.1　Introduction

11.8.1.1　Classification, structure and property

Tetracyclines produced by the *Actinomyces*, are hydrogenation naphthacene derivatives in chemical structure, so they are also called tetracycline (TCs). TCs nucleus consist of 4 rings of A, B, C and D. The main functional groups include dimethyl amino in position of C_4, amide in C_2-site and phenolic hydroxyl in C_{10}-site. TCs contain conjugated system formed by benzene, keto and enol and there are two chromophore strucutre (A ring and BCD ring) withe strong UV absorption and fluorescence near visible region (350 nm). Figure 11-13 and Table 11-11 list the chemical structure and properties for some TCs[64].

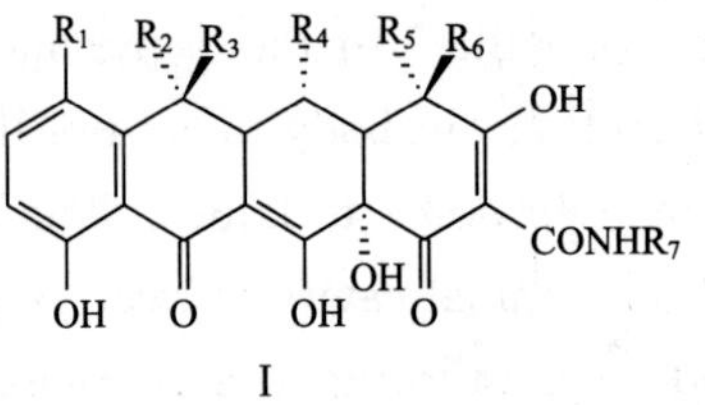

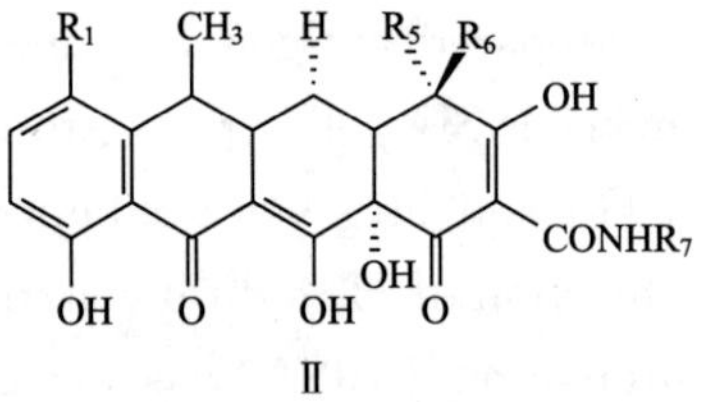

Figure 11-13　Chemical structure of TCs

Table 11-11　Physico-chemical properties of TCs

Compounds	Structure	R_1	R_2	R_3	R_4	R_5	R_6	R_7
Tetracycline(TC)	I	H	CH_3	OH	H	$N(CH_3)_2$	H	H
Oxytetracyclin(OTC)	I	H	CH_3	OH	OH	$N(CH_3)_2$	H	H
Chlortetracycline(CTC)	I	Cl	CH_3	OH	H	$N(CH_3)_2$	H	H
Doxycycline(DC)	I	H	CH_3	H	OH	$N(CH_3)_2$	H	H
Minocycline(MINO)	I	$N(CH_3)_2$	H	H	H	$N(CH_3)_2$	H	H
Metacycline(MTC)	I			CH_2	OH	$N(CH_3)_2$	H	H

Continued

Compounds	Structure	R_1	R_2	R_3	R_4	R_5	R_6	R_7
Dimethylchlortetracycline(DMCTC)	I	H	H	OH	H	$N(CH_3)_2$	H	H
pyrrolidinomethyltetracycline(PRMTC)	I	Cl	CH_3	OH	H	$N(CH_3)_2$	H	*
epitetracycline(ETC)	I	H	CH_3	OH	H	H	$N(CH_3)_2$	H
Anhydrotetraycline(ATC)	II	H	—	—	—	$N(CH_3)_2$	H	H
Epianhydrotetraycline(EATC)	II	H	—	—	—	H	$N(CH_3)_2$	H

* Represent [structure diagram: pyrrolidine ring] $N-\overset{H_2}{C}-$

TCs have similar physical and chemical properties. They are yellow crystalline powder, bitter taste and insoluble in water. Due to the presence of phenolic hydroxyl, enol and dimethyl amino in molecules, they are soluble in acid or alkaline solution. Hydrochloride salts of TCs are used widely in clinical treatment, which show good wa ter solubility and stability. TCs are relatively stable in weak acid solution, and are prone to degrade in strong acidic solution (pH <2), neutral or alkaline solution (pH> 7). TCs are stable in dry state, but easy to absorb water, and should be stored away from light.

11.8.1.2　Toxity mechanism

TCs are broad-spectrum antibiotics against Gram-positive and-negative bacteria and rickettsia. Their inhibitory effect mainly reflects as the binding of the end of 30S ribosomal subunit, and thus interferes with bacterial protein synthesis. Typical compound of TCs include tetracycline (TC), chlortetracycline (CTC), oxytetracycline (OTC), demeclocycline, doxycycline, methacycline, minocycline and so on. The C = O on C-2 site is necessary group for active center of tetracycline antibiotics. The changes of C-2 side chain can significantly increase the solubility of tetracycline in water. Changes of the C-6, C-7 can gain derivatives with stronger activity, such as doxycycline, methacycline, dimethyl tetracycline, which show stronger antibacterial effects than tetracycline, and are also effective against resistant bacteria[65].

11.8.1.3　Metabolism and residues

TCs have a good oral absorption, and are widely distributed in tissues. TCs are easily deposited in bones and teeth, leading to a lasting stain teeth ("tetracycline"). They can also enrich in the liver, causing liver damage. Their structures contain multiple active groups, which strongly bind with protein. TCs are mainly excreted renal and renal dysfunction lead to prolonged elimination period of TCs. Table 11-12 showed some MRLs in various tissues.

Table 11-12   MRLs for tetracyclines(mg/kg)

Drugs	Marked residual	Animal kinds	MALs	Tissue	Note
Tetracyclines	Prototype of drugs	All kinds of food animals	0. 6	kidney	The total residue level of tetracyclines can not exceed the regulation
			0. 3	liver	
			0. 2	egg	
			0. 1	muscle	
			0. 1	milk	

11. 8. 2   Immunoassay for TCs

Wu[66] used glutaraldehyde cross-linking method and mixed anhydride method to prepare complete antigen of tetracycline-BSA, respectively. The coupled tetracycline-BSA was used to immunize rabbit to obtain polyclonal antibody. The antibody was identified by ELISA with titers up to 1 : 320,000 and the IC_{50} value of 92. 6 μg/mL. The lowest detection limit of tetracycline was 6. 25 μg/mL. Cross-reaction rate for chlortetracycline was 78%.

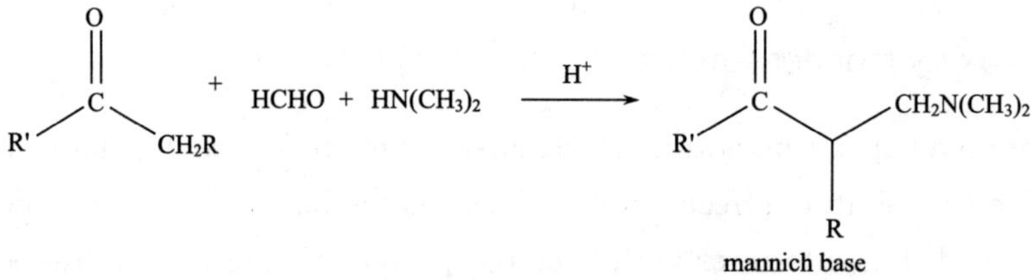

Figure 11-14   Hapten synthesis with mannich reaction

Gao[67] synthesized tetracycline (Tc) with carrier protein BSA (bovine serum albumin) and conjugates of TC with HSA (human serum albumin) by using Mannich reaction (Figure11-14). The prepared artificial antigens were identified by UV spectrophotometry and SDS-polyacrylamide gel electrophoresis (SDS-PAGE) with coupling ratio t of 23 : 1 (Tc-BSA) and 22 : 1 (Tc-HSA). A monoclonal antibody for tetracyclines was successfully obtained, which showed good specificity for Tc. An indirect ELISA was established with detection range 1-1,000 μg/mL and the minimum detection limit 1 μg/mL.

Xing[68] prepared complete antigen for oxytetaroycline (OTC) with carbodiimide method and glutaraldehyde method. The resulting conjugates of OTC-BSA was immunized mice. The titer of antisera reached the highest level 90 days later with the highest titer of 1 : 20,000. An indirect ELISA was established and the detection range of oxytetracycline was 5-200 ng/mL.

11.9 Anthelmintic (worm) agents

11.9.1 Introduction

Anthelmintic is a class of drugs which can eliminate parasitic worms in animals, so they are also called anthelminthin (worm) agents.

Worm is multi-cell parasite, and they are very different in their size, shape and physical structure. Worms can be divided into nematodes, flukes and tapeworms. Excepting of a minority of them, their worm eggs generally leave the host animals, and further develop in the soil or other intermediate hosts. According the type of targeted worms, vermifuges can be divided into antinematodal agent, anti-fluke drugs and antitapeworm drugs. The common anthelmintic additives in feed are mainly targeted gastrointestinal nematodes drugs[69].

11.9.1.1 Thiabendazole

It belongs to benzimidazole anthelmintic, with chemical name of 2-(4-thiazolyl) benzimidazole. As a stable white or yellow powder or crystalline powder, it is slightly bitter in taste. Thiabendazole is a broad-spectrum, high efficiency and low toxicity anthelmintic, which has efficient insecticide effect on a variety of gastrointestinal nematodes. Also, it is effective in control of lung nematodes and spear-shaped dual-chamber trematodes. Thiabendazole can be rapidly absorbed by animal digestive tract and widely distributed in body tissue, so it is good ate inhibition and killing larvae migrating in tissue and adults parasites. The mode of action of thiabendazole is inhibiting activity of polypide fumarate reductase of parasite.

11.9.1.2 Fenbendazole

It is benzimidazole anthelmintic, with the chemical name of 5-(phenylthio)-2-benzimidazole carbamate. It is a colourless powder, insoluble in water, and only soluble in dimethyl sub-alum. It is a broad spectrum benzimidazole anthelmintic used against gastrointestinal parasites including roundworms, hookworms, whipworms, the taenia species of tapeworms, pinworms, aelurostrongylus, paragonimiasis, strongyles and strongyloides and can be administered to sheep, cattle, horses, fish, dogs, cats, rabbits and seals. Drug interactions may occur if using bromsalan flukicides such as dibromsalan and tribromsalan. Abortions in cattle and death in sheep have been reported after using these medications together.

11.9.1.3 Levamisole hydrochloride

It is an imidazo-thiazole anthelmintic, with chemical name of L-2,3,5,6-tetrahydro-

6-phenyl-imidazole (2,1-b) thiazole hydrochloride, and also known as levamisole. It is white crystalline powder, stable, soluble in water. The drug shows high efficacy, low toxicity and similar insecticide spectrum to tetra-thiophene.

11.9.1.4　Pyrantel

It is often prescribed by veterinarians to treat and prevent the occurrence of intestinal parasites in small animal pets. Pyrantel is a nicotinic receptor agonist. Like levamisole and the related pyrimidine morantel, it can elicit spastic muscle paralysis in parasitic worms due to prolonged activation of the excitatory nicotinic acetylcholine (nACh) receptors on body wall muscle. This has been studied in body wall muscle preparation of the parasitic nematode Ascaris.

11.9.1.5　Phenothiazine

It is an organic compound that occurs in various antipsychotic and antihistaminic drugs. It has the formula $S(C_6H_4)_2NH$. This yellow tricyclic compound is soluble in acetic acid, benzene, and ether. The compound is related to the thiazine-class of heterocyclic compounds. Derivatives of the parent compound find wide use as drugs. It is effective on many nematodes of cattle, sheep, horses and poultry.

11.9.1.6　Hygromycin B and amphotericin A

They are the most used drugs in control of intestinal nematodes. These two antibiotics can damage the female reproductive function roundworm, nodular worms, whip worms, cecal worms, and capillaria in chicken or pig.

11.9.1.7　Aavermectin

The avermectins belong to a series 16-membered macrocyclic lactone derivatives with potent anthelmintic and insecticidal properties. These naturally occurring compounds are generated as fermentation products by Streptomyces avermitilis, a soil actinomycete. Eight different avermectins were isolated in 4 pairs of homologue compounds, with a major (a-component) and minor (b-component) component usually in ratios of 80 : 20 to 90 : 10. Other anthelmintics derived from the avermectins include ivermectin, selamectin, doramectin and abamectin.

11.9.2　Immunoassay for anthelmintics

A competitive, indirect enzyme-linked immunosorbent assay (ELISA) for thiabendazole has been developed by Antonio[70], which was applied to the analysis of fruit juices spiked with this fungicide. The immunoassay is based on a new monoclonal antibody derived from a hapten functionalized at the nitrogen atom in the 1-position of the thiabendazole structure (Figure 11-15). The IC_{50} value and the detection limit of the ELISA for standards were 0. 2 and 0. 05 ng/mL, respectively. Fruit juices were analyzed by diluting samples in assay buffer, without extraction or cleanup. Samples were not even centrifuged or filtered to remove fruit pulp. Under these conditions, the immuno-assay was able to accurately determine thiabendazole down to 1 ng/mL in orange and grapefruit juices, down to 5 ng/mL in banana juice, and down to 20 ng/mL in apple and pear juices. Sensitivity differences of the ELISA were caused by the minimum dilution required by each juice to minimize matrix effects: 1/10 for orange and grapefruit juices, 1/50 for banana juice, and 1/100 for apple and pear juices. In an attempt to further increase the sensitivity of the immunoassay for matrices showing the strongest interfe-rences, apple and pear juices spiked with thiabendazole at low levels (1-20 ng/mL) were extracted with ethyl acetate before analysis. This simple procedure entailed a significant reduction of matrix effects, which in fact allowed us to determine accurately as low as 5 ng/mL thiabendazole in apple and pear juices. Irrespective of whether samples were analyzed by the direct dilution method or after extraction, the simplicity, sensitivity, and sample throughput this monoclonal immunoassay makes it a very convenient method for the routine monitoring of thiabendazole residues in fruit juices.

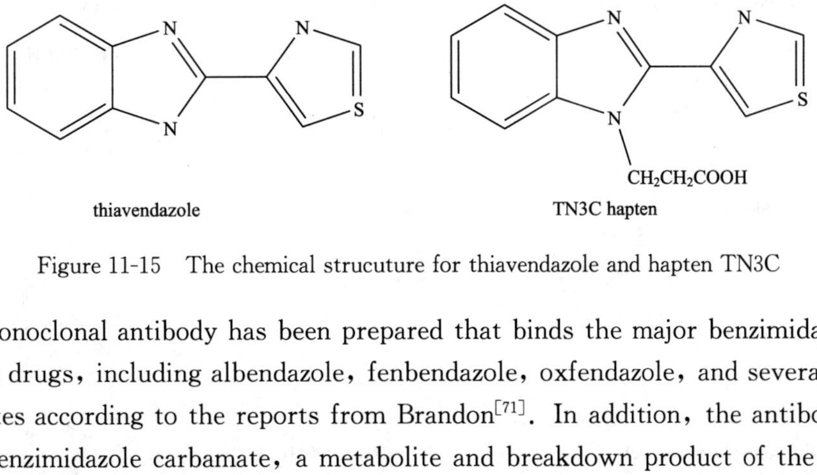

thiavendazole　　　　　　　　TN3C hapten

Figure 11-15　The chemical strucuture for thiavendazole and hapten TN3C

A monoclonal antibody has been prepared that binds the major benzimidazole ant-helmintic drugs, including albendazole, fenbendazole, oxfendazole, and several of their metabolites according to the reports from Brandon[71]. In addition, the antibody binds methyl benzimidazole carbamate, a metabolite and breakdown product of the pesticide benomyl. The antibody was elicited from mice using the novel hapten methyl 5(6)-[(carboxypentyl)-thio]-2-benzimidazolecarbamate (Table 11-13) and was used to develop an ELISA method that can detect multiple benzimidazole drug and pesticide residues at concentrations between 1 and 8 ppb. The ELISA provided the basis for quantification of drug residues in bovine liver using aqueous extraction. The sulfoxide and sulfone metab-olites of albendazole and fenbendazole were readily extractable and quantifiable by this method. ELISA of liver tissue from cows treated with fenbendazole produced excellent agreement with the results of HPLC analysis. In bovine liver samples fortified with equal amounts of benzimidazole drug and sulfoxide and sulfone metabolites, the limits of detection were 58 ppb for the albendazole group and 120 ppb for the fenbendazole compounds. This sensitivity enables rapid identification of samples requiring residue-specific quantitative analysis. Since the ELISA method employs stable nonhazardous materials and reagents, it could be performed in the field for rapid screening of meat products for undesired residues.

Table 11-13　Principal benzimidazole anthelmintics and related compounds, including hapten

Compound	R_1	R_2
methyl benzlmldazole carbamate	—NHCOOCH$_3$	H-
Albendozole	—NHCOOCH$_3$	CH$_3$CH$_2$CH$_2$S—
Albendozole Sulfoxide	—NHCOOCH$_3$	CH$_3$CH$_2$CH$_2$SO—
Albendozole Sulfone	—NHCOOCH$_3$	CH$_3$CH$_2$CH$_2$SO$_2$—
2-amino-5-propylsulfonyl bienzimidazole	—NH$_2$	CH$_3$CH$_2$CH$_2$SO$_2$—
2-amino-5-propylthio bienzimidazole	—NH$_2$	CH$_3$CH$_2$CH$_2$S—
oxibendozole	—NHCOOCH$_3$	CH$_3$CH$_2$CH$_2$O—
fenbendozole	—NHCOOCH$_3$	
Oxifendozole	—NHCOOCH$_3$	
Fenbendozole Sulfone	—NHCOOCH$_3$	
4'-Hydroxyl fenbendozole	—NHCOOCH$_3$	
Mebendazole	—NHCOOCH$_3$	
Flubendazole	—NHCOOCH$_3$	

Continued

Compound	R₁	R₂
	R_1	R_2
Thiabendazole		H–
Cambendazole		$(CH_3)_2CHOCN(OH)–$
5(6)-carboxypentylthlo-2-methyl benzimidazole carbamate (hapten, I)	$—NHCOOCH_3$	$HOOC(CH_2)_5S–$

11.10 Anticoccidial drugs

11. 10. 1 Introduction

Coccidiosis is an infection of the intestinal tract caused by a single cell parasite. All livestock species, as well as wild animals, can be infected and it is especially prevalent when animals or birds are grouped together in significant numbers. However, it can occur in less intensively managed situations, including outdoor flocks and herds. Each species of livestock has species-specific coccidia that cause infections in that species. Generally, there is no cross-infection between species. The disease is characterised by an invasion of the intestinal wall by the parasite. The parasite then undergoes several stages of growth and multiplication, during which there is damage to the mucosal and submucosal tissues. Severe haemorrhage may result and mortality in an unprotected poultry flock may be extensive. Although cattle, sheep and pigs can become infected causing significant depression of performance, it is usually in poultry species that the parasite can cause the most devastating losses[72]. For this reason, it is essential in most poultry rearing situations to use an anticoccidial agent period to prevent illness and control infections. The production of affordable, quality poultry meat owes much to the development of effective anticoccidial products used in the prevention, control and treatment of coccidiosis.

Several chemical compounds have remained to the present day and continue to provide a key role in the prevention and control of coccidiosis. A breakthrough came during the 1970s when a new class of compounds-the ionophores-was discovered and rapidly became established as the key component of coccidiosis control. The ionophores are unique in that they permit a small "leakage" of coccidia to enable the bird to develop a certain level of immunity. This allows a greater degree of protection against the parasite and is a much more efficient method of control. Resistance to ionophores develops very slowly and there is more of a tendency for increasing levels of tolerance. This means that the chicken producer is in a position to adjust the anticoccidial medication programme before there is an acute outbreak of disease[73].

11. 10. 2 Immunoassay for anticoccidial drugs

Halofuginone(Hal)is a widely used anticoccidiosis agent, which is usually used as a feed additive at the medicated concentration of 3mg/kg for the prevention and treatment of the coecidiosis in poultry. A hapten for halofuginone was developed by Wu[74]. The hydroxyl group on halofuginone was protected by N-trimethysilylimidazole (TMSI), and the amino molety readily reacted with succinic anhydride to produce the monosubstituted acid succinamide(Figure 11-16). Finally the trimethylsis moiety was removed by hydrolysis in acid condition to give a succiuic acid derivative of halofuginone. This compound was coupled to BSA and OVA by NHS ester procedure and the coupling ratios were 9. 5 : 1 and 5 : 1, respectively. Antisera from rabbits were prepared and an ELISA method for determination of halofuginone residue in chicken liver and muscle tissues was developed. The linear detection range of ELISA was 0. 1-10 ng/mL. The method detection limits in chicken liver tissue and muscle tissue were 23 ng/g and 15 ng/g, respectively. In this study, ELISA, and HPLC methods were compared for determination of halofuginone residues in chicken liver and muscle tissues and a good correlation was achieved.

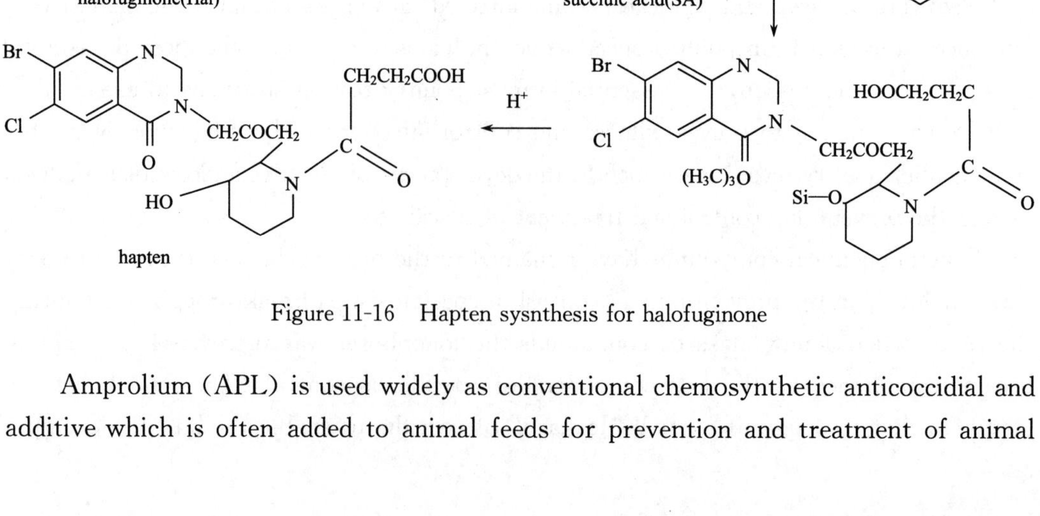

Figure 11-16   Hapten sysnthesis for halofuginone

Amprolium (APL) is used widely as conventional chemosynthetic anticoccidial and additive which is often added to animal feeds for prevention and treatment of animal deseases. Owing to its application improper or excessive, amprolium will accumulate in animal tissues. Therefore, the residues in the edible tissues of food-producing animal are paid more attention increasingly. Ma[75] coupled APL with carrier protein by glutaraldehyde method for preparation of immunogen, which was used to immunize Balb/C mice. The spleen cells and myeloma cells (SP2/0) were fused and screend by enzyme-linked immunosorbent assay (ELISA) and a hybridoma line secreting anti-APL monoclonal antibody was obtained. An indirect competitive ELISA method was established for APL determination and the cross reactivity rate between amprolium hydrochloride and thiamine was less than 5%.

Shen[76] synthesized 5 antigens of madubamycin using mixed anhydride method and activated ester method, two of which was used to immunize rabbits for the preparation of antibodies. Based on this antibody, Shen prepared specific immunoaffinity column for madubamycin with high capacity (the dynamic column capacity 3,750-7,160 ng/mL gel). This immunoaffinity chromatography column (IAC) was successfully applied in madubamycin analysis of chicken feed and chicken tissue.

11.11　Anabolic hormones

11.11.1　Introduction

Anabolic steroids, technically known as anabolic-androgen steroids (AAS) or colloquially simply "steroids", are drugs that mimic the effects of testosterone and dihydrotestosterone in the body. They increase protein synthesis within cells, which results in the buildup of cellular tissue (anabolism), especially in muscles[77]. Health risks can be produced by long-term use or excessive doses of anabolic steroids. These effects include harmful changes in cholesterol levels (increased low-density lipoprotein and decreased high-density lipoprotein), acne, high blood pressure, liver damage (mainly with oral steroids), dangerous changes in the structure of the left ventricle of the heart. Conditions pertaining to hormonal imbalances such as gynecomastia and testicular atrophy may also be caused by anabolic steroids.

11.11.1.1　Steroid hormones

Steroids are widely spread in plants and animals. In 1815, Chevreul isolated the first steroids cholesterol from human gallstones. In 1930s, estrone, estradiol, testosterone, corticosterone and other substances were gradually isolated from animal gonads, adrenal gland or urine, thus a far-reaching area of steroid chemistry appeared in biology and medicine. The role of steroid hormones is extensive, with important regulatory role on metabolism, organ development, reproduction and immune function.

Natural steroid hormones are produced by the gonads, adrenal cortex and secreted placental tissue. Due to the important value of steroid hormones in medicine and biology, industrial production of steroid hormones based on diosgenin in 1940s was achieved. Until now, more than 7,000 kinds of steroids through natural products isolation or artifical synthesis have been reported[78]. The steroid anabolic hormones (ASs) discussed here include sex hormones and adrenal cortex hormones such as androgens, estrogens, progestins and glucocorticods.

Structure of this class of drugs consist of 1,2-cyclopentane and basic nucleus of poly-hydrogen phenanthrene (Figure 11-17). Steroid nucleus are fused by the A, B, C and D rings together, A, B, C are 6-membered ring, D is 5-membered ring. Usually the presence of angular methyl is in the A/B and C/D fusion and D ring at C-17 have side chains. There often are double bonds or α, β-unsaturated ketones on rings and some are aromatic rings. The side chains are functional groups and halogen such as alkanes, hydroxyl, keto or ester

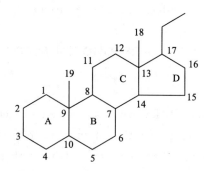

Figure 11-17   Structure of steroid hormone drugs

There are 4 main kinds of steroids with different structure and physical-chemical properties :① Glucocorticoids. They have 21 carbon atomswith $\triangle^4$-3-ketone in or $\triangle^{1,4}$-3-benone structure in A ring there is α-alcohol ketone structure and α-hydroxy at C-17 site of C ring, which is easily oxidized. B ring often contains halogen while D ring contains hydroxyl or keto at C-11 site and methyl group at C-10, C-13 sites. ② Progestins. They have 21 carbon atoms with $\triangle^4$-3-ketone in A ring. There is a ketone in C17 site and methyl group at C-10, C-13 sites. ③ Androgens. There are 19 or 18 (C-10 methyl) carbon atomswith $\triangle^4$-3-ketone in A ring and there often is a β-hydroxy esters in C-17 site. ④ Estrogen. There are 18 carbon atoms. A is benzene ring, with hydroxyl, ester or ether in C-3 site. Groups such as β-hydroxy, ketone, ester or acetylene present in C-17 site and there is no methyl at C-10 site. Some important steroids are shown in Table 11-14.

Table 11-14　Information for some typical steroids

Name	Molecular formula/wight	Melting point/℃	Chemcial constitution
Androgens			
Boldenone(BOL)	$C_{19}H_{25}O_2$/286.1933	—	
Chlorotestosterone (CTS)	$C_{19}H_{22}C_lO_2$/322.1700	—	
Dianabol (Metandienone) (DIA)	$C_{20}H_{28}O_2$/300.2089	163-167	
Fluotestin(FLO)	$C_{20}H_{28}FO_3$/336.2101	278	
Mesterolone(MES)	$C_{20}H_{22}O_2$/304.2402	—	
Methyltestosterone(MTS)	$C_{20}H_{23}O_2$/302.2246	163-167	
Nandrolone Phenylpropionate (PNT)	$C_{22}H_{24}O_3$/406.2058	93-99	
19-nortestosterone (Nandrolone) (17βNT)	$C_{19}H_{26}O_2$/274.1933	—	

Continued

Name	Molecular formula/wight	Melting point/℃	Chemcial constitution
Norethandrolone(NOE)	$C_{20}H_{24}O_2$/302.2246	130-136	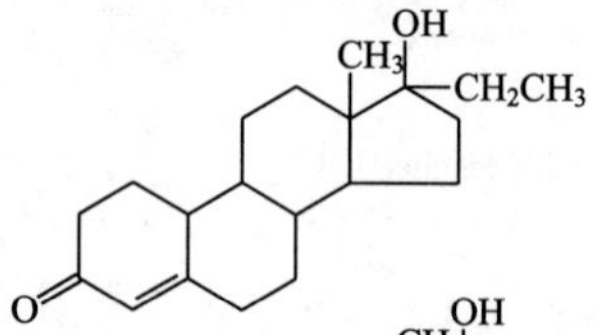
Oxymetholone(OXY)	$C_{21}H_{23}O_2$/332.2351	174-182	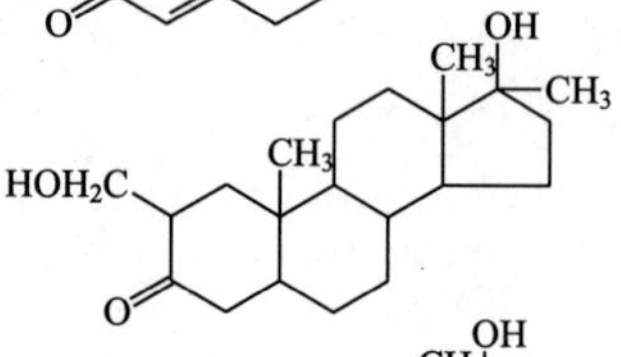
Stanozolol(STA)	$C_{23}H_{32}N_2O$/328.2545	153-156	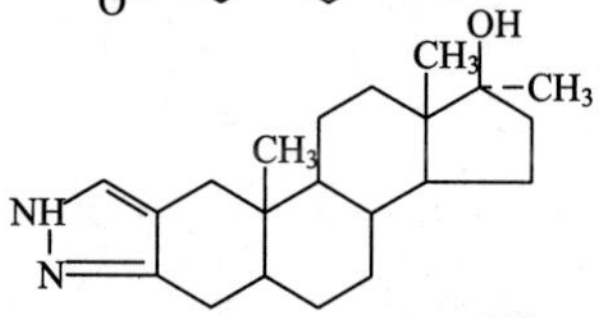
Trenbolone(TRE)	$C_{18}H_{22}O_2$/270.1620	183-186	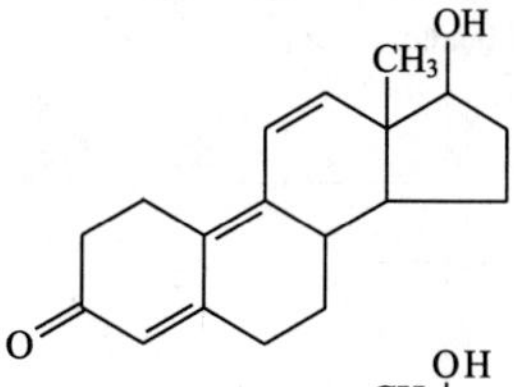
Testosterone(17βTS)	$C_{19}H_{24}O_2$/288.2089	155	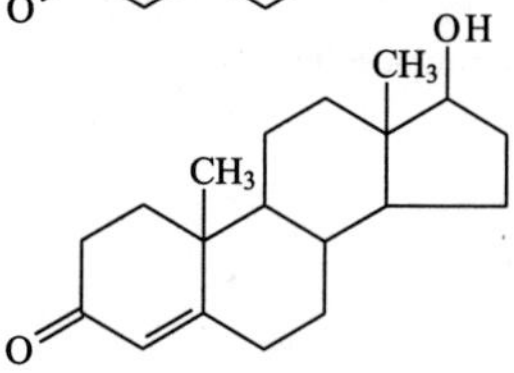
Epitestosterone (17αTS)	$C_{19}H_{24}O_2$/288.2089	—	17α—OH
Testosterone Propionate(PTS)	$C_{22}H_{32}O_3$/344.2351	118-123	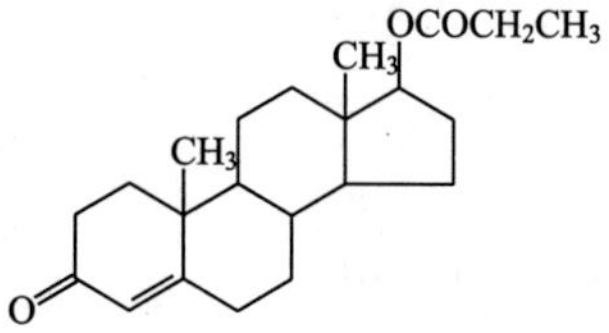

Estrogens

Estradiol(17βES)	$C_{18}H_{24}O_2$/272.1766	—	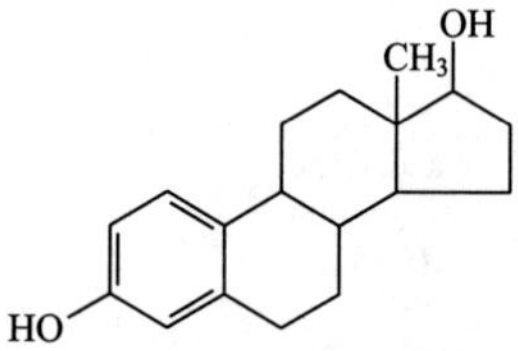
Estradiol(EST)	$C_{18}H_{24}O_3$/288.3808	282	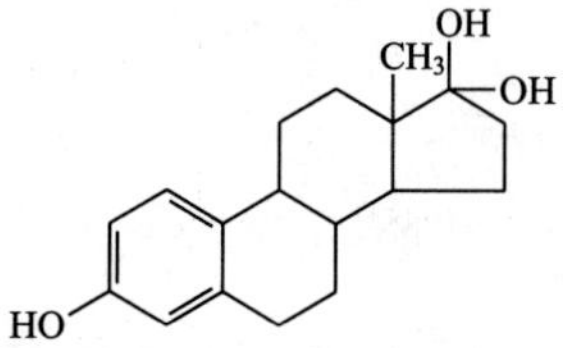

Continued

Name	Molecular formula/wight	Melting point/℃	Chemcial constitution
Ethinylestradiol(EES)	$C_{20}H_{21}O_2$/296.1776	180-186	
Estrone(ESN)	$C_{19}H_{22}O_2$/270.1620	256-262	
Estradiol Benzoate (BES)	$C_{25}II_{28}O_3$/376.2038	191 196	
Progestogens			
Chlormadinone(CHM)	$C_{23}H_{29}ClO_4$/404.1754	206-214.5	
Delmadinone(DEL)	$C_{23}H_{29}ClO_4$/404.1754	—	
Hydroxyprogesterone acetate(HPA)	$C_{22}H_{22}O_4$/372.2301	—	
Hydroxyprogesterone Caproate(HPC)	$C_{22}H_{40}O_4$/428.2927	119-122	

Continued

Name	Molecular formula/wight	Melting point/℃	Chemcial constitution
Medroxyprogesterone Acetate(MPA)	C$_{24}$H$_{34}$O$_4$/386.2457	202-208	
Melengestrol(MEL)	C$_{25}$H$_{32}$O$_4$/396.2301	—	
Megestrol(MEG)	C$_{24}$H$_{32}$O$_4$/384.2300	213-220	
Norethisterone(NOR)	C$_{20}$H$_{26}$O$_2$/298.1933	202-208	
Norgestrel(NOG)	C$_{21}$H$_{28}$O$_2$/312.2089	204-212	
Propylene algestone (PRO)	C$_{29}$H$_{36}$O$_4$/448.2614	—	

Continued

Name	Molecular formula/wight	Melting point/℃	Chemcial constitution
Progesterone(PG)	$C_{21}H_{30}O_2$/314.2246	121-137(α) 121(β)	
Glucocorticoid			
Cortisone(COR)	$C_{21}H_{28}O_2$/360.1937	217-224	
Dexamethasone (DEX)	$C_{22}H_{29}FO_5$/392.1999	262-264	
Betamethasone (BET)	$C_{22}H_{29}FO_2$/392.1999	—	16β—CH_3
Flumetasone (FLM)	$C_{22}H_{28}F_2O_5$/410.1905	—	
Hydrocortisone (HYC)	$C_{21}H_{30}O_5$/362.2093	212-222	
Methylprednisolone (MPP)	$C_{22}H_{30}O_5$/374.2093	—	

Continued

Name	Molecular formula/wight	Melting point/℃	Chemcial constitution
Prednisolone(PRE)	$C_{21}H_{28}O_5$/360.197.3	229	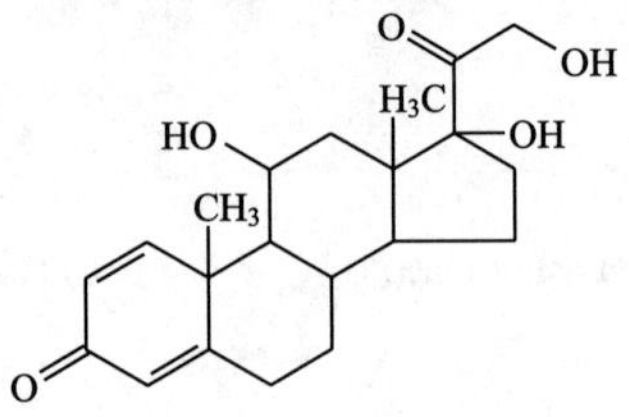
Triamcinolone (TRA)	$C_{21}H_{27}FO_5$/394.1792	264-268	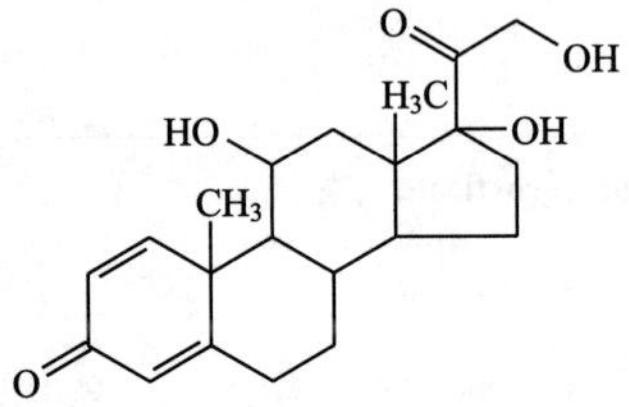

11.11.1.2　Non-steroidal anabolic hormones

Non-steroidal anabolic hormones are mainly non-steroid estrogens. Generally, compounds with the following structure may have estrogenic effects: a large bulk of rigid, inert skeleton; the distance of 8.55Å between two ends (ketone, phenolic hydroxyl or hydroxyl group) which can form hydrogen bonds[79]. Some natural or artificial non-steroidal anabolic hormones are listed in Table 11-15.

Table 11-15　Non-steroidal anabolic hormones

Name	Molecular formula/wight	Melting point/℃	Chemcial constitution
Diethylstilbestrol (DES)	$C_{18}H_{20}O_2$/268.1463	169-172	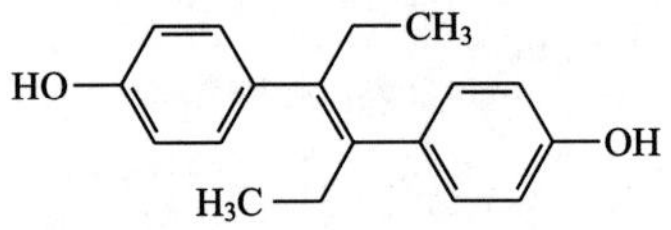
Dienestrol	$C_{18}H_{18}O_2$/266.1307	230-235	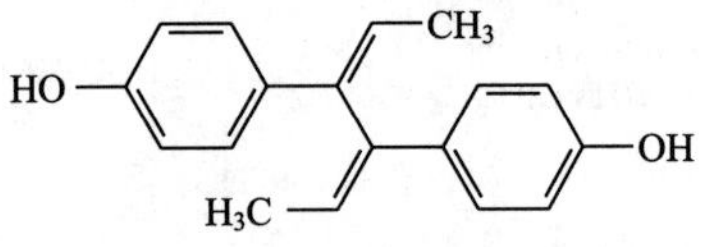
Butylenestrol	$C_{16}H_{16}O_2$/240.1150	—	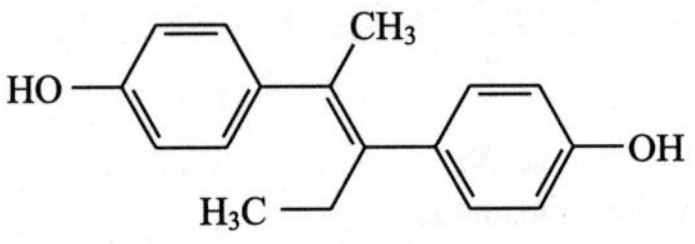
Hexestrol(HES)	$C_{18}H_{22}O_2$/270.1620	185-188	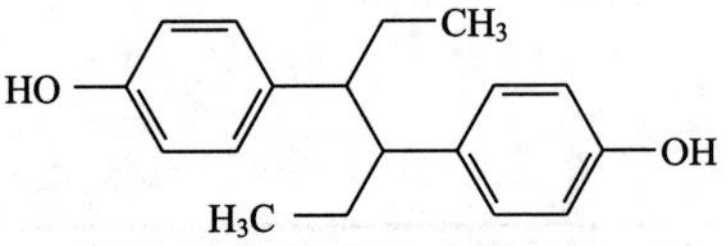

Continued

Name	Molecular formula/wight	Melting point/℃	Chemcial constitution
1- hydroxyl Diethylstilbestrol	$C_{18}H_{20}O_3$/284.1412	—	
Methoxyl Diethylstilbestrol	$C_{19}H_{22}O_2$/282.1620	—	
Taleranol	$C_{18}H_{26}O_5$/322.1780	181-185	
Zearanol	$C_{18}H_{26}O_5$/322.1780	—	7β—OH
Zearelone (ZLA)	$C_{18}H_{24}O_5$/320.1624	—	
Zearalenone (ZLE)	$C_{18}H_{22}O_5$/318.1467	—	

11.11.1.3　Pharmacology and toxicology

Receptors for steroid and thyroid hormones are located inside target cells, in the cytoplasm or nucleus, and function as ligand-dependent transcription factors. That is to say, the hormone-receptor complex binds to promoter regions of responsive genes and stimulates or sometimes inhibits transcription from those genes. Thus, the mechanism of action of steroid hormones is to modulate gene expression in target cells. By selectively affecting transcription from a battery of genes, the concentration of those respective proteins are altered, which clearly can change the phenotype of the cells. Steroids can affect levels of cholesterol. Athletes who use steroids show increased levels of LDL's (the "bad" cholesterol) and decreased levels of HDL's (the "good" cholesterol). This may also explain why anabolic steroids have been linked to cardiovascular problems as well. The liver can be affected with orally ingested steroids. This means that studies have shown the levels of liver enzymes to increase after ingesting oral steroids which is an indication of liver activity. Anabolic steroids increase the levels of estrogen in the body which can lead to female-like breast tissue in males. However, there are anti-estrogen drugs that can help reduce this risk. Likewise, females who use anabolic steroids might begin to develop male characteristics such as hirsuitism or deepening of the voice, clitoral enlargement.

Usually, the usage dose of ASs is very small and their metabolism and elimination are rapid. The metabolism products are complex with very limited parent the drug fractions. There are almost no parent drugs in urine, thus, the metabolites become markers in ASs detection. Most of ASs and its metabolites are excreted via biliary and renal. So most of drugs or its metabolites present in urine, bile or feces, which become important samples for ASs intaken check. Since 1980s, many countries and organizations have restricted or prohibit the use of ASs in food animals.

11.11.2　Immunoassay for ASs

Zeranol (ZER) is a non-steroidal estrogen agonist. It is a mycotoxin, derived from fungi in the *Fusarium* family. Zeranol is approved for use as a growth promoter in livestock, including beef cattle, in the United States. Liu[80] modified zeranol on hydroxyl group at C16 site(Figure 11-18) and the resulting ZER-16-carboxy propyl ether was coupled with the carrier protein BSA with the mixed anhydride method. New Zealand white rabbits were immunized and an indirect competitive ELISA was established with the minimum detection limit (IC_{10}) of 8.6 ng/mL.

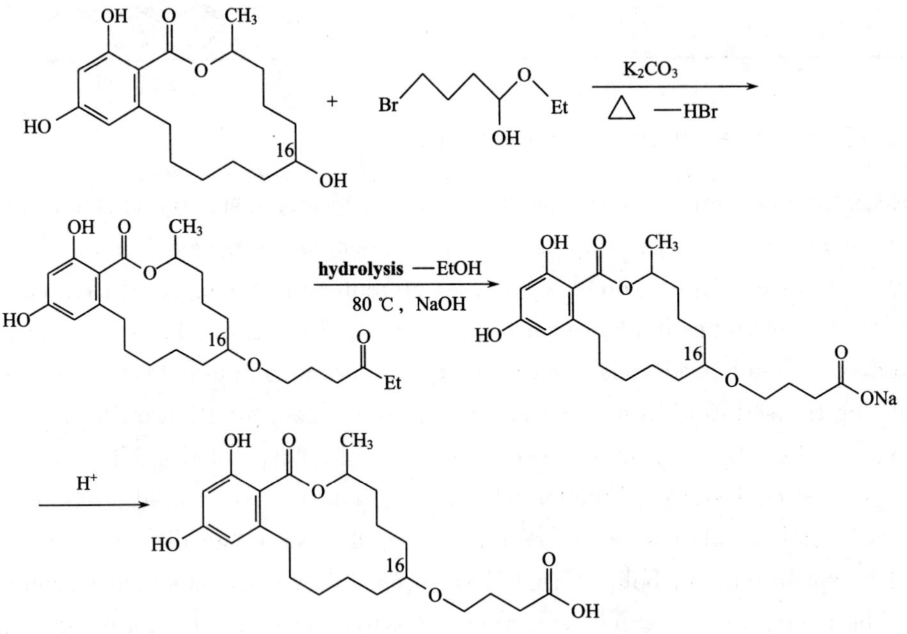

Figure 11-18　Synthesis hapten for zeranol

Wang[81] prepared hapten of zeranol (ZER) by modification of ZER with succinic anhydride (Figure 11-19), which was conjugated with BSA with mixed anhydride method. The ZER-BSA was used as immunogen to prepare monoclonal antibody. Two hybridomas producing anti-ZER monoclonal antibody were obtained by limited dilution. The characterization of antibody was performed by ELISA test. The results indicated that the subclass of antibody was IgG1 and the cross-reactivities of antibody to zeranol, β-zearalanol, α-zearalenol, β-zearalenol, zearalanone and zearalenone were 36. 25%, 52. 73%, 0. 67%, 74. 36% and 53. 70% respectively. The cross-reactivity rate with diethylstilbestrol, estradiol, testosterone, clenbuterol was less than 0. 1%. Indirect competitive ELISA method was used for detection of ZER. The IC_{50} value was 3. 0 ng/mL and calculated corresponding to 10% inhibition concentration, the method detection limit was ZER 0. 6 ng/mL.

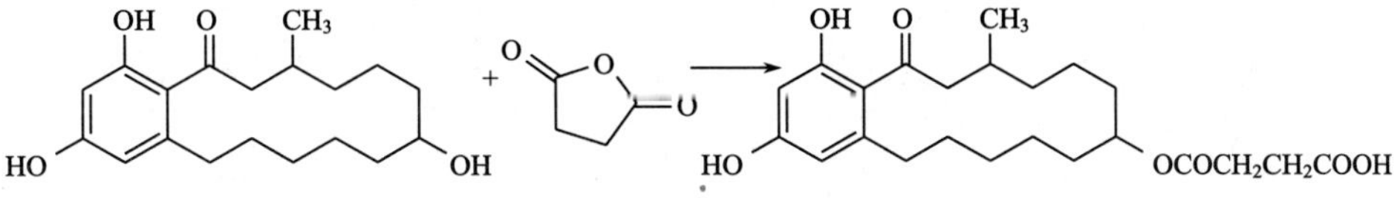

Figure 11-19　Synthesis of ZER-7-semi-succinate-BSA

Acetylgestagens are a class of synthetic steroid drugs and very important for the physiological function of humans and animals by regulating the growth and differentiation of cells. They are highly useful in oestrus regulation, growth-promoting and veterinary practice. Guo[82, 83] modified medroxyprogesterone acetate (MPA) with oximation reaction and the product was coupled with bovine serum albumin (BSA) and ovalbumin (OVA), respectively using mixed anhydride method. Rabbits were immunized with MPA-BSA. An indirect competitive ELISA method was established based on the antisera from rabbits. Under the optimized conditions, an ELISA kit was assemblied with IC_{50} of MPA 4. 95 ng/mL. The cross-reactivity rate to megestrol acetate was 4. 2% and no recognition with other analogues was found. This ELISA kit was successfully used for MPA monitoring of urine, milk and pork. Peng[84-97] synthesized and identified hapten 3-CMO-MPA and artificial antigen 3-CMO-MPA-BSA (Figure 11-20). The artificial antigen was used to immunize rabbits. An indirect competitive ELISA method was established with IC_{50} value of 1. 22 ng/g. The cross reactivity (CR) of MPA antisera with anabolic steroid analogswas carried out by this ELISA. CR of medroxyprogesterone was 96%. CRS of epitestosterone, nandrolon, 17α-methyltestosterone, testosterone 17-propionate, deoxycorticosterone, 17α-hydroxyprodesterone, prednisolone, cortisol, and dexamethasone were less than 0. 1%. CRS of 17β-estradiol 3-benzoate, progesterone (PG), and pregnenolone were less than 20%.

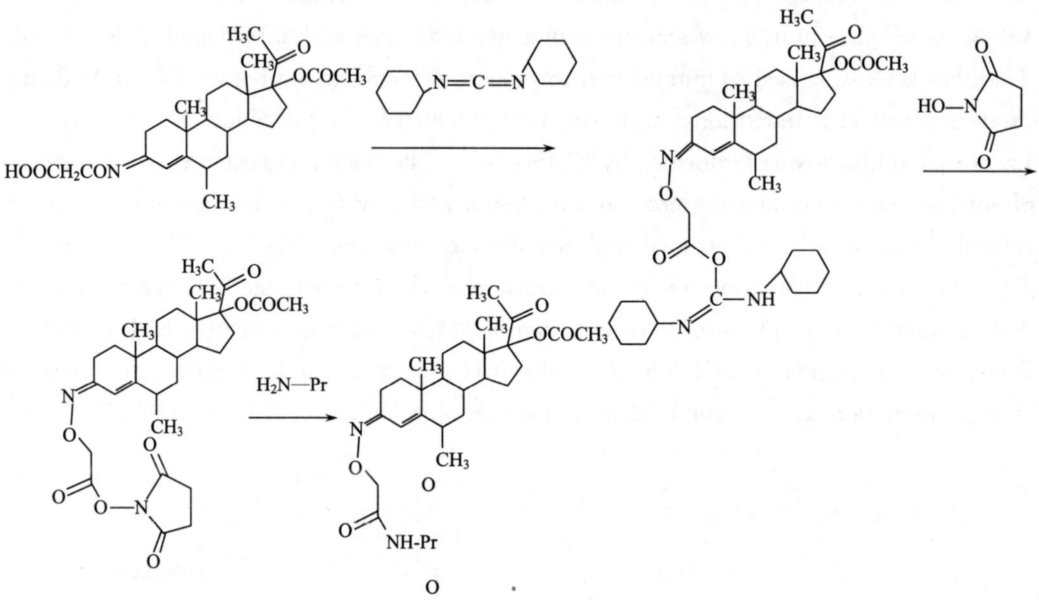

Figure 11-20　Complete antigen synthesis for MPA

According to the molecular structure of MPA, Yang[98] applied oxime reaction and acylation to insert spacer arm containing a carboxyl with 4C-atoms into MPA. The MPA reacted with hydroxylamine hydrochloride, followed by reaction of succinic anhydride to obtain MPA derivatives with carboxyl. Then, the resulting MPA derivatives containing carboxyl were coupled with carrier protein BSA and OVA, respectively with mixed anhydride method and carbodiimide method. Mice wer immunized and a monoclonal antibody was prepared with hybridoma technique. The antibody was purified with affinity chromatography-Protein G Sepharose 4 Fast Flow and used for ELISA of MPA development. The results showed that the minimum detection limits of indirect competitive ELISA were 0. 1-0. 5 ng/mL with IC_{50} 3. 39 ng/mL. The cross-reaction rate with megestrol was 25% and no cross-reaction with other veterinary drugs was found.

Diethylstilbestrol is a synthetic nonsteroidal estrogen that was first synthesized in 1938. In 1971, DES was shown to cause a rare vaginal tumor in girls and women who had been exposed to this drug in utero. The United States Food and Drug Administration subsequently withdrew DES from use in pregnant women. Follow-up studies have indicated that DES also has the potential to cause a variety of significant adverse medical complications during the lifetime of those exposed. Two haptens diethylstilbestrol-mono-carboxylic methyl ether (DES-MCME) and diethylstilbestrol-mono-carboxylic propyl ether (DES-MCPE) with different length of spacers were synthesized by Wang[99]. The haptens were covalently conjugated to carrier proteins BSA and OVA by the method of active ester or mixed anhydride to prepare artificial antigens DES-MCME-

BSA and DES-MCPE-BSA, and coating antigens DES-MCME-OVA and DES-MCPE-OVA. The UV scanning showed that the coupling ratio of hapten-carrier protein conjugates were 10 : 1-30 : 1. DES-MCME-BSA and DES-MCPE-BSA were immunized New Zealand white rabbits to prepare antiserum, respectively. An indirect competitive enzyme-linked immunosorbent assay for DES was successfully established. The best sensitivity was obtained using antibody and coating antigen from DES-MCPE with IC_{50} value of 1. 32 ng/mL. Chen[100] studied the preparation of derivatives of DES. Diethylstibestrol-4-carboxymethyl ether was synthesized and identified by nuclear magnetic resonance (NMR) and mass spectrometry (MS). The complete antigen was prepared with mixed anhydride and and carbodiimide method, respectively with coupling rate of hapten-carrier protein estimated to 25 : 1-32 : 1. Rabbits and mice were immunized for antisera production. The titer of antisera was determinated with agar double diffusion (1 : 6×10^4-1 : 7×10^7). Wang[101,102] synthesized oxime of DES by oxime reaction. The products were identified and used for the synthesis of DES-BSA, DES-OVA conjugate with mixed anhydride method. UV scanning analysis showed that the coupling ratio of DES-BSA and DES-OVA were calculated to be 1 : 13 and 1 : 10. New Zealand white rabbits were immunized with DES-BSA to produce antibodies against DES. The antiserum with highest titer was selected for development of ELISA of DES. The IC_{50} value of developed ELISA was 1. 89 ng/mL and cross-reactivity rate was 3. 1% to dienestrol. The ELISA was used to analyze DES spiked beef with recovery 50%-80%.

For the establishment of rapid detection methods of estradiol (E2) residues in animal products, Dai[103] prepared E_2 complete antigen by carbodiimide. Balb/C mice were immunized and antibody titers were measured by indirect enzyme linked immunosorbent assay (ELISA). Experimental results indicated that titer of anti-E_2 antibody was 1 : 3. 2 $\times$ 10^4. By hybridoma technique, two hybridoma cell lines $1E_3$ and $1B_4$ were obtained with secreting E_2 monoclonal antibody. The monoclonal antibody secreted by $1E_3$ cells were identified be IgG2b isoforms and IgG1 subtype of monoclonal antibody from 1B4 cell line. The ascites titer was detected as 1 : 1 $\times$ 10^5. The molecular weight of 1E3 monoclonal antibody was 179kDa, in which light chain molecular weight was 28kDa, heavy chain molecular mass was 61. 2kDa. The monoclonal antibody showed cross-reactivity less than 0. 5% with estriol, estrone and progesterone. Enzyme labelled estradiol antigens (E_2-BSA-HRP) and estradiol antibody ($1E_3$-HRP) were prepared by oxidation of sodium iodide. Indirect competitive ELISA method was used for development of immunoaasay for E_2 with IC_{50} value of 0. 54 ng/mL.

Yuan and others[104-106] modified dexamethasone, betamethasone and flumetasone-with succinic anhydridized and then conjugated the derivatives with bovine serum albumin and ovalbumin respectively. New Zealand white rabbits were immunized for preparation of antiserum. The antibody with higher titer was used for development of indirect competitive ELISA method. The IC_{50} values of optimized indirect competitive ELISA for dexamethasone, betamethasone and flumetasone were 2.66 ng/mL, 7.36 ng/mL and 6.12 ng/mL, respectively. The cross-reactivity tests for anti-dexamethasone antibody showed up to 120% to diflucortolone, 3% to betamethasone, 37% to triamcinolone, 21% to prednisolone, and less than 0.1% with several endogenous glucocorticoids. The chicken samples were analyzed with these ELISA and the results of the ELISA method were compared with LC-MS results, which had a good correlation.

A simple enzyme-linked immunosorbent assay for the detection of hexoestrol (HES) residue has been developed by Wang[107]. The HES derivatives, hexoestrol-mono-ether-butyrate-ethyl (HES-MEBE) and hexoestrol-mono-caroxyl-propyl-ethyl (HES-MCPE) were synthesised to enable the coupling of HES and proteins together. The immunogen HES-MCPE-BSA and enzyme-linked antigen HES-MCPE-HRP were prepared by mixed anhydride methods. After the antibody was obtained by immunisation of New Zealand rabbits, a direct enzyme-linked immunosorbent assay was developed for the determination of HES residue. Several parameters were optimised. Excellent linearity ($R^2 = 0.9905$) was demonstrated from 0.01 to 8.1 ng/mL. The 50% inhibition value (IC_{50}) and the limit of detection (LOD) were 0.23 and 0.01 ng/mL, respectively. The specificity of the immunoassay was evaluated by determining cross-reactivity of six hormones (diethylstilbestrol, dienestrol, 19-nortestosterone, medroxyprogesterone, testosterone and estrone), and all cross-reactivity rates were $<0.5\%$. The recoveries from beef and fish ranged from 61% to 102%. The stabilisation of the coated plates was also studied. The ELISA developed was used to detect the residues in chicken muscle tissues and confirmed by high-performance liquid chromatography and tandem mass spectrometry (LC/MS/MS).

Yu[108,109] modified 19-nortestosterone(19-NT) with carboxymethyl hydroxylamine. The oximation product was coupled with bovine serum albumin (BSA) and ovalbumin (OVA)by mixed anhydride method with coupling ratios 22 : 1 and 19 : 1, respectively. The polyclonal antibodies against 19-nortestosterone was obtained from immunized rabbits. By determination of indirect ELISA, the titer of antiserum could be up to 1 : 102,400. The conditions of ELISA were optimized with linear range of 0.1-25 ng/mL and IC_{50} value of 3.5 ng/mL. No significant cross-reaction was found with some analogues. Spiked pork samples (at 1 ng/g and 5 ng/g spiked level) were analyzed and the recoveries were 76.9%-104.7%. Based on this antibody, Liu[110-112] developed test strip for 19-NT detection. The detection limit was 200 ng/mL when it used to analyze pork urine samples.

11.12　β-Agonist

11.12.1　Introduction

Beta adrenergic drugs (β-agonist) are the most potent bronchodilators currently availablefor clinical use in asthma and obstructive lung disease. Among the beta agonists, the individual agents vary in their rapidity of onset and duration of action. Inhaled, short-acting, selective beta-2 adrenergic agonists are the mainstay of acute asthma therapy, while inhaled, long-acting, selective beta-2 adrenergic agonists (in combination with inhaled glucocorticoids) play a role in long-term control of moderate to severe asthma. β-Agonists act by impeding the uptake of adrenal hormones by nerve cells and through the stimulation of the cardiac system. They alter body composition by redistributing fat from muscle tissue, resulting in higher production efficiencies. Since 1980s, β-agonist was firstly applied in livestock and poultry breeding in the United States. β-Agonists, also called repartitioning agents, can promote the re-distribution of animal nutrition. Moreover, β-agonists can increase the proportion of animal muscle and lower fat content.

Since the beginning of 1990s, β-agonists have drawn attention of research due to foodsafe problems caused by its residues. The potential risk for human health posed by the presence of β-agonists is high, due to the severity of the possible adverse effects. The β-agonist clenbuterol has been implicated in many poisoning cases in European and Asian countries. The monitoring of raw meat and animal feed for drug and chemical residues is necessary to ascertain that these compounds are not misused and do not present a danger to consumers[113].

11.12.2　Immunoassay for β-agonist

He[114-116] coupled clenbuterol (CL) with bovine serum albumin and ovalbumin, respectively with diazotization reaction. The conjugation of CL-BSA was used to immunize rabbits. The high titer(3.2×10^5) of antiserum was obtained. An indirect ELISA and chemiluminescent enzyme immunoassay (CLEIA) was developed based on this antibody. Under the optimized conditons, the detection limit of ELISA was 0.1 ng/mL for CL while lower detection limit of 0.01 ng/mL for CLEIA was achieved. Shao[117] also prepared conjugates of CL-BSA by diazotization method. The antigen was immunized mice and the positive spleen cells were fused with myeloma cells (SP2/0) for prepration of hybridoma cells. The anti-CL monoclonal antibody showed higher specificity to CL. An indirect ELISA was established with linear detection range of 0. 1-10 ng/mL nd the minimum detection limit of 0. 1 ng/mL for CL. 540 clinical samples were analyzed with this ELISA and the results were compared with detections from marketed ELISA kit(from Netherlands E-D company). The positive results were futher confirmed with GC-MS, with coincidence of 96. 39%.

Yu[118] prepared conjugates of ractopamine-carrier protein with mixed anhydride. Firstly, ractopamine was activated with glutaric anhydride into ractopamine glutaric anhydride semialdehyde. Then, KLH and BSA were used as carrier protein to conjugate with activated ractopamine. The immunogen was used for immunization of rabbits for preparation of polyclonal antibody. The titer of antibody was above 6,000. Indirect inhibit ELISA was established based on the antibody with IC_{50} value of 10 ng/mL. No cross-reaction to clenbuterol, salbutamol and terbutaline were found.

Shi[119] conjugated salbutamol (Sal) to keyhole hemocyanin (KLH) and ovalbumin (OVA) respectively by carbodiimide method. The immunogen (Sal-KLH) was used for the immunization of mice. With hybridoma technique, anti-Sal monoclonal antibody was produced. An indirect ELISA was established with IC_{50} value of 40. 5 ng/mL for Sal. The cross-reaction(CR) to clenbuterol was 100% and the CRs to adrenal, norepinephrine, isoprenaline, dopamine, ephedrine and aminophylline were less than 1%.

11.13 Other drugs

11. 13. 1 Benzodiazepines

11.13.1.1 Physico-chemical properties for benzodiazepines

Benzodiazepines (BZD), psychoactive drugs, are mostly derivatives of 1,4-benzobenzodiazepines. The core chemical structure is the fusion of a benzene ring and a diazepine ring (Figure 11-21). The first benzodiazepine, chlordiazepoxide (librium), was discovered accidentally by Leo Sternbach in 1955, and made available in 1960 by Hoffmann-La Roche, which has also marketed diazepam (valium) since 1963.

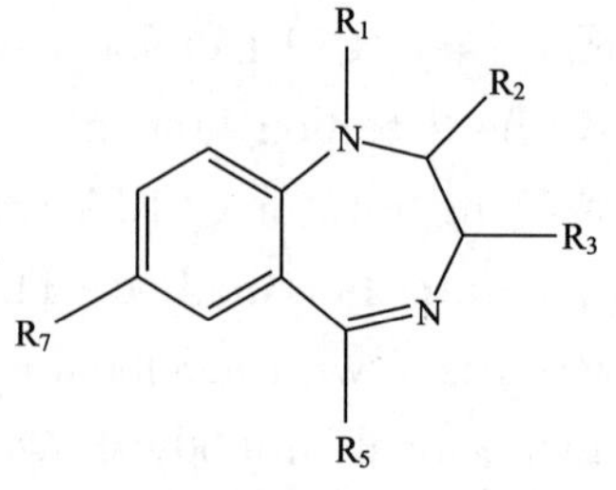

Figure 11-21 Structure of benzodiazepines

11.13.1.2   Pharmacological properties of BZD

Benzodiazepines enhance the effect of the neurotransmitter gamma-aminobutyric acid (GABA), which results in sedative, hypnotic (sleep-inducing), anxiolytic (anti-anxiety), anticonvulsant, muscle relaxant and amnesic action. These properties make benzodiazepines useful in treating anxiety, insomnia, agitation, seizures, muscle spasms, alcohol withdrawal and as a premedication for medical or dental procedures. Benzodiazepines are categorized as either short-, intermediate-or long-acting. Short-and intermediate-acting benzodiazepines are preferred for the treatment of insomnia; longeracting benzodiazepines are recommended for the treatment of anxiety.

11.13.1.3   Immunoassay for BZD

Illegal use of benzodiazepines in animal production was reported in recent years, which pose great risk on health of human. Li[120-122] prepared antigen for nitrazepam (NZP), clonazepam (CZP) and diazepam (DZP). The obtained immunogen was used to immunize New Zealand white rabbits. The results showed that the antibody from NZP conjugates recognized clonazepam, diazepam and two metabolites of 7-amino-nitrazepam and 7-amino-clonazepam well. Indirect competitive ELISA method for NZP detection was developed with IC_{50} value of 1. 53 ng/mL. Also, a colloidal gold competitive immunochromatographic test strip was developed for NZP based on this antibody. The particle size of colloidal gold was 18. 6nm. The test line and control line were coated with 0. 5 mg/mL of NZP coating antigen and 0. 2 mg/mL goat anti-rabbit IgG antibody, respectively. The visual detection limit of this strip for NZP was 150 ng/mL.

Barbiturate drugs are sedative hypnotic drugs, which was used as sedative and chemical diazepam in clinical of veterinary drugs (Figure 11-22). An overdose could easily lead to acute death, with lethal time of 10 h in general after taking drugs. Long-term use is easy to produce drug accumulation, leading to death. Barbiturate drugs are strictly prohibited in many countries for human use, and they are also prohibited as feed additives[123].

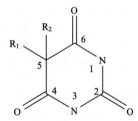

Figure 11-22   The common stutruce of barbiturate drugs

Wang[124] prepared artificial antigen of phenobarbital (PB). PB is chemically modified with nitrification and reduction reaction. An active amine group was inserted on benzene ring of PB. The hapten was used to couple with bovine serum albumin (BSA) and ovalbumin (OVA) by diazotization method for preparation of artificial immunogens (Figure 11-23).

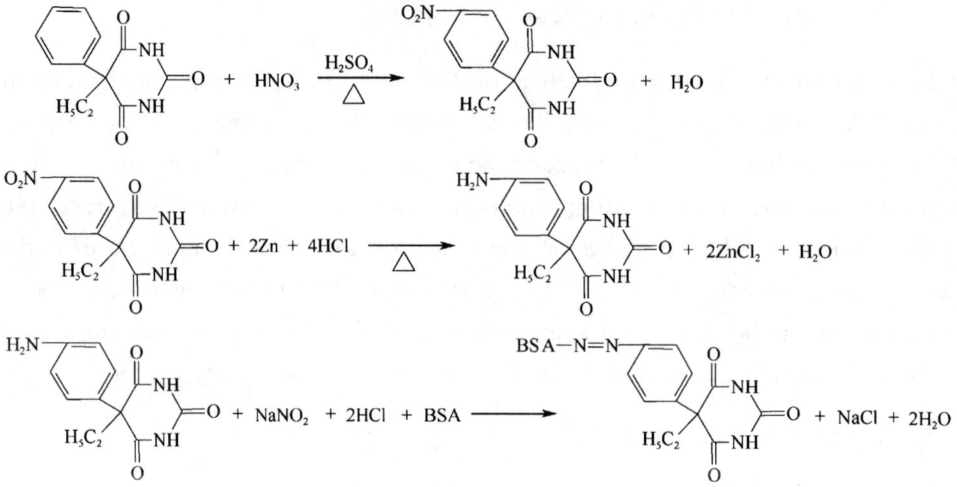

Figure 11-23　Synthesis route for PB antigen by diazotization

The polyclonal antibody (pAb) and monoclonal antibody (mAb) against PB were prepared. The pAb was characterized with IC_{50} value of 14.8 $\mu g/L$ for PB and cross-reaction to pentobarbital 9.7%. The mAb was characterized with IC_{50} of 11.35 $\mu g/L$ for PB and 8.4% cross-reaction to pentobarbital. A rapid detection of PB ELISA was established based on mAb. The detection limit was 1.0 $\mu g/L$, and the linear detection range was 1.0-128.0 $\mu g/L$. The average recoveries in feed and pig urine were 83.85% and 95.53%. F Based on the principle of colloidal gold immunochromatography test (GICA), anti-PB mAb was applied for the development of test strip for PB detection. The best colloidal gold particle size was 25±1.0 nm. Under the optimized conditions, the visual detection limit of this test strip was 2.0 $\mu g/L$. GC-MS method, ELISA and strip were used to analyze pork urine samples with coincidence of 100%.

11.13.2　Retention analysis for polyettler antibiotics

Polyether antibiotics (PEs) are a class of ionophore antibiotics developed from 70s of the 20th century. Most of them are secondary metabolites of streptomyces which belong to actinomycete. PEs has been used all over the world as feed additives to prevent coccidiosis and to increase bioability of the feed for poultry. Since the residues of PEs in edible animal tissues entering human food chain represents potential hazard to human health, It is necessary to develop a precise, specific, accurate, and sensitive technique for the rapid detection and identification of PEs in food supply. PEs can be divided into glucoside (β-configuration) PEs and non-glucoside PEs. According to the existence of spiral ketal structure, they can be further divided into single spiral ketal, spiral II ketal, Polyspiral Ⅱ ketal PEs, or non-spiral ketal PEs (Figure 11-24).

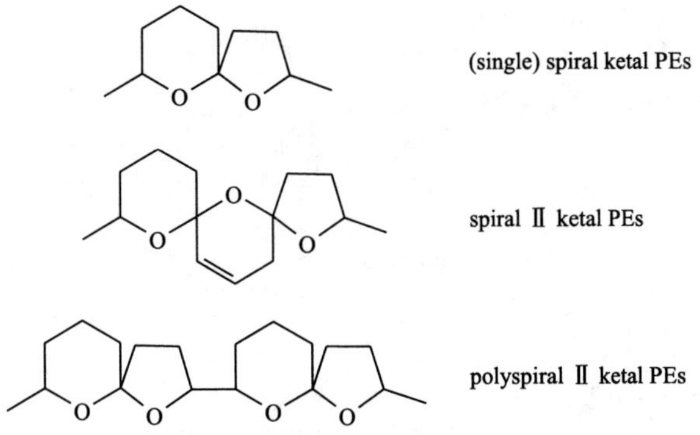

Figure 11-24　Structual categorization of polyether antibiotics

PEs consist of carboxyl and hydroxyl groups and can form salts or esters. PEs are widely distributed in the body after intake. The highest accumulation of PEs is in liver and fat tissues. Some typical PEs are showen in Figure 11-25.

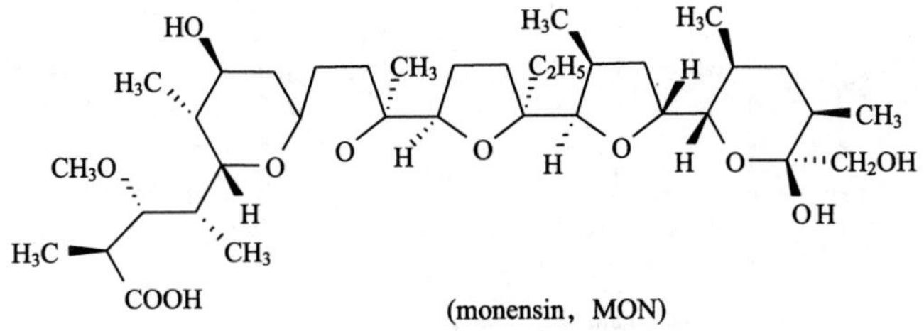

(monensin, MON)
$C_{46}H_{62}O_{11}$, MW 670.9242

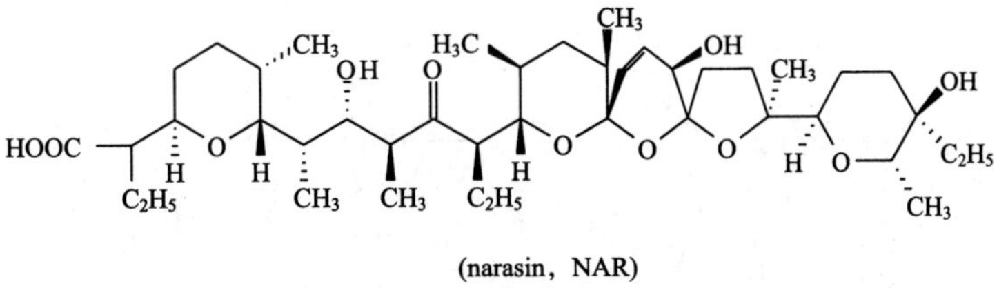

(narasin, NAR)
$C_{42}H_{70}O_{11}$, MW 750.4918

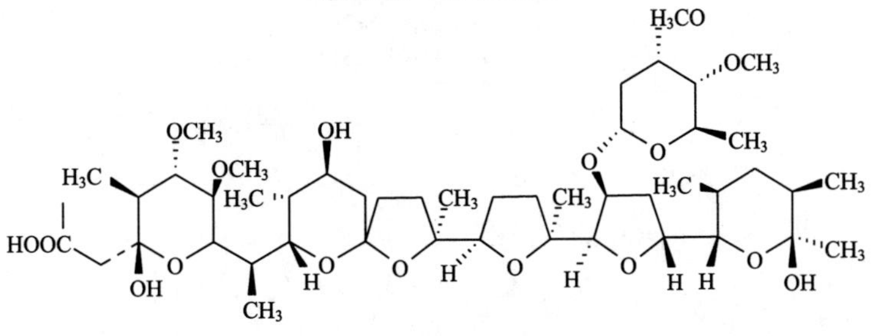

(maduramicin, MAD)
$C_{47}H_{79}O_{17}$, MW 916.5379

Figure 11-25　The chemical structure of some typical PEs(Continued)

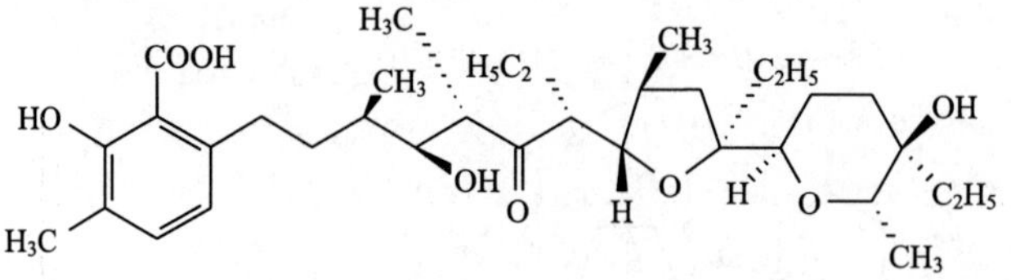

(lasalocid, LAS)
$C_{44}H_{54}O_8$, MW 590.3819

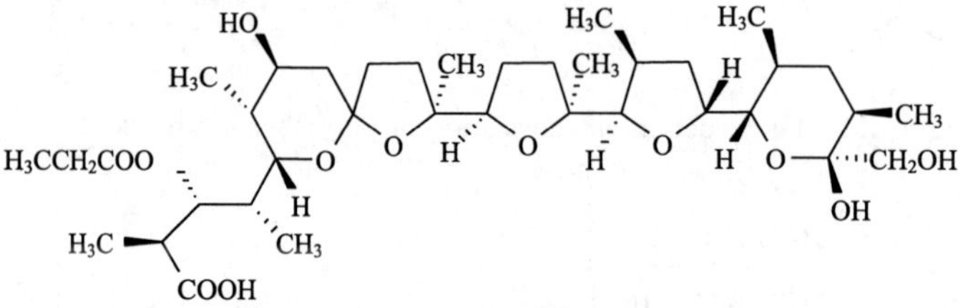

(laidlomycin, LAL)
$C_{47}H_{62}O_{12}$, MW 698.4241

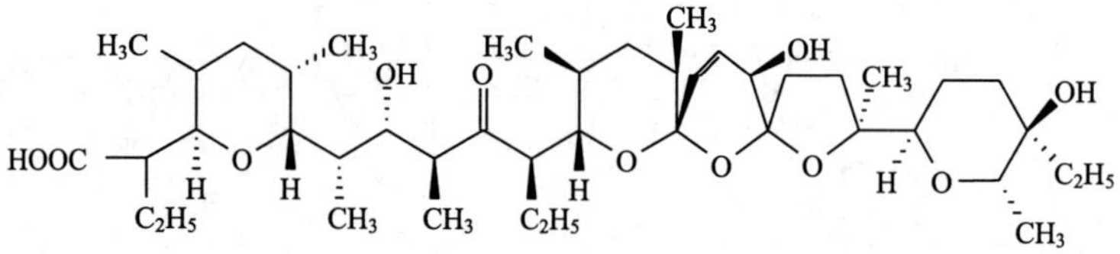

(narasin, NAR)
$C_{43}H_{72}O_{11}$, MW 764.5075

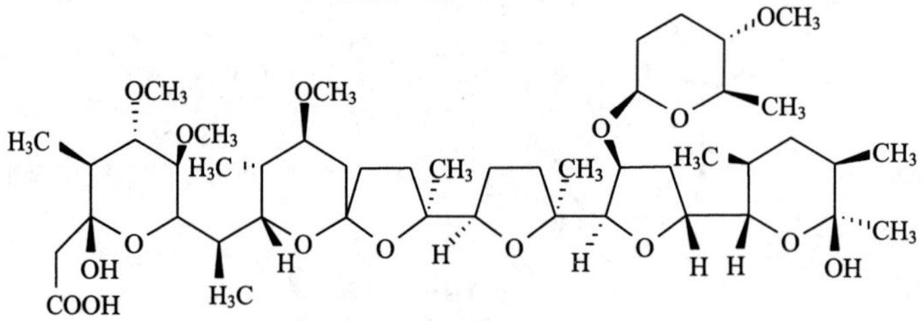

(semduramicin, SEM)
$C_{45}H_{76}O_{16}$, MW 872.5133

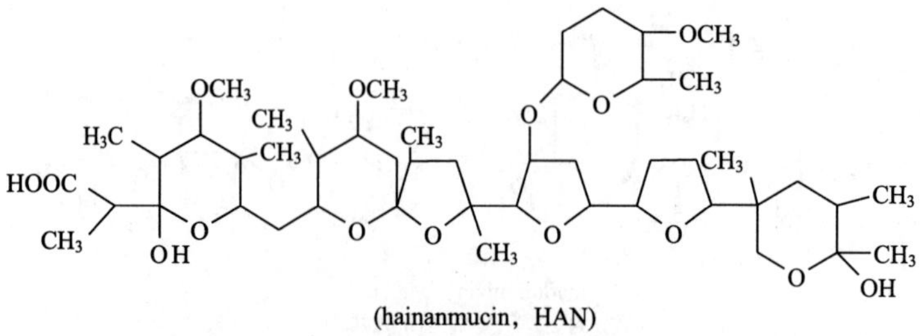

(hainanmucin, HAN)
$C_{47}H_{80}O_{15}$, MW 884.5479

Figure 11-25　The chemical structure of some typical PEs(Continued)

Wei[125] synthesized complete antigens for monensin (MON), which was used for production of polyclonal antibodies. An indirect competitive ELISA was developed with IC$_{50}$ value of 32. 4 ng/mL and the detection limit of 2. 75 ng/mL. no corr-reaction to salinomycin, hainanmycin, and sulfamonomethoxine were found. This proposed ELISA was successfully used for determination of MON in eggs.

11. 13. 3   Immunoassay for olaquindox and its metabolites

Olaquindox is a class of veterinary drugs which is of limited for use, and its main metabolite in animals was 3-methyl-quinoline-2-carboxylic acid. Olaquindox are known as synthetic antimicrobial agents against several gram-positive and gram-negative bacilli. Due to the beneficial effects in animal husbandry, these compounds have been added to cattle, swine, fish and poultry feeds as growth promoters. Carbadox and olaquindox are well known members in this family and there are a number of reports about their toxicity and potential adverse effects[126-129]. China has published relevant policies, regulations and testing standards for limiting their use[130,131].

Ji[132-134] synthesized hapten of olaquindox and its metabolites 3-methylquinoline-2-carboxylic acid, which were then coupled with with bovine serum albumin (BSA) and ovalbumin (OVA) for preparation of immunogen and coated antigen. By immunizing of New Zealand white rabbits to obtain antiserum, highly specific antibody was achieved, with titer 32,000. The results from indirect competitive ELISA showed that IC$_{50}$ value of olaquindox was 151. 9 ng/mL, IC$_{50}$ value of 3-methyl-quinoline-2-carboxylic acid was 3. 84 ng/mL. Based on the antibody, a biosensor was assembled based on nanomaterials and sensitive fluorescence quenching technique. The signal was amplified and detected by fluorescence quenching with the detection limit 7. 52 pg/mL, and detection range from 8. 5 pg/mL to 60 pg/mL in detection of 3-methyl-quinoline-2-carboxylic acid.

11.14   Multi-residal immunoassay for veterinary drugs

He[135] prepared haptens for avermectin (AVM), ivermectin (IVM) and doramectin (DOR) including 4'-o-succinoylAVM, 4'-o-succinoyIVM, and 4'-o-succinoylDOR. The molecular weights of products were 995. 79 (M+Na$^+$), 997. 8 (M+Na$^+$), and 998. 6 (M-H$^+$) through mass spectrum detection. The haptens were coupled to BSA as immunogen, and to OVA as coated antigen. The conjugates were identified by UV-VIS spectrophotometer, and the coupling ratio of hapten and carrier protein was calculated as ranging from 3. 4 to 31. 6. Polyclonal antibodies were prepared from 4'-o-succinoy-lAVM-BSA, showed high cross-reaction with eprinomectin (EP, 145. 4%), AVM (100%), IVM (25. 0%), and DOR (12. 3%). The working conditions for indirect competitive ELISA of AVM was optimized with the 50% inhibition concentration (IC$_{50}$)

4. 8 ng/mL and linear detection range 1. 1 ng/mL to 21. 9 ng/mL. The spiked bovine liver and beef at 5 ng/g, 10 ng/g, and 20 ng/g levels were analyzed with this ELISA. The recovery of AVMs was 53. 8%-108. 6% with coefficients of variation (CV) of 3. 4%-17. 9%. Immunoaffinity chromatography columns were prepared by coupling purified the antibody to CNBr activated Sepharose 4B. The dynamic column capacities of AVM, IVM, DOR, and EP were 3,531, 3,542, 3,543, and 3,284 ng/mL gel, and the specific column capacities were 495, 499, 493, and 460 ng/mg IgG, respectively. After 15 cycles in a month, the capacity of IAC also was above 1,300 ng/mL gel for every drug. The IAC-HPLC-FLD method was developed to detect the residue of AVM, IVM, EP, and DOR in animal tissue. Recoveries ranged from 79. 4% to 113. 8% with CV of 1. 1%-19. 4% when AVM, IVM, DOR, and EP were spiked in bovine liver at levels of 5-100 ng/g. Recoveries ranged from 77. 3% to 119. 5% with CV of 1. 5%-18. 9% when AVM, IVM, DOR, and EP were spiked in bovine muscle at levels of 5-100 ng/g. The limit of detection of the method was 0. 8 ng/g in bovine tissue.

Li[137-139] prepared norfloxacin (NOR) derivatives by alkylation of norfloxacin (Figure 11-26) and the NOR derivative was characterized by UV, LC-MS and H[1]NMR. The NOR derivative was coupled to bovine serum albumin (BSA) and ovalbumin (OVA) to synthesize immunogen respectively by glutaraldehyde method and active ester method. The polyclonal antibodies were acquired from animal immunization with tier of 1 : 102,400. Based on this antiserum, enzyme-linked immunosorbent assay was developed for detection of QNs. After optimization, the standard curve was obtained and the detection limit of the assay is as low as 0. 013 ng/mL of NOR. Cross-reactivity studies demonstrated that 13 other quinolones such as norfloxacin, ciprofloxacin, ofloxacin, were recognized to different extents with the limit of detection (LOD) 0. 01-10 ng/mL.

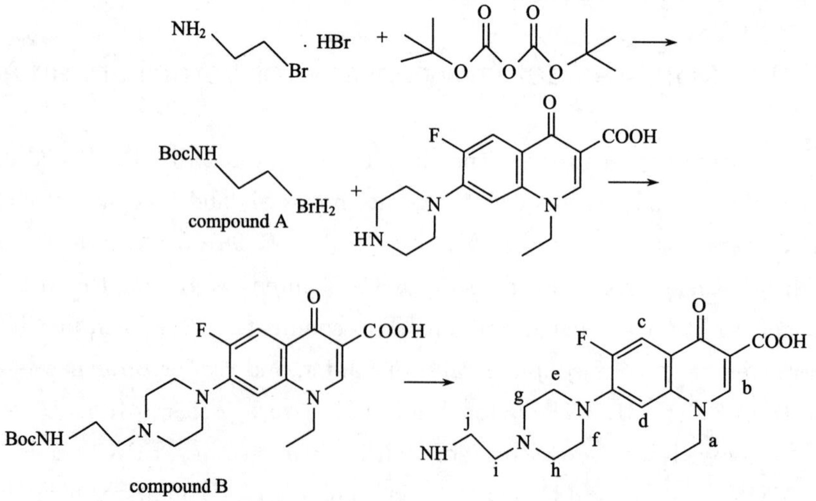

Figure 11-26　The synthesis route of norfloxacin derivative

According to the structural characteristics of fluoroquinolones (FQN), Heng[140] prepared ciprofloxacin hapten, which was coupled with carrier protein for immunogens. The antisera from rabbits were collected. The IC_{50} value of antibody for ciprofloxacin 31. 3 ng/mL was detected with an indirect competitive ELISA. Cross-reaction percentage(CR%) with danofloxacin mesylate was 67%, 62% for pefloxacin mesylate, 17. 61% for ofloxacin and 7. 87% for lomefloxacin hydrochloride. A colloidal gold immunochromatography test strip was assembled based on this antibody and colloidal gold particles 40 nm in diameter. The results showed that the detection sensitivity was 50 ng/mL of ciprofloxacin.

References

[1]　Cao FX. Production technology of secondary metabolites and their products. Beijing: National Defense University Press, 2003:208

[2]　Li D. Pharmacology. Beijing: People's Medical Publishing House, 2003:381

[3]　Zu JJ. Tetracycline polyclonal antibody preparation and amikacin ELISA preliminary study[Dissertation]. Zhengzhou:Zhengzhou University,2007

[4]　Tang N. ELISA kit for detection of streptomycin residues in milk[Dissertation]. Yangzhou: Yangzhou University,2006

[5]　Fan GY. Immunological rapid detection of streptomycin residues[Dissertation]. LingYang:Northwest A & F University,2007

[6]　Zhang GX. Streptomycin monoclonal antibody and its preliminary application study[Dissertation]. Chongqing: Southwest China University, 2006

[7]　Sha C. Neomycin artificial antigen construction, polyclonal antibody preparation and test methods preliminary study[Dissertation]. Chengdu:Sichuan University, 2007

[8]　Xu T. Gentamicin monoclonal antibodies and preliminary applications [Dissertation]. Yangzhou:Yangzhou University, 2005

[9]　Li JS and others. Veterinary drug residue analysis. Shanghai: Shanghai Science and Technology Press, 2002: 413

[10]　Huang DB, Tian ZG, Pang GS. Integrated traditional and western medicine pharmacology and clinical. Wuhan: Hubei Science and Technology Press, 2006:445

[11]　Cui HH. Ivermectin immunoassay technique [Dissertation]. Yangzhou:Yangzhou University,2007

[12]　Lu J. Development and application of ivermectin monoclonal antibodies[Dissertation]. Yangzhou: Yangzhou University, 2007

[13]　Xiang JZ. Pharmacology. Beijing: Science Press, 2002:482

[14]　Chen Y, Zhou M. Radical medicine. Beijing: People's Medical Publishing House, 1991, 179

[15]　Zhao J and others. Chloramphenicol ELISA detection method. Biotechnology, 2005, 15 (1): 56-59

[16]　Wang ZL and others. Development of gold standard rapid test strip for chloramphenicol monoclonal antibody and its performance measurement. China agricultural University, 2006, 21 (5): 127-131

[17]　Lin S. Establishment of chemiluminescence enzyme immunoassay method for Chloramphenicol [Dissertation]. Beijing:National Institute of Atomic Energy, 2005

[18]　Huang YL. Chloramphenicol ELISA method and kits development [Dissertation]. Hangzhou:Zhejiang University, 2005

[19] Hao K, Guo SD, Xu CL. CELISA detection method for Thiamphenicol residues in aquatic products. Food Science and Technology, 2006, 2: 186-189

[20] Hao K, Guo SD, Xu CL. Analysis method of thiamphenicol residues in fish. Food Science, 2006, 27 (3): 169-172

[21] Hao K, Guo SD, Xu CL. CELISA detection method for thiamphenicol residues in aquatic products. Food Science and Technology 2006, 27 (2): 186-189

[22] Hao K, Guo S, Xu C. Development and optimization of an Indirect enzyme-linked immunosorbent assay for thiamphenicol. Analytical Letters 2006, 39: 1087-1100

[23] G. Lanqi Ni. Antibiotics multidisciplinary research. 1995. The main translation by Yiguang Wang. Beijing: People's Medical Publishing House, 1998: 89

[24] Yan BX. The clinical application of anti-infectives. Beijing: Chemical Industry Press, 2002: 15

[25] He D. Ampicillin residues in ELISA method and Optimization[Dissertation]. Hangzhou: Zhejiang University, 2004

[26] Yan L, Wu GJ, Wang L and others. Ampicillin monoclonal antibody identification and preliminary research of enzyme-linked immunosorbent assay. Animal Husbandry and Veterinary Medicine, 2005, 37 (10): 1-4

[27] Zhang ZY. Penicillin G polyclonal antibody and the preliminary study of its rapid detection of [Dissertation]. Hohhot: Inner Mongolia Agricultural University, 2007

[28] Yang Y. Penicillins drug residues and ELISA detection[Dissertation]. Hangzhou: Zhejiang University, 2006

[29] Hao ZH. Ceftiofur monoclonal antibody and the preliminary study of ELISA and HPLC [Dissertation]. Chongqing: Southwest China Agricultural University, 2005

[30] Xie HL. Cephalosporin antibiotics in milk and rapid immunological muliti-residue detection [Dissertation]. Wuxi: Jiangnan University, 2009

[31] Xie HL, Chen W, Peng CF and others. β-lactam antibiotic residues Immunoassay Methods in Foodstuffs of animal origin. Food Science. 2008, 29 (7): 490-494

[32] Xie HL, Liu LQ, Xu CL. Development and validation of an immunochromatographic assay for rapid multi-residues detection of cephems in milk. Analytica Chimica Acta. 2009, 634(1): 129-133

[33] Ni YX, Hong XH. Bacterial resistance monitoring and anti-infection treatment. Beijing: People's Medical Publishing House, 2002: 22

[34] Li ZH. Medicinal Chemistry. Beijing: People Health Press, 1987: 374

[35] Xu B, Xu ZX, Wang FX and others. Furazolidone metabolites ELISA detection method in animal foods. Food Research and Development, 2007, 28 (12): 145-149

[36] Xu YP, Jin ZY, Xu CL. ELISA detection for nitrofurantoin metabolite 1-amino-hydantoin. Food Science and Technology, 2008, 29 (8): 272-275

[37] Xu YP, Xu CL. Detecion technology of nitrofurans substances and their metabolites residues in food of animal origin. Food science, 2007, 28 (10): 590-593

[38] Xu YP, Liu LQ, Li QS and others. Development of an immunochromatographic assay for rapid detection of 1-Aminohydantoin in urine specimens rapid detection of 1-Aminohydantoin in urine specimens. Biomedical Chromatography. 2009; 23: 308-314

[39] Luo J. Furazolidone residues in aquatic products by indirect competitive enzyme-linked immunosorbent assay (ciELISA) and antiserum preparation of chloramphenicol [Dissertation]. Qingdao: China Ocean University, 2004

[40] Xia KH, Zhou L, Hao LF. Quinolone antibiotics research. Foreign pharmaceutical antibiotics, 2004, 25 (3): 138-141

[41] Jiang J, Liu ML, Guo HY. New classification and rational use of quinolone antibacterial drugs. Foreign phar-

maceutical Antibiotics, 2003, 24 (6): 272-275

[42]　Liu LJ and others. Quinolone drug adverse reactions and drug interactions. Medicine, 2005, 24 (10): 959-960

[43]　Liu CN. Enrofloxacin antibody and immunoassay preliminary study[Dissertation]. Qingdao: China Ocean University, 2005

[44]　Lin BF. Enrofloxacin monoclonal antibodies preparation, identification and immunological detection of[Dissertation]. Changchun: Jilin University, 2005

[45]　Li H. Anti-enrofloxacin monoclonal antibodies Production and application [Dissertation]. Chongqing: Southwest China Agricultural University, 2005

[46]　Wang Y. Hybrid cell line secreting monoclonal antibodies against norfloxacin and its Preliminary Application [Dissertation]. Zhengzhou: Henan Agricultural University, 2006

[47]　Zhan CX. Norfloxacin monoclonal antibody screening, preparation and preliminary identification[Dissertation]. Guangzhou: Jinan University, 2007

[48]　Liu XH. Lomefloxacin and amoxicillin and polyclone antibody detection by Ci-ELISA and preliminary study [Dissertation]. Zhengzhou: Zhengzhou University, 2006

[49]　Du FB. Pefloxacin monoclonal antibody and its application in ELISA [Dissertation]. Changchun: Jilin University, 2006

[50]　Pan XC. Ofloxacin and difloxacin antibody preparation and ELISA detection [Dissertation]. Hefei: Anhui Agricultural University, 2006

[51]　Yu HF. Ofloxacin monoclonal antibody and the establishment of ELISA method [Dissertation]. Changchun: Jilin University, 2007

[52]　Li ZH. Fluoroquinolones screening and danofloxacin indirect competitive ELISA method [Dissertation]. Urumqi: Xinjiang Agricultural University, 2007

[53]　Lu SX. Danofloxacin, pefloxacin and lomefloxacin by enzyme-linked immunosorbent assay [dissertation]. Jinan: Shandong University, 2007

[54]　Li D. Pharmacology. Beijing: People's Medical Publishing House, 2003: 349

[55]　Wang XQ, Wu YN and others. Chromatography application in food safety analysis. Beijing: Chemical Industry Press, 2004: 104

[56]　Xu MD, Mao GN. Food safety and analysis detection. Beijing: Chemical Industry Press, 2003: 178

[57]　Liu BH and others. Direct competitive ELISA method for rapid detection of sulfadiazine residue in animal food. China Veterinary Medicine, 2008, 42 (2): 16-19

[58]　Xu WG. Anti-SD antibody and initial establishment of enzyme-linked immunoassay for sulfadiazine[Dissertation]. Beijing: China Institute of Atomic Energy, 2007

[59]　Chen LY, Wang HD, Wang ZY. Enzyme-linked immunosorbent assay for sulfametoxydiazine residues. Chinese Journal of Veterinary Science, 2004, 24 (4): 375-378

[60]　Yu Y. Sulfamethazine in aquatic by indirect competitive ELISA (ciELISA) Detection and Its Primary Application[Dissertation]. Qingdao: China Ocean University, 2005

[61]　Liu C. Sulfaquinoxaline antibody preparation and enzyme-linked immunosorbent assay development[Dissertation]. Beijing: China Institute of Atomic Energy, 2006

[62]　Yang CH. Development of enzyme-linked immunosorbent assay kit for Sulfadoxine-5-methoxy pyrimidine residues[Dissertation]. Yangzhou: Yangzhou University, 2006

[63]　Shen H. Anti sulfa drugs antibody preparation and immune analysis[Dissertation]. Beijing: China Agricultural University, 2005

[64]　Gu JF. Rational use of antibiotics. Shanghai: Shanghai Science and Technology Press, 2004: 107

[65]　Feng QH. Veterinary Clinical Pharmacology. Beijing: Science Press, 1983: 86

[66]　Wu YX and others. Tetracycline polyclonal antibody preparation, identification and ELISA detection. Chinese Veterinary Medicine, 2007, 41 (9): 13-16

[67]　Gao ZB. Preparation and application of hybridoma strains secreting monoclonal antibodies against tetracycline [Dissertation]. Zhengzhou: Henan Agricultural University, 2006

[68]　Xing SJ. Oxytetracycline residues in honey by ELISA. Wuhan: Huazhong Agricultural University, 2005

[69]　Kong FJ. Practical medical manual. Jinan: Shandong Science and Technology Press, 1985:209

[70]　Antonio A, Juan JM, Maria JM and others. Determination of thiabendazole in fruit juices by a new monoclonal enzyme immunoassay: Pesticide analysis using immunoassay. Journal of AOAC International, 2001, 84(1): 156-161

[71]　Brandon DL, Binder RG, Bates AH and others. Monoclonal antibody for multiresidue ELISA of benzimidazole anthelmintics in liver. Journal of agricultural and food chemistry, 1994, 42(7): 1588-1594

[72]　Wu TX, Lin SH and others. Essence of additive premix feed production and preparation techniques. Beijing: Chemical Industry Press, 1993:54

[73]　Shen JZ, Xie LJ. Veterinary Pharmacology. Beijing: China Agricultural University Press, 2000:239

[74]　Wu NP. Halofuginone residues in chicken tissues by ELISA detection method and HPLC detection method [Dissertation]. Beijing: China Agricultural University, 2004

[75]　Ma LN. Development of Amprolium hydrochloride monoclonal antibodies and initial establishment of ELISA detection [Dissertation]. Yangzhou: Yangzhou University, 2006

[76]　Shen JZ, Qian CF, Yang HC and others. Residue of madura ADM in broiler tissue. China Agricultural Science, 1998, 31 (6): 1-5

[77]　Jin YY. Pharmacology. Beijing: Chinese Peace Press, 1996:141

[78]　Yang XW and others. Chinese medicine metabolic analysis. Beijing: Chinese Medical Science and Technology Press, 2003:504

[79]　Mao GN and others. Hazardous substances in environment and detection. Beijing: Chemical Industry Press, 2004:226

[80]　Liu Y. Corn zeranol and oxime compounds Immunoassay Research [Dissertation]. Nanjing Normal University, 2005

[81]　Wang HJ. Residual zeranol in cow urine by ELISA detection method [Dissertation]. Beijing: China Agricultural University, 2004

[82]　Guo JH. Food medroxyprogesterone acetate residues by enzyme immunoassay method [Dissertation]. Wuxi: Jiangnan University, 2006

[83]　Guo JH, Xu CL, Jin ZY and others. Anti-medroxyprogesterone acetate antibodies and identification. Cellular and Molecular Immunology, 2006, 22 (2): 266-267

[84]　Peng CF, Shen CY, An KJ and others. High performance liquid chromatography-tandem mass spectrometry for determination of acetyl progesterone residues in fats. Analytical Chemistry. 2008, 36 (8): 1117-1120

[85]　Xu CL, Peng CF, Hao K and others. Chemiluminescence enzyme immunoassay method for the determination of chloramphenicol residues in aquatic products. Analytical Chemistry, 2005, 33 (12): 1809

[86]　Peng CF, Xu CL, Jin ZY. Diethylstilbestrol residues in food and preparation of artificial antigens, analysis and identification. Food Science, 2005, 26 (10): 46-49

[87]　Peng CF, Chen YW, Chen W and others. Development of a sensitive heterologous ELISA method for analysis of acetylgestagen residues in animal fat. Food Chemistry. 2008, 109: 647-653

[88]　Peng CF, Chen YW, Chen HQ and others. A rapid and sensitive enzyme-linked immunosorbent assay (ELISA) method and validation for progestogen multi-residues in feed. Journal of Animal and Feed Sciences, 2008, 17: 434-441

［89］ Peng C, Huo T, Liu L and others. Determination of Medroxyprogesterone Acetate Residues by Capillary Electrophoresis Immunoassay with Chemiluminescence Detection. Electrophoresis 2007, 28: 970-974

［90］ Xu C, Peng C, Liu L and others. Determination of hexoestrol residues in animal tissues based on enzyme-linked immunosorbent assay and comparison with liquid chromatography-tandem mass spectrometry. Journal of Pharmaceutical and Biomedical Analysis 2006, 41 (3): 1029-1036

［91］ Xu C, Peng C, Hao K and others. Chemiluminescence enzyme immunoassay (CLEIA) for the determination of chloramphenicol residues in aquatic tissues. Luminescence 2006, 21(2): 126-128

［92］ Xu C, Peng C, Hao K and others. Simultaneous Determination 9 Anabolic Steroids Residues in Animal Muscle Tissues by Gas Chromatography-Mass Spectrometry. Analytical Letters 2006, 39: 555-568

［93］ Peng C, Xu C, Jin Z and others. Determination of anabolic steroid residues (medroxyprogesterone acetate) in pork by ELISA and comparison with liquid chromatography tandem mass spectrometry. Journal of Food Science 2006, 71(1): C44-C50

［94］ Peng C, Xu C, Jin Z. Comparative Analysis of Medroxyprogesterone Acetate Residue in Animal Tissues by ELISA and GC-MS Analytical Letters,2006, 39(9): 186-1873

［95］ Xu C, Peng C, Wang L and others. Separation and identification of synthetic antigens of hexoestrol residue in animal derived food by HPLC-MS. Food and Agricultural Immunology,2006, 17(1): 21-27

［96］ Xu C, Peng C, Hao K and others. Synthesis and identification of immunogen medroxyprogesterone acetate residues in edible foods and preparation of the antisera. Chemical Papers-Chemicke Zvesti,2006, 60 (1): 61-64

［97］ Xu C, Peng C, Hao K and others. Determination of clenbuterol residual by chemiluminescent enzyme immunossay. Chinese Journal of Analytical Chemistry,2005, 33(5): 699-702

［98］ Yang ZX. Acetate progesterone monoclonal antibody and ELISA Detection [Dissertation]. Yangling: Northwest A & F University, 2008

［99］ Wang H. Diethylstilbestrol chemistry immunoassay [Dissertation]. Yangzhou: Yangzhou University,2006

［100］ Chen DK. Diethylstilbestrol Immunoassay [Dissertation]. Yangling: Northwest A & F University, 2004

［101］ Wang B, Xu CL, Peng CF. Diethylstilbestrol conditions in indirect enzyme-linked immunosorbent assay. Food Science. 2008, 29 (5): 273-278

［102］ Wang B, Peng CF, Xu CL. Diethylstilbestrol complete antigen synthesis and analysis. Chemical Industry and Engineering, 2007, 58 (6): 1523-1528

［103］ Dai XH. Estradiol monoclonal antibody preparation and ELISA kit preliminary study. Urumqi: Xinjiang Agricultural University, 2005

［104］ YuanY. Glucocorticoid residues of Immunity and quantum dots in the immune Analysis [Dissertation]. Wuxi: Jiangnan University, 2009

［105］ Yuan Y, Peng CF, Xu CL. What blocks steroid abuse "black hole"? -Glucocorticoids residue detection methods and standards improvement. Animal health in China, 2007, 9: 95-97

［106］ Yuan Y, Xu C, Peng C. Analytical methods for the detection of corticosteroids-residues in animal-derived foodstuffs. Critical Reviews in Analytical Chemistry, 2008, 38(4): 227-241

［107］ Wang L, Xu C, Peng C. Development and optimization of an indirect enzyme-linked immunosorbent assay for the determination of Hexoestrol. Food and Agricultural Immunology 2006, 17(3): 157-171

［108］ Zhang Y, Yu DH, Xu CL and others. Nortestosterone and its metabolite residues detection in foodstuffs of animal origin. Food Science, 2008, 29 (8): 643-646

［109］ Yu DH, Xu CL, Peng CF and others. 19-nortestosterone residue in food and immunogen synthesis and identification. Food Science, 2006, 27 (5): 195-198

［110］ Liu LQ, Peng CF, Jin ZY and others. Nanoparticles gold technological development and application in food safety rapid detection. Food Science, 2007, 28 (5): 348-352

[111] Li K, Liu L, Xu C and others. Rapid determination of chloramphenicol residues in aquaculture tissues by immunochromatographic assay. Analytical Sciences 2007, 23: 1281-1284

[112] Liu L, Xu C. Development and evaluation of a rapid lateral flow immunochromatographic strip assay for screening 19-nortestosterone. Biomedical Chromatography, 2007, 21: 861-866

[113] Yang SM. Feed safety testing and evaluation. Beijing: China Agricultural Science and Technology Press, 2003, 196

[114] He TM. Clenbuterol residues in food by enzyme immunoassay [Dissertation]. Wuxi: Jiangnan University, 2005

[115] Huo T, Peng C, Xu C and others. Immumochromatographic assay for determination of hexoestrol residues. European Food Research and Technology 2007, 225(5-6): 743-747

[116] Huo T, Peng C, Xu C and others. Development of colloidal gold-based immunochromatographic assay for the rapid detection of medroxyprogesterone acetate residues. Food and Agricultural Immunology, 2006, 17(3): 183-190

[117] Shao JD. Development of enzyme-linked immunosorbent assay kit for clenbuterol monoclonal antibody [Dissertation]. Suzhou: Suzhou University, 2006

[118] Yu HX. Ractopamine enzyme-linked immunosorbent assay (ELISA) [Dissertation]. Beijing: Chinese Academy of Agricultural Sciences, 2005

[119] Shi LJ. Salbutamol monoclonal antibodies preparation, identification and immunological detection of [Dissertation]. Changchun: Quartermaster University of PLA, 2003

[120] Li QS, Shen CJ, Xu JZ and others. High performance liquid chromatography-electrospray ionization tandem mass spectrometry for determination of metabolites of nitrazepam of 7-amino-nitrazepam in urinary. Food Science, 2008, 29 (8): 510-512

[121] Li QS, Xu CL, Peng CF and others. Research Advances of benzodiazepine residue in foodstuffs of animal origin, food science, 2007, 28 (08): 521-524

[122] Chen W, Peng C, Xu C and others. Ultrasensitive immunoassay of 7-aminoclonazepam in human urine based on CdTe nanoparticle bioconjugations by fabricated microfluidic chip. Biosensors and Bioelectronics. 2009, 24 (7): 2051-2056

[123] Lu YQ, Gong QY. Pharmacology. Shanghai: Fudan University Press, 2003: 62

[124] Wang ZL. Phenobarbital residues of rapid detection by immunology [Dissertation]. Yangling: Northwest A & F University, 2007

[125] Wei RC. Monensin antibody preparation and application [Dissertation]. Nanjing: Nanjing Agricultural University, 2006

[126] Commission ReguLation (EC). No. 2788/98, Official journal of european communities, 1998, L347: 3-32

[127] FAO/WHO. Joint expert committee on food additives: evaluation of certain veterinary drug residues in food. Technical Series, 1990, 799: 45

[128] FAO/WHO. Joint expert committee on food additives: toxicologyical evaluation of certain veterinary drug residues in food, Additives Series, 1991, 27: 175

[129] FAO/WHO. Joint expert committee on food additives: evaluation of certain veterinary drug residues in food, Technical Series, 1995, 851: 19

[130] AQSIQ. Animal derived food residue monitoring, compilation of documents and regulations. 2003

[131] Ministry of Agriculture documents. Pastoral issued 38 veterinary drug residues detection in food of animal origin. 2001

[132] Zhang Y, Ji BQ, Xu CL. Analysis methods of quinoxaline and its metabolites in foodstuffs of animal origin. Food Science, 2008, 29 (9): 668-671

[133]　Qing JB, Xu CL, Jin ZY and others. Synthesis and characterization of olaquindox metabolites 3-methyl-quino-line-2-carboxylic acid. Chinese feed. 2008, 12: 33-35

[134]　Shim BS, Chen W, Doty C and others. Smart electronic yarns and wearable fabrics for human biomonitoring made by carbon nanotube coating with polyelectrolytes. Nano Letters, 2008, 8 (12): 4151-4157

[135]　He JH. Avermectin multi-residue analytical methods [Dissertation]. Beijing: China Agricultural University, 2005

[136]　Li YL. Quinolones veterinary drug residues by enzyme-linked immunosorbent assay method [Dissertation]. Wuxi: Jiangnan University, 2008

[137]　Li YL, Hao XL, Ji BQ and others. HPLC-ESI-MS/MS determination of 19 quinolone residues in animal food. Food Science, 2008, 29 (8): 502-506

[138]　Li YL, Xu CL. Quinolone drug residues and detection methods. Food Science, 2007, 28 (11): 628-633

[139]　Li Y, Ji B, Chen W and others. Production of new class-specific polyclonal antibody for determination of fluo-roquinolones antibiotics by indirect competitive ELISA. Food and Agricultural Immunology, 2008, 19(4): 251-264

[140]　Hu H. Development of fluoroquinolone multi-residues test strip [Dissertation]. Changchun: Jilin University, 2008

Chapter 12   Biological Toxin Immunoassay

12.1   Animal toxins

12. 1. 1   Overview

Animal toxin is self-defense weapons formed in the struggle for survival, which is also used to capture prey. Animal toxins can be divided into "active toxicants" and "passive toxicants". Active toxin results from the divergent evolution and is produced from a special organ. Passive toxin can be divided into "primary poison" and "secondary poison". The former is made by animal's special apparatus and the latter is mainly derived from food. For example, some poisonous fish get toxins from the toxic algae. In recent years; many significant progresses in research of animal toxins, especially in marine animal toxins have been made. Some marine toxins, which are very complex in chemical structure, such as palytoxin (PTX), have been clearly identified. A variety of new toxins from cyanophyta and pyrrophyta have been isolated successfully. It has been found that these micro-algae toxins can be transferred to fish, or shellfish along the food chain. The people and livestock will be poisoned when they took these contaminated foods carelessly. Notorious ciguatera poisoning, paralytic shellfish poisoning (PSP), diarrhetic shellfish poisoning (DSP) and neurotoxic shellfish poisoning (NSP) belongs to this case.

12.1.1.1   Paralytic neurotoxin

Tetrodotoxin, frequently abbreviated as TTX, has been isolated from widely differing animal species, including western newts of the genus taricha, pufferfish, toads of the genus Atelopus, several species of blue-ringed octopuses of the genus hapalochlaena, several sea stars, certain angelfish, a polyclad flatworm, several species of chaetognatha, several nemerteans (ribbonworms) and several species of xanthid crabs. The chemical structure of TTX has been identified and was successfully synthesized in 1972. Tetrodotoxin blocks action potentials in nerves by binding to the voltage-gated, fast sodium channels in nerve cell membranes, essentially preventing any affected nerve cells from firing by blocking the channels used in the process, which has been an important tool in the research of membrane channels. In the past, Japan was in exclusive production of tetrodotoxin. Now, China is capable of producing tetrodotoxin, which has been marketed. Recently, natural tetrodotoxin analogues including tetrodonic acid, dehydrogenation tetrodotoxin and 4-tetrodotoxin have been firstly been isolated from the fugu niphobles. In addition, saxitoxin was isolated from the liver of fugu pardalis. These findings subverted the conventional idea that there was only one kind of tetrodotoxin in puffer fish.

Shellfish toxin often results in paralytic poisoning, and it has been identified that these toxins stem from protogonyaulax. If protogonyaulax heavily multiplys in appropriate environment, red tide will appear. Filter-feeding shellfish enriched toxins by feeding protogonyaulax. At some places such as Ningbo, Zhoushan in China, food poisoning took place due to exposure of contaminated massariidae. The symptoms were similar to paralytic shellfish poisoning. Up to now, over 10 kinds of chemicals, analogues of saxitoxin(STX), have been isolated and identified from shellfish and algae. The nuclear structure of these toxins is hydrogen purine, containing two guanidine groups. STX is the typical representative and displays similar pharmacological effects to TTX by specifically blocking the sodium channels.

12.1.1.2   Polyether toxins

In recent years, series of highly toxic polyether toxins with larger molecular have been found, such as palytoxin, gymnodinium breve toxin, dinophysistoxins (DTX), pectenotoxins (PTX) and ciguatoxins (CTX). Scientists from the US. and Japan isolated and identified 7 PTXs from palytoxin in Hawaii, Tahiti, the Ryukyu Islands and the Caribbean area. Up to now, PTX is the most toxic in non-protein toxin family. PTX also indicates strong cardiac toxicity and cell toxicity, which make it become a new measure in the research of cell membrane. Ciguatera poisoning is common in fish poisoning, and it is said that 400-500 species of fish can cause ciguatera poisoning. Some poisoning accidents have happened in Taiwan and sea islands of South China. Ciguatoxin is sourced from new species of gambier-discus toxicus. Ciguatera poisoning is mainly caused by the following toxins: ciguatoxin (CTX), maitotoxin (MTX), scaritoxin (SG) and cigua phallotoxin. The most important toxin is CTX. A recent study on diarrhetic shellfish poisoning (DSP) showed that this toxin stemmed from dinophysistoxins and prorocentrum Lima, which belong to fat-soluble polyether. These algae do not form red tide. However, the toxicity of shellfish will exceed the limit, resulting in DSP poisoning if the density of these algae in sea was close to 200 per liter.

12.1.1.3   Peptide toxins

The peptide toxin is widely distributed in bacteria, animals and some plants, which has become an issue in toxin research because of its great varieties and strong toxicity. There are a number of common important toxicological mechanisms in this toxin family[1]. ① The chemical structure includes an effector chain (chain A) and a binding chain (chain B). Only chain A is toxic and the role of chain B is binding with membrane receptors and to help the chain A entry into the cell. Individual chain A or chain B doesn't show any toxicity. ② The activity mechanism of chain A is similar to enzyme's catalysis. ③ The ganglioside receptors in the cell membrane are the potential targeted effectors almost for all peptide toxins. The primary action mode is based on the change of ion permeability. Identification of these basic action modes and molecular interaction is very necessary for understanding of poisoning mechanism. In recent years, insect toxin has attracted much attention, which is produced by the insects for self-defense, attacking, hunt or communication among the same species. Now, over 600 species of insects from at least 10 orders produce toxins. The universal existence of insect toxin ensures the subsistence, reproduction and evolution of insects. The composition of insect toxin is very complex. According to the chemical structure, it can be classified into three groups. ① Small molecule toxins. There are more than 110 kinds of chemicals including hydrocarbons, alcohols, aldehydes, ketones, organic acids, terpenes and so on. Their chemical structures are generally simple containing 7-28 carbons. Over 50 alcohol toxins belong to defensive toxins and half of them stem from hymenoptera insects. ② Peptide and protein toxins. ③ Enzyme toxins. The typical toxins include phospholipase (PLA), hyaluronidase, acidic phosphatase and lipase. Phospholipase A2 toxins, which are important allergen, can be produced by almost all insects.

12. 1. 2   Immunoassy of biological toxins

Zhou[2] coupled tetrodotoxin with bovine serum albumin (BSA) and keyhole limpet hemocyanin (KLH) respectively. The synthesis of antigen is shown in Figure 12-1. Mice were immunized with KLH conjugate using routine immunization combination spleen immunization. Spleen cell from immunized mouse was fused with SP2/0 cancer cell and 3 hybridoma strains (5B7, 5B9 and 6B9) which can stably secrete monoclonal antibody against tetrodotoxin(TTX) were obtained. An indirect ELISA was established based on the monoclonal antibody from 5B7 hybridoma. The detection range was 3. 16-1,000 ng/mL and limit of detection was estimated as 3. 16 ng/mL.

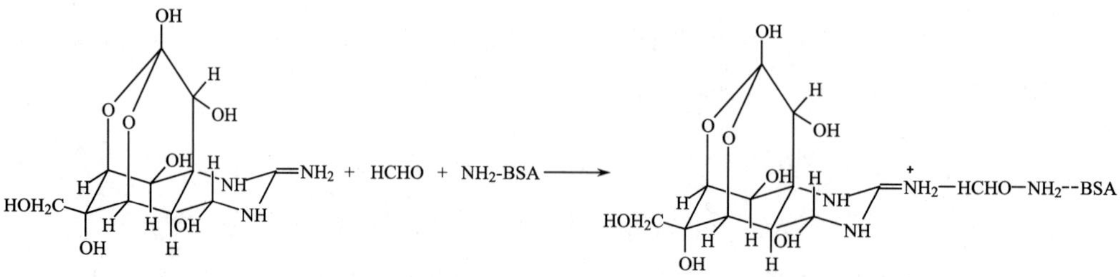

Figure 12-1   The coupling way of TTX with carrier protein BSA

Yang[3] developed a piezoelectric immunosensor through immobilization of antibody (IgY) against viperotoxin on the surface of piezoelectric quartz crystal. The conventional surface immobilization method, glutaraldehyde cross-linking was used here. The linear range of this method for viperotoxin detection was 0. 5-100 μg/mL.

12.2   Fungi and mycotoxins

12. 2. 1   Overview

Fungal contaminations are widespread in various types of food. Fungi not only makes food go bad but also produce mycotoxins, which result in poisoning. Some species of filamentous fungi, especially Penicillium, Aspergillus and Byssochlamis, are considered the main producer of mycotoxins. Maize, rice, peanuts and wheat are the most easily polluted foods by mycotoxins. The clinical symptoms of mycotoxin poisoning is complicated such as acute poisoning, chronic poisoning and carcinogenic, teratogenic and mutagenic[4].

12.2.1.1   Classification of mycotoxins

There are two types of fungal toxins that cause human poisoning. One is mycotoxins such as aflatoxins, the other is the mushroom toxins, such as Amanita toxins. In the growth of harmful mould process, it can produce toxic metabolites and remain in foods. People eat the foods that contain mycotoxins causing poisoning. And the toxic mushroom is much like domestic fungus, but they can produce lethal toxins. Once people eating these mushrooms severe symptoms of poisoning are caused. Toxin-producing fungi is only a small part of fungi family and some have been listed as follows[5].

(1) Eurotium

Aspergillus flavus, *Aspergillus ochraceus*, *Aspergillus versicolor*, *Aspergillus fumigatus*, *Aspergillus nidulans* and *Aspergillus parasiticus*.

(2) Penicillium

Penicillium islandicum, *Penicillium citrinum*, *Penicillium citreoviride*, *Penicillium rubrum*, *Penicillium expansum*, *Penicillium cyclopium*, pure green Penicillium, *Penicillium decumbens*, etc.

(3) Fusarium

Fusarium graminearum, *Gibberella zeae*, *Fusarium poae*, *Fusarium oxysporum*, *Fusarium nivale*, *Fusarium moniliforme*, *Fusarium sporotrichioides*, *Fusarium equiseti*, *Fusarium solani* and *Fusarium roseum*.

(4) Other fungi

Claviceps, amanita, helvella and alternaria.

12.2.1.2 Structure and Properties

(1) Aflatoxin

Aflatoxins are substances that are chemically related to difuronocoumarin and classified in two broad groups according to their chemical structure. One is the difurocoumarocyclopentenone (including AFB1, AFB2) and the difurocoumarolactone series(including AFG1, AFG2, AFM1, AFM2). AFB1 is the important toxin in aflatoxin family, which may be related to mutagenic and carcinogenic effects in humans. Some chemical structures of typical aflatoxin derivatives or isomers are shown in Figure 12-2.

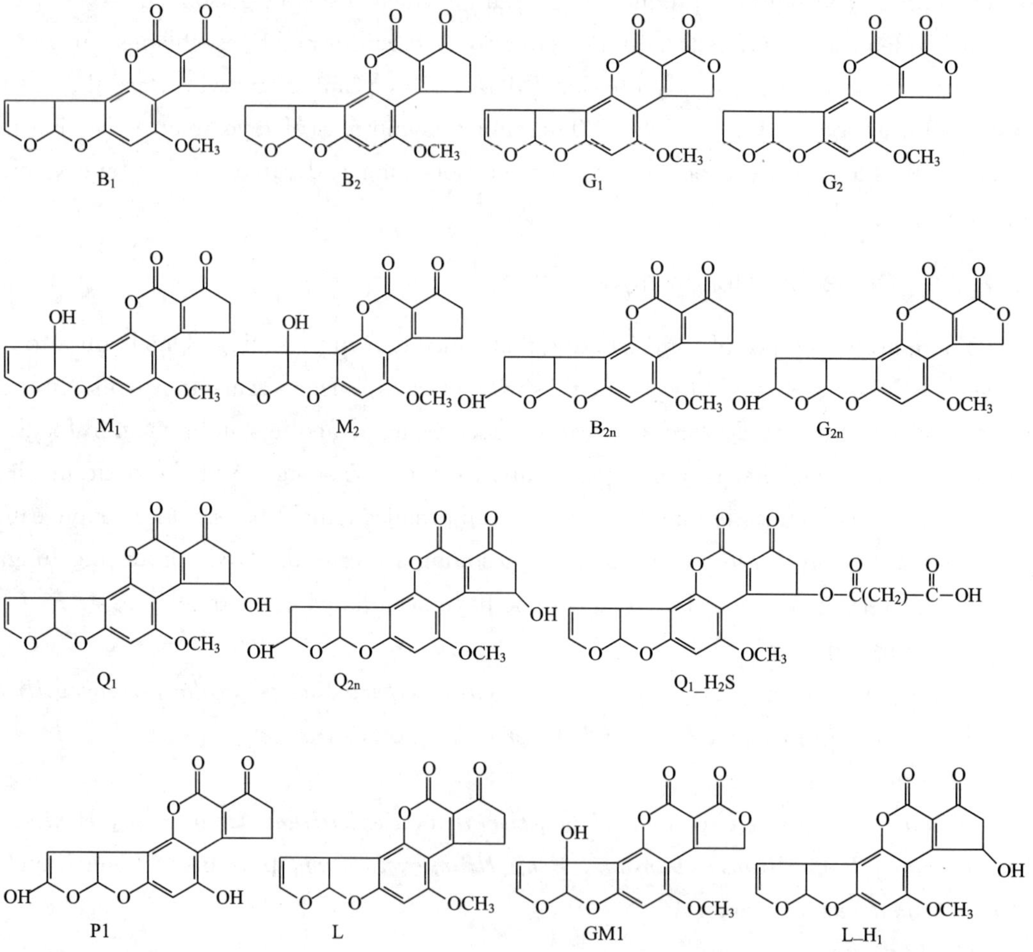

Figure 12-2   Typical Structures of various aflatoxins

The main target organ of aflatoxin is animal liver, which can cause liver damage and occurrence of liver cancer. Some animals are very sensitive to carcinogenic effects of aflatoxin and even a very small amount of toxin would induce tumors. However, other animals are relatively of strong resistance to aflatoxin, though still some serious symptoms of poisoning would appear.

(2) Ochratoxin

Ochratoxin is produced by both Aspergillus ochraceus and Penicillium viridicatum. Ochratoxin has a good heat resistance and can not be destroyed by conventional heating. Ochratoxin A (OTA), is a colourless crystalline compound, the toxin of utmost concern compared to ochratoxin B and C and all of them containing dihydroisocoumarin. OTA is soluble in polar organic solvents and dissolves in dilute organic bicarbonate or water. There is a chlorine atom on the aromatic ring, which accounts for its toxicity (Figure 12-3). There is no chlorine atom in the structure of ochratoxin B, which results in its low toxicity.

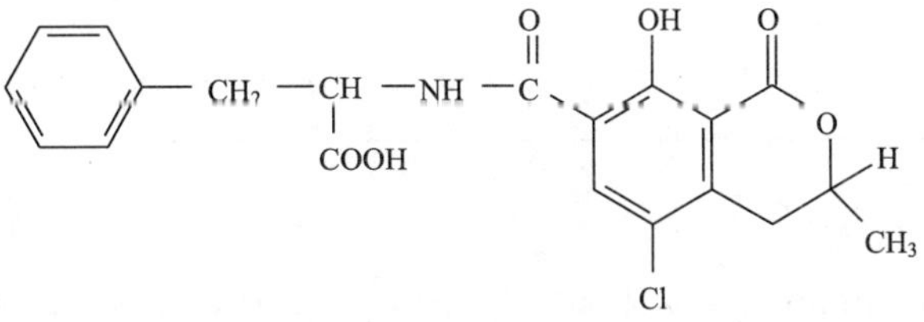

Figure 12-3   Structure of OTA

Ochratoxin has a strong toxicity on the kidneys and liver. Human and animal ingesting of toxin contaminated food and feed would course acute or chronic poisoning. Characteristics of ochratoxin poisoning are to cause structural changes in renal interstitial fibrosis and tubular dysfunction, and inflammation. It can cause serious kidney disease and acute liver dysfunction, fatty degeneration, hyaline degeneration and localized necrosis, and is also carcinogenic in long-term exposure.

(3) Sterigmatocystin

Sterigmatocystin is mainly produced by *Aspergillus versicolor*, *Aspergillus nidulans* and *Aspergillus ustus*. It is soluble in chloroform, benzene, pyridine, acetonitrile and dimethyl sulfoxide, slightly soluble in methanol and ethanol but insoluble in water and alkaline solution. Sterigmatocystin is strong carcinogenic next to aflatoxin. Sterigmatocystin can cause acute poisoning, which is characterized by liver and kidney necrosis. It can also cause subacute and chronic poisoning, leading to lobular focal necrosis, chronic hepatitis, fibrosis and cirrhosis. Sterigmatocystin can form complex with uracil through the double bond at the end of the second furan ring, affecting some enzyme activity in the blood, liver and kidney. Some typical sterigmatocystins are shown in Table 12-1.

Table 12-1　Chemical structure of some sterigmatocystins

		R_1	R_2	R_3		R	R_1
Sterigmatocystin		—H	—H	—H	Dihydro-O-methyl-sterigmatocystin	—CH₃	—CH₃
O-methyl-sterigmatocystin		—CH₃	—H	—H	Dihydro-demethoxy-sterigmatocystin	—H	—H
5-methoxy- sterigmatocystin		—H	—H	—OCH₃	Dihydro-sterigmatocystin	—CH₃	—H

(4) Trichothecenes

T-2 toxin (CAS number 21259-20-1) belongs to trichothecenes family, and can be formed by Fusarium (in particular, *Fusarium sparotrichioides*, *Fusarium poae* and *Fusarium langsethiae*), *Myrothecium*, *Stachybotrys*, *Trichoderma*, *Cephalosporium*, *Trichothecium* and *Verticimonosporium*. T-2 toxins (Figure 12-4) are chemically stable enoughstored at room temperature over 6-7 years or even heated at 200 ℃ for 2 h. T-2 toxins are highly soluble in ethyl acetate, acetone, chloroform, dichloromethane and diethyl ether. T-2 toxin is teratogenic and weak carcinogenic and is very toxic to active proliferation cells, especially to lymphocyte through inhibition of protein or DNA synthesis.

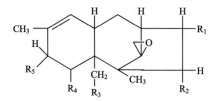

Figure 12-4　Structure of T-2 toxin
R_1=OH　R_2=OAc　R_3=OAc　R_4=H
R_5=OCOCH₂CH(CH₂)

Deoxynivalenol (DON, CAS number 51481-10-8), also known as vomiting toxins, is mainly produced by certain *Fusarium*, resulting in pollution of barley, wheat, oats, corn and other crops. It is mainly synthesized by *Neurospora* in lack of nutrients. DON (Figure 12-5) has a strong cyto-toxicity. Its chemical structure is similar to T-2 toxin where R_1, R_3 and R_4 are hydroxyl group while $R_2 = H$. Due to its highly hygroxylated chemical structure, DON is soluble in water and polar solvents.

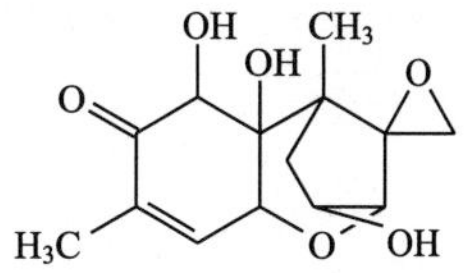

Figure 12-5   The structure of DON

(5) Citrinin

Citrinin (CAS number : 518-75-2) is originally isolated from *Penicillium citrinum* and can also be produced by *Aspergillus niveus*, *Aspergillus ochraceus*, *Aspergillus oryzae*, *Aspergillus terreus*, *Monascus ruber*, *Monascus purpureus*, *Penicillium camemberti*. Pure citrinin (Figure 12-6) is difficult to dissolve in water and can be thermaly decomposed whether in acidic or alkaline solution.

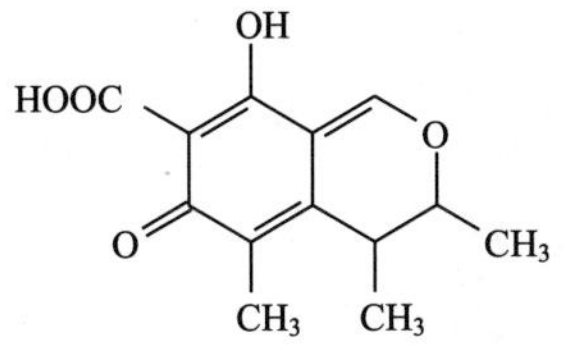

Figure 12-6   The structure of citrinin

Citrinin acts as a nephrotoxin in all species in which it has been tested, but its acute toxicity varies. It causes mycotoxic nephropathy in livestock and has been implicated as a cause of Balkan nephropathy and yellow rice fever in humans.

(6) Patulin

Patulin(CAS number 149-29-1) is produced mainly by Penicillium patulum. Other fungi such as *Penicillium expansum*, *Penicillium cyclopium*, *Penicillium papaya*, *Aspergillus terreus*, *Aspergillus clavatus*, and *Byssochlamys fulva* can also generate patulin. Patulin (Figure 12-7) is a neutral substance and is now known to be present in apple and apple products. Patulin is soluble in water, ether, acetone, ethyl acetate and chloroform, and less soluble in benzene. Patulin is more stable in

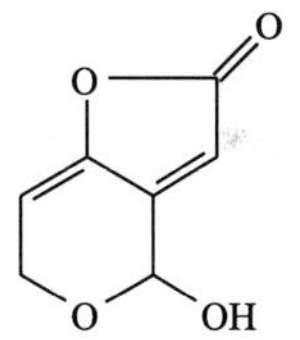

Figure 12-7   The structure of patulin

acidic solution and reduced toxicity was found under alkaline conditions. Patulin is found to be carcinogenic, and obviously teratogenic to embryo.

(7) Other mycotoxins

Luteoekyrin(Figure 12-8) is produced mainly by Penicillium islandicum. It has been shown to be carcinogenic in animals. The main symptom of Luteoekyrin poisoning is liver damages, such as lobular focal necrosis and fatty degeneration.

Citreoviridin (CIT) is produced mainly by *Penicillium citreoviride*. It is chemically stable and can persist in high temperature up to 270 ℃. The chemical structure of CIT is shown in Figure 12-9. It is soluble in acetone, chloroform, acetic acid, methanol and ethanol, slightly soluble in benzene, ether, carbon disulphidl and carbon tetrachloride, but insoluble in petroleum, ether and water. CIT was thought to be the cause of yellow rice poisoning, manifested in humans and experimental animals by respiratory and circulatory failure, paralysis, convulsions and death.

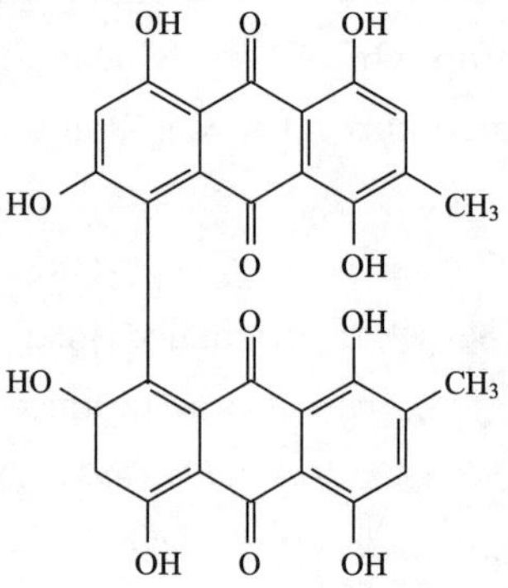

Figure 12-8　The structure of luteoekyrin

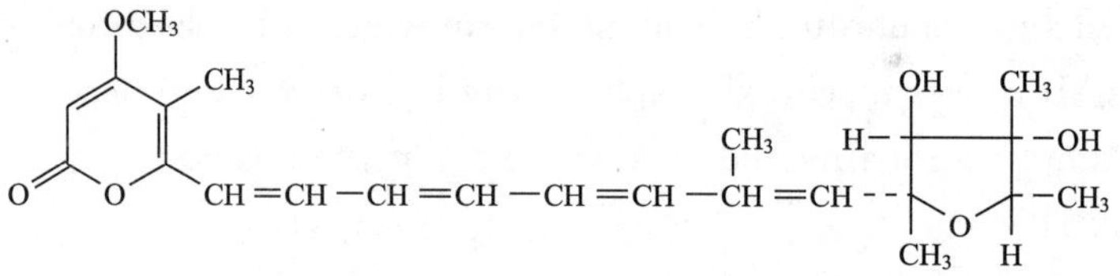

Figure 12-9　The structure of citreoviridin

12.2.2　Analytical methods

Liu[6] prepared artificial antigen of ochratoxin A using active ester method and immunized mice to screen monoclonal antibody. The cross reactivity rate of monoclonal antibody against Ochratoxin B was 35% and the cross reactivity data was less than 0.01% with citrinin, aflatoxin B1 and patulin. Competitive ELISA was 200-6,000pg/mL and detection limit was 150 pg/mL. The spiked experiments in wheat samples showed that recoveries ranged from 83% to 116% with with coefficient of variation of 9.4%-19.8%. 23 samples were analysed with this proposed method and OTA was found in 0.6-2.56 μg/kg range. Wu[7] prepared anti ochratoxin A monoclonal antibody and the antibody fragments were used to immunize rabbits to obtain anti-idiotypic antibody against ochratoxin A. The anti-idiotypic antibodies simulate the characteristics of ochratoxin A standards in antibody identification and can be used as alternative of toxin standard in immunoassay. Thus, an atoxic ELISA kits for ochratoxin A detection in cereals was successfully developed, which manifested highly sensitive, specific, safe and cheap advantages in application.

An immunochromatography strip for aflatoxins was developed by Sun[8]. The conjugate of BSA and AFB1 was synthesized by carbodiimide method and the coupling ratio was 20 : 1. The complete antigen AFB1-O-BSA was immunized in Balb/C mice. Two cell lines secreting of specific anti AFB1 monoclonal antibodies (3A7 and 5D3) were gotten through the process of cell fusion, subcloning and expansion of culture. The affinity constants were 3. 81×107 mol and 3. 43×107 mol, respectively for two cell strain. Both monoclonal antibodies belong to IgG2a family. Gold nanoparticles (the average particle size 10. 7 nm, 14. 4 nm, 25. 1 nm, 36. 9nm and 67. 4nm) was prepared. Direct competition Gold immunochromatographic assay (GICA) was established where gold nanoparticles labelled AFB1 antibody acted as probes and AFB1-O-BSA was used as competitive antigen. The optimized GICA has a linear range from 0. 01 to 2. 5 μg/kg and the minimum detection limit was 0. 019 μg/kg based on 25. 1 nm gold nanoparticles. 67 contaminated natural samples were analysed with GICA and ELISA, and good linear correlation was found with these two methods. It was shown that the salt content, metal ions (iron, copper, lead, arsenic) content and fat content in food affected the detection of GICA when it was used for feed and fermented food analysis. The secondary extraction with chloroform and petroleum ether can eliminate these impacts. The research above provided theoretical reference for the detection of AFB1 in food by GICA. Xie[9] transformed AFB1 into AFB2a, and then coupled it with protein for the preparation of complete antigen in order to obtain a wider cross-reaction range against various aflatoxin. A monoclonal antibody was successfully prepared and competitive ELISA method was established with the detection limit of 2 ng/mL. The colloidal gold (particle size 20 nm) was labelled the purified monoclonal antibody to assemble strip. Three aflatoxins including AFB2, M1 and M2 can also be detected except AFB1 (the minimum detection limit of 5 ng/mL for AFB1). An immunonephelometry based on carboxylated polystyrene latex particles (CPLP) was developed by Wang[10]. The diameters of CPLP were 240 nm and the latex was stable even when it was stored at room temperature for 15 days. The specific antibody against AFB1 was prepared by immunization of mice with AFB1-carrier protein conjugates. The sensitized conditions on latex agglutination were optimized and the minimum detection limit was 5 ng/mL.

For preparation of polyclonal antibody against patulin (PAT), Yang[11] prepared complete antigen of by carbodiimide method. In order to increase the coupling ratio between PAT and carrier protein, PAT was modified with succinic anhydride to increase its carboxyl group, while the carrier protein bovine serum albumin (BSA) was modified with ethylenediamine to increase the amine group in its structure. The derivative of PAT (PAT-BSA) was used to immunize mice to prepare antibody. The conditions for ELISA including coating buffer, blocking buffer, reaction time or temperature were optimized, which laid a basis for the preparation of patulin monoclonal antibodies.

Li[12] has studied the rapid detection of DON, including hapten design, antibody preparation and the establishment of ELISA. The conversion of DON to 3-HS-DON was facilitated by protecting two of the three available hydroxyls with a cyclic boronate ester. The C3 hydroxyl of DON reacted with succine anhydride to get hapten 3-HS-DON. Preparative TLC was used to purify 3-HS-DON and high performance liquid chromatography tandam mass spectrometry confirmed the sucsesful synthesis. The derivative products were coupled with the carrier protein BSA or OVA by carbodiimide method. Balb/C mice and cavies were immunized to prepare polyclonal antibody. The cross-reactivity for T-2 toxin was lower than 6.3%. The linear detection rang of optimized ELISA is 0.1-100 μg/mL. The proposed method was applied for DON analysis in scabofwheat seeds. Deng[13] prepared 3-hemiglutarate-deoxynivalenol (3-HG-DON) by protection of the C7-and C15-hydroxyls with a cyclic boronate ester, and then esterified with glutarate at the C3-hydroxyl. 3-HG-DON was activated by active ester method, and conjugated to ovalbumin (OVA) as detecting antigen (DON-HG-OVA) while DON was activated by 1,1-Dicyclohexylcarbodiimide (CDI) to couple with carrier protein for immunogen. A monoclonal antibody of DON was obtained and the sub-class of antibody was IgG2a. The cross reactivity between antibody and zearalenone and citrinin were less than 1%. An indirect ELISA was established with linear detection range 20-460 ng/mL. Distilled water was used to extract DON spiked wheat and corn samples (250-4,000 ng/g) and the extracts were detected by ELISA. Recoveries were 82.1%-96.6% and 85.5%-95.6%, respectively. An immuno-affinity column was prepared based on this monoclonal antibody using CNBr activated Sepharose 4B gel. Peptide from random peptide library was screened using solid phase panning method. A heptad CMRPWLQ was selected as the minic epitops of DON to develop ELISA for DON detection. The linear range of DON-phage-ELISA was 20-400 ng/mL. There was good consistency for analysis of spiked wheat and corn samples between chromatography methods and DON-phage-ELISA.

　　Zheng[14] screened stable secreting anti-T-2 toxin monoclonal antibody with cell fusion hybridoma technique. The obtained antibodies were used for development of rapid preparation of F (ab') 2 fragments by ficin digestion and the F (ab') 2 fragments were identified to have the same immunoreactivity to intact antibody. F (ab') 2 fragments were used as immunogen for the production of idiotypic antibodies. Anti-idiotypic antibodies can take the place of T-2 Toxin in non-toxic assay development. Based on the idiotypic antibodies, non-toxic ELISA was developed with the minimum detection limit 2.5 μg/kg.

12.3　Bacteria and bacterial toxins

12.3.1　Overview

Bacteria are large domain of prokaryotic microorganisms. There are typically 40 million bacterial cells in a gram of soil and a million bacterial cells in a milliliter of fresh water. Although the vast majority of bacteria are harmless or beneficial, quite a few bacteria are pathogenic. Endotoxin or exotoxin formed in or elaborated by some bacterial cells do harm to human health in different degrees.

12.3.1.1 Categories

In the process of growth and metabolism, certain bacteria can produce some toxic substances released into the surrounding environment, which was named as exotoxin. Most of bacteria producing exotoxin belong to G+ bacterial strains and these exotoxins usally are proteins, extremely sensitive to heat and certain chemicals. Exotoxin is known to be the most toxic substance in all of toxic substances, which can specifically damage certain elements of the body cells or inhibit certain metabolic pathways. Some information of bacteria producing exotoxins is shown in Table 12-2.

Table 12-2   Some exotoxins produced by bacteria

Bacteria	Habitat	Toxin	Disease
Bacillus anthracis	Soil	Complex	Anthrax
Clostridium botulinum	Soil	Neurotoxin	Botulinum
Staphylococcus aureus	Human skin	Enterotoxin	Vomiting
Diphtheria bacilli	Human	Diphtheria toxin	Diphtheria
Vibrio cholerae	Human gastrointestinal tract	Toxin	Cholera
Clostridium tetanus	Soil	Neurotoxin	Tetanus
Shigella dysentery	Human gastrointestinal tract	Neurotoxin, Neurotoxin	Dysentery
Yersinia pestis	Mice, flea	Plague toxin	Plague

Compared with exotoxin, endotoxin has the following characteristics. Firstly, endotoxin is produced in living cells. It isn't released to the surrounding medium but the cells are broken down. Mostly, endotoxin is located on the outer layer of celluar wall of G-bacterial. Secondly, most endotoxins are not proteins but lipopolysaccharides (LPS). Third, although endotoxins may be produced by different bacterials, the symptoms caused by endotoxin are similar, including fever, diarrhea or hemorrhagic shock. Fourth, the antigenicity of endotoxin is weak and shows low toxicity. The bacteria such as *Salmonella*, *Shigella*, *E. coli* and *Neisseria perflava and* so on can secret endotoxin. These bacteria are widely present in the intestinal tract, faeces or faecal polluted environment.

12.3.1.2 Common toxins

(1) Clostridium botulinum toxin

Clostridium botulinum exotoxin is known to be one of the strongest toxins. The minimum lethal dose of injecting subcutaneously in mice was 0. 000,12 mg/kg. The different types of Clostridium botulinum produce five kinds of Clostridium botulinum neurotoxin (from A to E), which have similar function on the nerve endings, but the sensitivity of different types of Clostridium botulinum neurotoxin for various animals are obviously different. Each antitoxin could counteract the same type of toxin.

The occurrence of animal botulism poisoning is related with the intake of feed containing this toxin. Ducks are sensitive to this toxin, showing depression, weight loss and muscle paralysis ("soft neck"). The sick ducks often die of their inability to drink, and, therefore, dehydration. The cattle, sheep deficient in calcium and phosphorus often intake toxins after they eat rotten corpse bones. Clostridium botulinum toxin is produced in the anaerobic environment while Clostridium botulinum located in intestines seldom produces toxin.

(2) Tetanus toxin

Tetanus toxin is a potent neurotoxin (150,000 Da) produced by the anaerobic bacterium Clostridium tetani. It is also called spasmogenic toxin, tetanospasmin or abbreviated to TeTx or TeNT. The LD50 of this toxin has been measured to be approximately 1 ng/kg, making it second only to Botulinin toxin D as the deadliest toxin in the world. Titanic spasms can occur in a distinctive form called opisthotonos and be sufficiently severe to fracture long bones. The shorter nerves are the first to be inhibited, which leads to the characteristic early symptoms in the face and jaw, risus sardonicus and lockjaw. The toxin bind to the neurons is irreversible and nerve function can only be returned by the growth of new terminals and synapses.

(3) Gas gangrene toxins

Gas gangrene toxin (GGT) is a toxin produced by the bacterium *Clostridium*. It is a zinc metalloenzyme and binds to the membrane in the presence of calcium. It is a phospholipase. Gas gangrene is caused by exotoxin-producing Clostridial species (most often *Clostridium perfringens*, and *C. novyi* but less commonly *C. septicum* or *C. ramnosum*), which are mostly found in soil but also found as normal gut flora, and other anaerobes (e. g. bacteroides and anaerobic streptococci). The exotoxin is commonly found in *C. perfringens* type A strain and is known as alpha toxin. These environmental bacteria may enter the muscle through a wound and go on to proliferate in necrotic tissue and secrete powerful toxins. These toxins destroy nearby tissue, generating gas at the same time. A gas composition of 5. 9% hydrogen, 3. 4% carbon dioxide, 74. 5% nitrogen and 16. 1% oxygen was reported in one clinical case.

12. 3. 2　Immunoassay for bacterial toxins

Li[15] used colloidal gold immunochromatographic technology in sandwich format to detect bacillus anthracis. The sensitivity and specificity of the method were evaluated.

"White powder" made of flour, starch, milk powder were used as a matrix and different levels of bacillus anthracis were mixed with this white powder, which verified the optential application of this proposed method in anthrax detection. The detection limit of bacillus anthracis was 1×10^6 cfu/mL.

Lv[16] established an indirect competitive ELISA to detect avian cholera. Three coating antigens including products of supersonic schizolysis, lipopolysaccharides and killed cholera were optimized and the best sensitivity was achieved using killed cholera as coating antigen in ELISA.

Duan[17] prepared polyclonal antibodies of staphylococcal enterotoxin B (SEB) and developed an indirect competitive ELISA for detection SEB in milk samples. SEB was extracted from fermentation broth of Staphylococcus aureus and purified with ultrafiltration, ammonium sulphate precipitation and anion exchange chromatography. The SDS-PAGE electrophoresis was applied to confirm its purity. The antibody was obtainal by immunization of mice. Under the optimized conditions, the detection range of this proposed ELISA was 1-1,000 ng/mL.

Escherichia coli O_{157} is the main strain in shiga-producing. Shiga toxin (Stxs) is the main toxin of *Escherichia coli* O_{157}. There are two types of Stxs including Stxs1 and Stxs2. Stxs2 is secreted outside of cell and causes diseases at very low dosage. Thus, the capacity of secreting Stxs2 represents the strength of toxicity of bacterial strain. Gao[18] transferred the recombinant plasmid Stx2B subunit (pGEX-Stx2B) into *E. coli* BL21. After identification by PCR and enzyme digestion, the positive transformation bacteria was chosen for expression of recombinant protein. The recombinant Stx2B was used to immunize rabbits to prepare polyclonal antibody. At the same time, the *E. coli* O_{157} ATCC43889 was used superantigen to immunize rabbits. The antiserum was purified with Protein G affinity chromatography. Average size of 20 nm of colloidal gold nanoparticles was prepared, which was used to label antibody. A lateral flow test strip in sandwich format was assemblied. The established strip was used to test 58 strains and the results were compared with immuno-agglutination. The consistence rate of two methods was 87.5% and the detection limit of strip was 1×10^5 cfu/mL of *E. coli* O_{157}.

12.4   Plant toxins

12.4.1   Introduction

Plants synthesize a broad range of secondary metabolites, including alkaloids and terpenoids that are toxic to herbivores and pathogens, and so are believed to act as defense compounds. Classical examples of plants that are poisonous to humans, such as poison hemlock, foxglove, and aconite, demonstrate how well natural products can defend plants, at least against mammalian herbivores[19].

All plant compounds that have negative effects on the growth, development or survival of another organism can be regarded as toxins. Plant-associated toxins (PATs), present a cost to agriculture and a risk to animal and human consumers of food that may be contaminated by such chemicals. The mechanisms of action of some plant toxins are well known. For example, saponins disrupt cellular membranes, hydrogen cyanide released from cyanogenic glycosides inhibits cellular respiration, and cardenolides are specific inhibitors of the Na^+/K^+-ATPase. But the modes of action of many other toxins still wait discovery.

12.4.1.1 Plant phenols

Phenolic molecules produced by higher plants in response to biotic and abiotic stresses exert numerous effects in their life. The typical phenolic compounds including tannic acid, salicylates, rotenone, dicoumarol, and tetrahydrocannabinol (THC) and so on show acute toxicity to mammals, especially to predators. Phenolic compounds may affect the bioengergetics in action of oxidative phosphorylation uncoupler. Some polyphenols with better fat-solubility do harm to the permeability of the membrane and they also conjugate with some metals. Plants such as ginkgo, cotton seeds, pinellia and sumac are rich in phenolic compounds. Cottonseed contains gossypol, which results in slow growth, poor appetite and weight loss of animals. Gossypol inhibits the synthesis of prothrombin, causes anemia by formationof iron complex, and also affects the application of some amino acids.

12.4.1.2 Cyanogenic compounds

The cyanogenic glycosides belong to the products of secondary metabolism, to the natural products of plants. These compounds are composed of an a-hydroxynitrile typeaglycone and of a sugar moiety (mostly D-glucose). The distribution of the cyanogenic glycosides (CGs) in the plant kingdom is relatively wide, the number of CG-containing taxais at least 2,500, and a lot of such taxa belong to families Fabaceae, Rosaceae, Linaceae, Compositae and others.

12.4.1.3 Alkaloids

One of the largest groups of chemical arsenals produced by plants is the alkaloids. Many of these metabolic by-products are derived from amino acids and include an enormous number of bitter, nitrogenous compounds. More than 10,000 different alkaloids have been discovered in species from over 300 plant families. Alkaloids often contain one or more rings of carbon atoms, usually with a nitrogen atom in the ring. The position of the nitrogen atom in the carbon ring varies with different alkaloids and with different plant families. In some alkaloids, such as mescaline, the nitrogen atom is not within a carbon ring. In fact, it is the precise position of the nitrogen atom that affects the properties of these alkaloids. Although they undoubtedly existed long before humans, some alkaloids have remarkable structural similarities with neurotransmitters in the central nervous system of humans, including dopamine, serotonin and acetylcholine. The remarkable signs of alkaloid poisoning include nervousness, weakness, salivation, nausea, bloating, rapid heart rate, death. Excitement and physical exercise after ingesting large amounts can intensify all signs of poisoning. The amazing effect of these alkaloids on humans has led to the development of powerful pain-killer medications, spiritual drugs, and serious addictions by people who are ignorant of the properties of these powerful chemicals.

12.4.1.4 Terpenoids

Terpenoid phytoalexins are low molecular-weight, antimicrobial compounds that are synthesized by, and accumulate in, plants after exposure to microorganisms. It has long been known that the basic unit of most secondary plant metabolites, including terpenes, consists of isoprene, a simple hydrocarbon molecule.

Terpenoids (terpenes) occur in all plants and represent the largest class of secondary metabolites with over 22,000 compounds described. The simplest terpenoid is the hydrocarbon isoprene (C_5H_8), a volatile gas emitted during photosynthesis in large quantities by leaves that may protect cell membranes from damage caused by high temperature or light. Terpenoids are classified by the number of isoprene units used to construct them. For example, monoterpenoids consist of two isoprene units, sesquiterpenoids (three units), diterpenoids (four units), and triterpenoids (six units). Terpenoids exist in various forms of resin, bitter elements, latex, oil, pigments. Sesquiterpene lactones, diterpenoids and triterpenoids are the most toxic terpenoids by causing allergic contact dermatitis, cell poisoning and even death. Grayanotoxins in diterpenoids family is highly toxic to heart and nervous system of mammals. Triterpenes in forms of resin, latex, and bitter elements and so on, will cause liver necrosis or death.

12.4.1.5 Other toxic compounds from plants

Plant seeds are rich sources of lectins especially those of the Leguminosae. This explains why eating raw peas or beans may lead to severe poisoning in man. The toxicity of lectins, however, is different from substance to substance. Ricin for example, is extremely toxic. Other shows a mild toxicity only. Some vegetable oils containing certain specific fatty acids interfere with metabolic processes of animals after consumption of this oil.

12. 4. 2　Immunoassay for plant toxins

The tertiary structure of ricin was shown to be a globular, glycosylated heterodimer of approximately 60-65 kDa. Ricin toxin A chain and ricin toxin B chain are of similar molecular weight, approximately 32 kDa and 34 kDa respectively. Hao[20] used ricin pretreated with formaldehyde to immunized mice and prepared anti-ricin monoclonal antibody. The antibody was characterized to react to epitops from chian A. Gold nanoparticles at 20 nm were prepared and used to label antibody for developing immunochromatographic strip. The sensitivity of the test strip in sandwich format was 1 $\mu g/mL$.

Zhao[21] prepared a hapten, 4-O-succinyl-podophyllotoxin for podophyllotoxin(Figure 12-10). Artificial antigens of podophyllotoxin were successfully synthesized with mixed anhydride method and the carbodiimide method. Balb/C mice were immunized for monoclonal antibody development. An indirect competitive ELISA method was established based on this antibody with IC_{50} value of 6. 97 ng/mL. The cross-reactivity rates with deoxypodophyllotoxin, β-Ahpodophyllotoxin and podophyllotoxin hydrazide were detected (11. 6%, 0. 67%, and less than 0. 07%, respectively). Under the optimum conditions, vegetable samples were analyzed with this immunoassay with linear detection range 0. 13-390 ng/mL. Good correlation was obtained compared with HPLC analysis. Xu[22] also used the conjugates of 4-O-succinyl podophyllotoxin and BSA to prepare polyclonal antibody by immunization of rabbits. The IC_{50} value of the antibody was 5. 6 ng/mL and low cross-reactivity between etoposide glycosides, picropodophyllotoxin and podophyllotoxin acid hydrazine were found.

podophyllotoxin　　　　　　　　　　　　4-O-succinyl podophyllotoxin

Figure 12-10　Chemical structure of podophyllotoxin and its hapten

A toxic protein (P1) was purified by ammonium sulphate precipitation and chromatography from the seeds of *Abrus precautorius L.* The P1(molecular weight 64,300 Da) can be split into two polypeptide chains with beta-mercaptoethanol: chain A(35,400 Da) and chain B(29,300 Da). The purity 97% of P1 was obtained and used to prepare antibody after treatment with formaldehyde. Song[23] prepared monoclonal antibody using P1 protein. Very low cross-reaction with ricin, ricinus communis agglutinin or abrus agglutinin was found. A colloidal gold immunochromatography strip was developed with detection limits 0.1-0.5 μg/mL.

Swainsnine (SW) is an indolizine alkaloid. It is a potent inhibitor of Golgi alpha-mannosidase II, an immunomodulator, and a potential chemotherapy drug. As a toxin in locoweed (likely its primary toxin) it also is a significant cause of economic losses in livestock industries. Gou[24] extracted SW from locoweed and purified the products with chromatography. The produced SW was identified with mass spectrum, infrared spectrum and melting point analysis. Based on the SW products, Gou prepared complete antigens by coupling it with carrier protein (OVA, BSA and HSA). Mice were immunized by antigens and polyconal antibody against SW was achieved.

12.5   Algae and phycotoxins

12.5.1   Overview

Phycotoxins which are produced by a tiny single-cell microalgae, are accumulated into the marine food along the food chains. Algae toxins may produce different toxic effects, such as paralysis, diarrhea, short-time memory impairment, nerve poisoning. Humans poisoning cases caused by algal toxins are common throughout the world, mainly due to the accumulation of these toxins through consumption of marine food. Also, algae toxins from freshwater have led to poisoning of livestock.

Some shellfish, especially mussels, scallops, oysters and clams are major vectors for toxins that are implicated in several human toxic syndromes. These toxins include amnesic shellfish poisoning (ASP), diarrheic shellfish poisoning (DSP), paralytic shellfish poisoning (PSP), neurotoxic shellfish poisoning(NSP) and azaspiracid poisoning (AZP). Some toxins accumulated in fish such as ciguatoxin has led to human toxicosis and animal death.

12.5.1.1   PSP toxins

The PSP toxins are basic, water-solutable compounds and potent neutrotoxins with great harm to human health. Marine dinoflagellates have been identified as the progenitors of PSP toxins that contaminate bivalve shellfish. To date, there are over 20 known kinds of PSP toxins, which are mainly distributed in tropical and temperate climate zone. The parent structure of PSP toxin is shown in Figure 12-11. The PSP toxins vary by differing combinations of hydroxyl and sulphate substituents at four sites (R_1-R_4). These toxins can classified into caramates, sulfocarbamoyl, decarbamoyl and deoxyde-carbamoyl toxins, which demonstrate different toxicity, with the carbamate toxins being the most toxic.

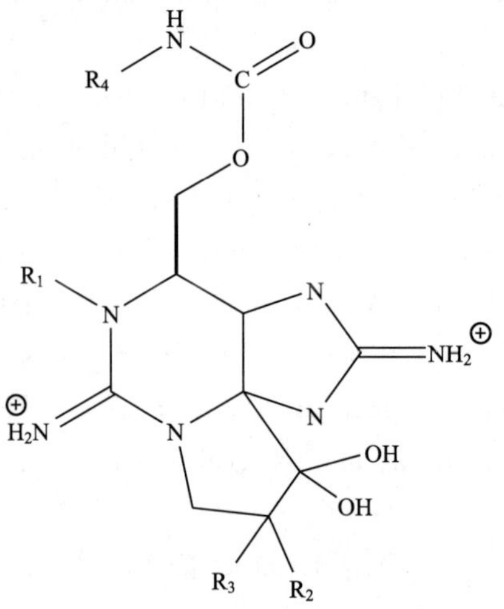

Figure 12-11　The chemical structure of the most common PSP toxins

$R_4=H$; carbamoyl;STX　$R_1=R_2=R_3=H$;GTXII　$R_1=R_3=H$, $R_2=OSO_3^-$;GTXIII　$R_1=R_2=H$, $R_3=OSO_3^-$;NEO　$R_1=OH$, $R_2=R_3=H$;GTXI　$R_1=OH$, $R_2=OSO_3^-$, $R_3=H$

12.5.1.2　The DSP toxins

DSP was first reported in Japan in 1978, which is a widely distributed seafood contamination and is now recognized as an important threat to public health all around the world. Three classes of DSP include ① dinophysistoxins ② pectenotoxins(PTX) and ③yessotoxins (YTXs). Bivalve shellfish can accumulate DSP toxins through filter feeding on the dinoflagellate, *Dinophysis fortii*. Most incidents of DSP involved the demethyl analogue, okadaic acid(OA), which arising from *Dinophysis* sp. Serious diarrhetic effecs have been proved for DTXs, especially OA, DTX2 and DTX1 are known potent inhibitors of protein phosphatases and tumour promoters. Some identified DTX toxins are shown in Figure 12-12.

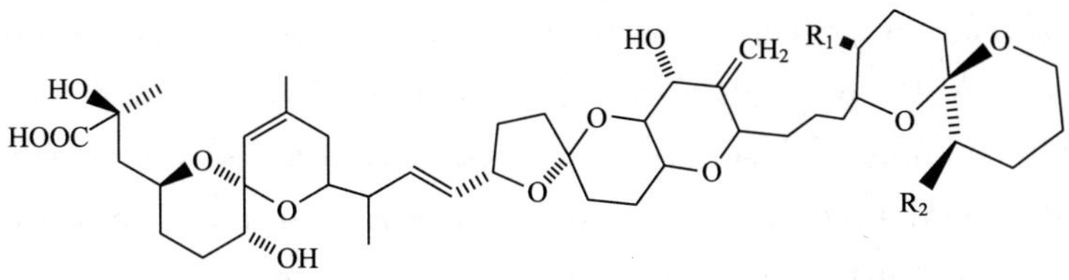

Figure 12-12   The chemical structure of DTXs

structure of DTXs: OA, R_1 = methyl; R_2 = H; DTX1, R_1 = R_2 = methyl; DTX2, R_1 = H, R_2 = methyl

12.5.1.3   AZP toxins

The dinoflagellate, *Protoperdidinium crassipes*, was identified as the responsible organism for the production of azaspiracids, which results in toxic syndrome of humans such as nausea, vomiting, diarrhoea and abdominal cramps. Besides, azaspiracids shows toxicity to liver, spleen, the small intestine and even cause cancer.

12.5.1.4   NSP toxins

The polycyclic ether toxins, brevetoxins(PbTx), are responsible for NSP causing massive kills of marine animals and threaten human health. Brevetoxins are depolarizing substances that open voltage gated sodium ion channels in cell membranes, leading to uncontrolled sodium ion influx into the cell. The effects of NSP are felt in 30-60 min, including chills, headache, diarrhoea, muscle weakness, muscle and joint pain, nausea and vomiting, double vision and difficulty in swallowing. NSP has not been known to be fatal and the extinction of these symptoms will be resolved in a few days.

12.5.2   Immunoassay

A competitive indirect ELISA was developed for measurement of okadaic acid (OA) by Liu[25]. OA was coupled to BSA and OVA by carbodiimide reaction. OA-BSA was used as immunogen to immunize mice. Titres of the antisera against OA were determinated using OA-OVA as coating antigen by ELISA method. The spleen cells of immunized mice were fused with Sp2/0 cells. Four hybridoma cell strains stably produced ant-OA monoclonal antibody were obtained. Under optimal condition, the competitive indirect ELISA for OA in shellfish was established with the detection limit 31.2 μg/L. Lu[26] also prepared anti-OA monoclonal antibody. Two formats including direct and indirect ELISA were used to detect OA in seafood. The minimum detection limits were 0.6 μg/L and 0.18 μg/L, respectively. Jiao[27] prepared antiserum against pectenotoxins (PTX), which was purified and labelled with horseradish peroxidase for the establishment of a direct enzyme-linked immunosorbent assay for PTX detection. The detection range of this proposed ELISA for PTX was 5-10 μg/mL.

Xu[28] established a competitive enzyme immunoassay method of domoic acid (DA), which belongs to ASP toxins. DA was directly coupled to carrier proteins and immunized mice. After immune for several times, polyclonal antiserum against DA was obtained. DA-OVA were used as coating antigen to establish an ELISA for DA analysis. The results showed that the minimum detection limit of seawater was 10 μg/L.

To obtain hybridoma cell line which can secrete anti-gonyautoxin II (GTX II) and anti-gonyautoxin III (GTX III) monoclonal antibodies(mAb), the immunogen (GTX-BSA) and solid phase immunogen (GTX-KLH) were prepared with aldehyde conjugation method by Tang[29], in which, GTX II and GTX III were conjugated with BSA and KLH respectively. The spleen cells from immunized BALB/c mice were fused with murine Sp2/0 myeloma cells. The specific hybridoma cells were selected to prepare the ascites and the titer of purified monoclonal antibody was determinated as 1.4×10^5. Based on this monoclonal antibody, Luo[30] developed direct and indirect competitive ELISA for detection of GTX II and GTX III. The IC_{50} values of this proposed two ELISA were 10 μg/mL and 4 μg/mL.

Yuan[31] prepared various antisera with various red tide algae including *Chaetoceros*, *Heterosigma akashiwo*, *Prorocentrum micans*, and *Gymnodiniumsanguineum hirasaka*. These monoplast algae were directly immunized animals. Three animal models including a mouse, an egg-laying hen and a rabbit were used and best titer of antisera were obtained from rabbit. The anisera were found with serious cross-reactions among other algae.

References

[1]　Chen NQ. Practical biological toxins. Beijing:Science and Technology Press, 2001:477

[2]　Zhou XC. Preparation of monoclonal antibodies of tetrodotoxin and establishment of ELISA method [Dissertation]. Changchun:Jilin University, 2008

[3]　Yang YY, Pu J H, Liu ZM. Development of a piezoelectric immunosensor array for viper venom. Beijing Biomedical Engineering, 2008, 27 (3): 289-291

[4]　Shi X M. Food safety and hygiene. Beijing:Agriculture Press, 2003:56

[5]　Zhao DY. Food additives and contaminants. Beijing:Standard Press, 2003:632

[6]　Liu RR, Yu Z, He QH. Study on competitive enzyme-linked immunosorbent assay for ochratoxin a monoclonal antibody determination. Food Science, 2005, 26 (11): 174-177

[7]　Wu Y. Study of substitution technology for ochratoxin a in immunoassay and development of innocuous ELISA-Kit quantitative analysis [Dissertation]. Changchun:Jilin University, 2006

[8]　Sun XL. Study on gold labelled immunochromatographic assay for detection of aflatoxin B1 in Foods [Dissertation]. Wuxi:Jiangnan University, 2005

[9]　Xie GH. Development of gold immunochromatography assay for rapid detection of aflatoxin B1 [Dissertation]. Changchun:Jilin University, 2008

[10]　Wang P. Application of carboxylated polystyrene latex particles in the detection of aflatoxin B1 [Dissertation]. Wuhan:Huazhong Agricultural University, 2007

[11] Yang XQ. Synthesis of complete hapten and the preparation of polyclonal against patulin [Dissertation]. Wuhan: Huazhong Agricultural University, 2006

[12] Li H. The artificial antigen synthesis and detection of deoxynivalenol using indirect competitive ELISA [Dissertation]. Nanjing: Nanjing Agricultural University, 2003

[13] Deng SZ. Development of phage enzyme-linked immunosorbent assay for detecting deoxynivalenol and preliminary study on immunoassay of patulin [Dissertation]. Nanchang: Nanchang University, 2006

[14] Zheng J. Research on preparation of multi-cloning anti-ldiotype antibody of T-2 toxin and development of nontoxic ELISA quantitative detection reagent cases [Dissertation]. Zhengzhou: Zhengzhou University, 2006

[15] Li W. To Build the fast detection of bacillus anthracis by using the technology of gold-immunochromatography. Chinese Journal of Frontier Health and Quarantine 2004, 27 (6): 329-331

[16] Lv XJ. Establishment of ELISA with whole cell antigen for detection anti-cholera antibody. Animal Husbandry and Veterinary Medicine, 2008, 40 (5): 85-87

[17] Duan HB. Enzyme linked immunosorbent assay staphylococcal enterotoxin B from milk [Dissertation]. Yangling: Northwest A & F University, 2008

[18] Gao CX. Rapid gold immunochromatography assays for detection of E. Coli O157 and shiga toxin [Dissertation]. Shanghai: Shanghai Jiao Tong University, 2008

[19] Wittstock U, Gershenzon J. Constitutive plant toxins and their role in defense against herbivores and pathogens. Current Opinion in Plant Biology, 2002, 5(4): 300-307

[20] Hao LQ. The application and preparation of monoclonal antibody of ricin. Chinese Journal of Cellular and Molecular Immunology, 2003, 19 (6): 517-518

[21] Zhao P. Preparation and application of anti-podophyllotoxin monoclonal antibody [Dissertation]. Yangling: Northwest A & F University, 2006

[22] Xu DM. Studies on the Immunochemistry for residue analysis and safety assessment of botanical podophyllotoxin [Dissertation]. Yangling: Northwest A & F University, 2005

[23] Song WX. Development of the immunochromatograpic colloidal gold dipsticks for abrins [Dissertation]. Changchun: Jilin University, 2008

[24] Gou C. Studies on synthesis of three artificial antigens of swainsonine and their immunogenicity [Dissertation]. Yangling: Northwest A & F University, 2008

[25] Liu RY. Preparation and application of the monoclonal antibody against okadaic acid Journal of Hygiene Research, 2008, 37 (4): 443-445

[26] Lu SY. The study on rapid detection assays for okadaic acid from the shellfish [Dissertation]. Changchun: Jilin University, 2007

[27] Jiao H, Liu BY, Wang J J. Direct ELISA detection of palytoxin. Inspection and Quarantine Science, 2005, 15: 8-10

[28] Xu DY. Study on detection method of amnesic shellfish poisoning(ASP). Marine Environmental Science, 2006, 26 (3): 237-240

[29] Tang Q. Preparation and character analysis of monoclonal antibodies against GTX2,3 [Dissertation]. Guangzhou: Jinan University, 2005

[30] Luo HW. Production and characterization of monoclonal antibodies against GTX2,3 and the primary development of enzyme-linked immunosorbent assay for the detection of GTX2,3 [Dissertation]. Guangzhou: Jinan University, 2006

[31] Yuan HG. Antibody preparation and ELISA development for some algae from red tide [Dissertation]. Qingdao China Ocean University, 2005

Chapter 13　Immunoassay for Other Residues

13.1　Dioxins

13. 1. 1　Introduction

Dioxins represent a class of dioxin compounds. In chemistry, a dioxin is a heterocyclic 6-membered ring, where 2 carbon atoms have been substituted by oxygen atoms. Dioxin-like compounds include a diverse range of chemical compounds which are known to exhibit "dioxin-like" toxicity. They are made of two six carbon-atom aromatic rings connected together by one or two bridging oxygen atoms, including polychlorinated dibenzo-p-dioxins (PCDDs) and polychlorinated dibenzofurans (PCDFs). PCDDs and PCDFs are two groups of planar tricyclic compounds that have similar structures, physical-chemical properties and toxicity. Some Co-PCBs (coplanar polychlorinated biphenyls) have similar chemical structures, properties and toxicity to dioxins. Since "The Law Concerning Special Measures against Dioxins" took effect in Japan in July 1999 and executed in January 2000, dioxins are known including three groups of substances: PCDDs, PCDFs as well as Co-PCBs[1,2].

13.1.1.1　Structures and properties

Figure 13-1 showed the structures of dioxins[3]. Numbers of carbon on aromatic rings represent potential substitute positions for chlorine atoms and there are potential 8 sites in case of PCDDs, PCDFs and 10 for PCBs. Depending on the number and position of chlorine atoms on aromatic rings, there are 75 dioxins compounds, 135 furans and 209 PCBs representing 419 congeners. In the case of co-PCBs, the toxic configuration usually has 4-7 chlorine atoms in its rings, representing 13 allotropes.

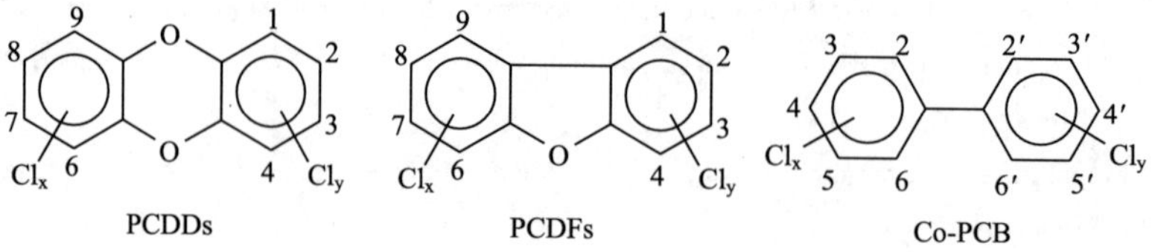

Figure 13-1　Structures of dioxins

Dioxins are highly stable and lipophilic compounds, which are extremely persistent in the environments. They are almost insoluble in water, but soluble in many kinds of organic solvents. They showed intense affinity to soil and sediments, easily accumulate in fat tissue of living organisms. Dioxins are also resistant to microbiological deterioration, hydrolization and photochemical degradation, and resist degradation[4].

13.1.1.2 Toxicity

Dioxins are a class of chemical contaminants that are formed during combustion processes such as waste incineration, forest fires, and backyard trash burning, as well as during some industrial processes such as paper pulp bleaching and herbicide manufacturing. The most toxic chemical in the class is 2,3,7,8-tetrachlorodibenzo-para-dioxin (TCDD). The highest environmental concentrations of dioxin are usually found in soil and sediment, with much lower levels found in air and water[5].

Humans are primarily exposed to dioxins by eating food contaminated by these chemicals. Dioxin accumulates in the fatty tissues, where they may persist for months or years[6]. People who have been exposed to high levels of dioxin have developed chloracne, a skin disease marked by severe acne-like pimples. Studies have also shown that chemical workers who are exposed to high levels of dioxins have an increased risk of cancer. Other studies of highly exposed populations show that dioxins can cause reproductive and developmental problems, and an increased risk of heart disease and diabetes. More research is needed to determine the long-term effects of low-level dioxin exposures on cancer risk, immune function, and reproduction and development.

13.1.2 Immunoassay for doxins

A quartz crystal microbalance (QCM) immunosensor was developed for the detection of 2,3,7,8-tetrachlorodibenzo-p-dioxins (TCDD) in environmental pollutants by Shigeru[7]. An anti-TCDD antibody was immobilized on the gold surface of the QCM *via* chemical coupling, and its immunologic activity was then maintained by treatment with an artificial stabilizing reagent such as poly(2-methacryloyloxyethyl phosphorylcholine-co-n-butyl methacrylate). A competitive immunoreaction with TCDD conjugated ovalbumin (TCDD-ovalbumin) was used to detect TCDD. A calibration curve was obtained through the competitive immunoreaction, and linearity was shown from 100 ng/mL to 0.1 ng/mL. Also, the cross-reactivities of the anti-TCDD monoclonal antibody were thoroughly evaluated with several TCDD derivatives. The relationships between GC-MS, ELISA, and QCM were compared using fly ash samples from a municipal solid waste, which were prepared using an accelerated solvent extractor. For 23 samples, the experimental relationship between the TCDD concentration by QCM *vs.* the TCDD concentration by ELISA was $y = 1.07x + 2.70$, $r = 0.99$, and the TCDD concentration by QCM *vs.* the toxic equivalent quantity (TEQ) value by GC-MS was $y = 2.46x - 14.98$, $r = 0.89$.

The development of an enzyme-linked immunosorbent assay (ELISA) based on polyclonal antibodies for the polychlorinated dibenzo-p-dioxins is described by Sugawar[8]. The hapten of 2,3,7,8-tetrachlorodibenzo-p-dioxin (TCDD) was prepared and used to produce antibody. Antisera were screened with seven different coating antigens (hapten-protein conjugates), including trans-3-(7,8-dichlorodibenzo-p-dioxin-2-yl)-cis-2-methylpropeno ic acid (Ⅶ) and 5-(3,7,8-trichlorodibenzo-p-dioxin-2-yl)penta-trans, trans-2,4-dien oic acid (Ⅹ). All inhibition screening and optimization studies were conducted using a less toxic surrogate standard for TCDD [2,3,7-trichloro-8-methyl-dibenzo-p-dioxin (TMDD; ⅩⅦ)] which responded similarly to 2,3,7,8-TCDD in the ELISA. The most sensitive assay from the screening studies [coating antigen Ⅶ-BSA, 0.1 microgram/mL, and antiserum 7,598 (anti-X-LPH), 1 : 10,000] was further optimized and characterized. It exhibited an IC_{50} value of 12 pg/well (240 pg/mL), with working range from 2 to 240 pg/well (40 to 4,800 pg/mL). The influence of various physical and chemical factors (time, solvent, detergent) was investigated. The optimized assay was then used to assess cross-reactivity by congeners of halogenated dioxins and related structures. DMSO up to concentrations of 37.5% decreased the IC_{50} value in the assay, whereas methanol to concentrations of 30% did not lead to improved IC_{50} values.

13.2　Polybrominated biphenyl

13.2.1　Introduction

Polybrominated diphenyl ethers (PBDEs) are a class of organic compound based on bromine atoms, its chemical formula are $C_{12}H_{(0-9)}Br_{(1-10)}O$. Chemical structure of PBDEs was shown in Figure 13-2.

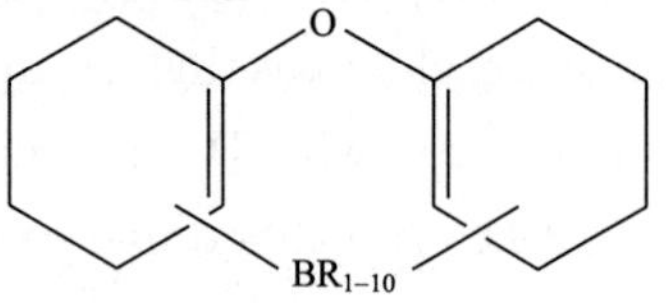

Figure 13-2　Chemical structure of PBDEs

PBDEs are widely used as flame retardants in a variety of consumer products such as textiles, furniture, building material, vehicle and electronics. As non-reactive flame retardants, PBDEs can easily enter into environment through various routes. At present, PBDEs were found in various environmental matrices (such as air, sediment, soil, indoor air and different organisms) and human body. Moreover, polybrominated dibenzo-p-dioxin (PBDDs) and polybrominated dibenzofuran (PBDFs) were formed during the preparation, combustion and thermal decomposition of PBDEs. The persistence and potential hazard of PBDEs have heightened concerns about their immigration in environment and control[9]. PBDEs are classified according to the average number of bromine atoms in the molecule There are a total of 209 PBDE congeners. The typical PBDEs were tetra-BDEs, penta-BDEs, hexa-BDEs, octa-BDEs and deca-BDEs.

PBDEs shared similar in their chemical structure, persistence and environmental distribution. PBDEs have low vapour pressures and high fat-solubility at ambient temperature. They are structurally akin to the PCBs and other polyhalogenated compounds, consisting of two halogenated aromatic rings. The health hazards of these chemicals have attracted increasing scrutiny, and they have been shown to reduce fertility in humans at levels found in households[10].

13. 2. 2　Immunoassay for PBDEs

A sensitive magnetic particle enzyme-linked immunoassay (ELISA) was developed by Weilin[11] to analyze polybrominated diphenyl ethers (PBDEs) in water, milk, fish, and soil samples. The assay was rapid and can be used to analyze fifty samples in about 1h after sample cleanup. The assay has a limit of detection (LOD) below 0. 1 ppb towards the following brominated diphenyl ether (BDE) congeners: BDE-47, BDE-99, BDE-28, BDE-100, and BDE-153, with the LOD approximately the same as GC-NCI-MS. The congeners most readily recognized in the ELISA were BDE-47 and BDE-99 with the cross-reactivities of BDE-28, BDE-100, and BDE-153 being less than 15% relative to BDE-47. As anticipated, the sensitivities are proportional to the similarities between the hapten structure and the BDE congener structure. Some oxygenated congeners with structural similarity to the hapten showed high to moderate cross-reactivities. Very low cross-reactivity was observed for other PBDEs or chlorinated environmental contaminants. The assay gave good recoveries of PBDEs from spiked water samples and a very small within and between day variance. Comparison with GC-NCI-MS demonstrated the ELISA method showed equivalent precision and sensitivity, with better recovery. The lower recovery of the GC-NCI-MS method could be caused by the use of an internal standard other than an isotopically substituted material that could not be used because of the fragmentation pattern observed by this method. The cleanup methods prior to ELISA were matrix dependent, no pretreatment was needed for environmental water samples, while fish, milk, and soil samples required various degrees of cleanup. Analysis of this wide variety of environmental samples by both ELISA and GC-MS demonstrated ELISA provides a timely and cost-effective method to screen for PBDEs in a variety of samples.

13.3　Sudan red and para red

13. 3. 1　Introduction

The red dyes Sudan Ⅰ, Ⅱ, Ⅲ and Ⅳ are oil soluble, azo dyes used legally in leather and fabric industries. They are fairly inexpensive and readily available. However, they are not approved at any level for use in foods. The International Agency for Research on Cancer (IARC), a part of the World Health Organization, has assessed the Sudan dyes as Group 3 genotoxic carcinogens. The Group 3 category is used most commonly for agents, mixtures and exposure circumstances for which the evidence of carcinogenicity is inadequate in humans and inadequate or limited in experimental animals. Statements issued by the Food Standards Agency (FSA) in the United Kingdom have indicated that the risk is very small from consuming items contaminated with the dyes. However, their use in food products is illegal. The industrial dye Para Red is chemically similar to Sudan I and is also an industrial dye not permitted for use in food.

13.3.1.1　Sudan red

Repeated notifications were disseminated via the EU rapid alert system in about the detection of the non-authorised dyes Sudan red I-IV in foods in 2003. These dyes were mainly found in chilli powder and in foods prepared with it. On 9 May 2003 France passed on information via the rapid alert system that it had detected the dye Sudan red I in hot chilli products from India. There is a suspicion that Sudan red I may be a genotoxic carcinogen which means that it is not possible to establish a tolerable daily intake. The European Commission, therefore, issued a decision on emergency measures (2003/460/ EC) whereby the Member States prohibit the import of hot chilli and hot chilli products unless an analytical report accompanying the consignment demonstrates that the product does not contain any Sudan red I (CAS No. 842-07-9) (CEC, 2003). In the opinion of the European Commission "the findings reported by France point to an adulteration constituting a serious health risk". In February 2005, British Food Standard Office stated the recall of 400 species of food contaminated with Sudan red I. In March 2005, Sudan red was also found in pepper spices, egg and lipstick in China[12].

Sudan red I is classified as a category 3 carcinogens and as a category 3 mutagen in Annex I of the Directive 67/548/EC. The main targeted organ of carcinogenicity for Sudan red I is liver. Animal tests showed that Sudan red I can also cause the tumor of mouse, rat and rabbit in bladder and spleen. However, there is still lack of evidence for its carcinogenicity in human body. Sudan red is highly liposoluble, with appearance of lamellar, kermesinus or deep yellow. The information of four Sudan red is listed in Table 13-1.

Table 13-1　Categorization and property of Sudan red

Categorization	Name/Synonym (Examples)	Formula	Molecular weight
Sudan red I	1-phenylazo-2-naphthol	$C_{16}H_{12}N_2O$	248.28
	1-phenylazo-β-naphthol		
	2-hydroxy-1-phenylazonaphthalene		
	2-hydroxynaphthyl-1-azobenzene		

Continued

Categorization	Name/Synonym (Examples)	Formula	Molecular weight
	Solvent Yellow 14		
	Sudan Gelb		
	Dispersol Yellow PP		
	Ölorange E		
	Scharlach B		
Sudan red Ⅱ	1-(2,4-dimethylbenzolazo)-2-naphthol	$C_{18}H_{16}N_2O$	276.33
	Solvent Orange 7		
	D&C Red No. 14. Ext.		
	Sudan orange		
	CEN-C2		
Sudan red Ⅲ	1-[94-benzolazo)-benzolazo]-2-naphthol	$C_{22}H_{16}N_4O$	352.38
	Solvent Red 23		
	D&C Red No. 17		
	Ölrot 3G		
	C-Ext. Rot 56		
Sudan red Ⅳ	2',3-dimethyl-4-(2-hydroxy-1-naphthylazo)-azobenzene	$C_{24}H_{20}N_4O$	380.44
	o-tolylazo-o-tolylazo-β-naphthol		
	Solvent Red 24		
	ölrot 2B		
	Scharlachrot		
	CEN-C5		

13.3.1.2 Para red

Para red also called contra-nitro aniline red is widely used as azo dyes in various industrial products, including machine oil, wax, paint and petrol. Among the red azo-dyes, para red is the least diaphanous and the cheapest. As shown in Figure 13-3, the structure of para red is very similar to Sudan red Ⅰ.

Similar to Sudan Red, para Red is not allowed to be used in food and it could cause cancer. On 21 April 2005, the Food Standards Agency of UK announced that some batches of Old El Paso dinner kits had been contaminated with the dye, and issued an alert. Also, reported on the 5 May 2005, the dye was found in 35 products which have been taken off supermarket shelves. The products were mainly cooking sauces, though some were also spices. Chinese government also banned the use of dyes in food production.

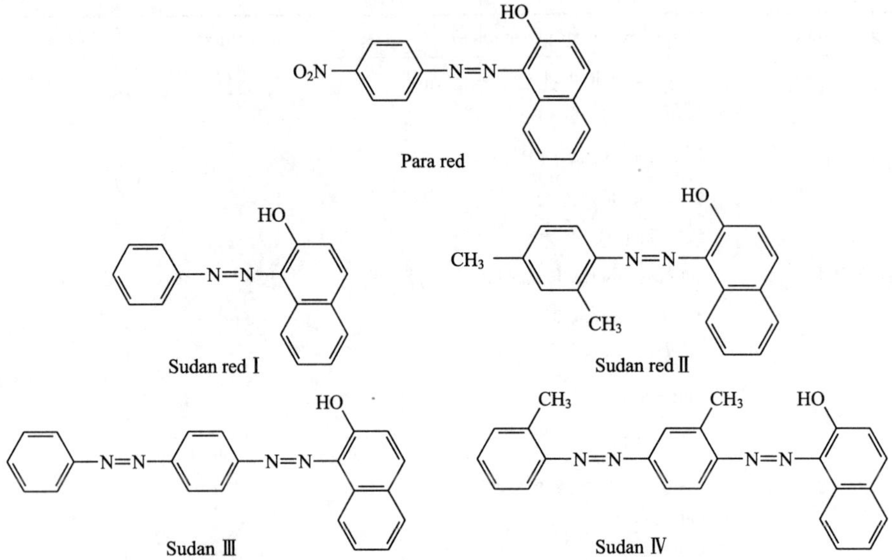

Figure 13-3　Molecular structure of para red and its analogs of Sudan red Ⅰ-Ⅳ

13.3.2　Immunoassay for Sudan red

Two hapten derivatives with different lengths of carboxylic spacer at the azobound para-position were synthesized and coupled to carrier proteins by Han Dan[13] (Figure 13-4). The hapten-bovine serum albumin (BSA) conjugates were used as immunogens, while the hapten-ovalbumin (OVA) conjugates were applied as coating antigens. The antisera obtained from four immunized rabbits were characterized in terms of sensitivity and specificity. At optimal experimental conditions, it was found that IC_{50} value of 0.3 ng/mL for Sudan Red Ⅰ. Cross-reaction rates of 0.1%-14.3% between antibody and other Sudan red dyes (Sudan red Ⅱ, Ⅲ and Ⅳ) was achieved. There were nearly no cross-reaction between these antibodies and other dyes such as sunset yellow, lemon yellow, amaranth, carmine red, bright orchid and indigo blue. Six kinds of food from market were analyzed with ELISA. The recovery of Sudan red Ⅰ ranged from 92.4% to 114%, with variation coefficients of 5.9%-24.8%. Analysis by high performance liquid chromatograph(HPLC)and compared the results with that of ELISA, showing a strong correlation(r=0.84; n=7). An analogue of Sudan red Ⅰ was synthesized by Ren[14]. Compared to the molecular structure of Sudan red Ⅰ, a carboxy l group was induced to the counter point of chemical group-N= N. Anti-Sudan red Ⅰ monoclonal antibody was prepared by hybridoma techniques. The ascties from hybridoma showed highly specificity to Sudan red Ⅰ and also recognized Sudan red Ⅱ.

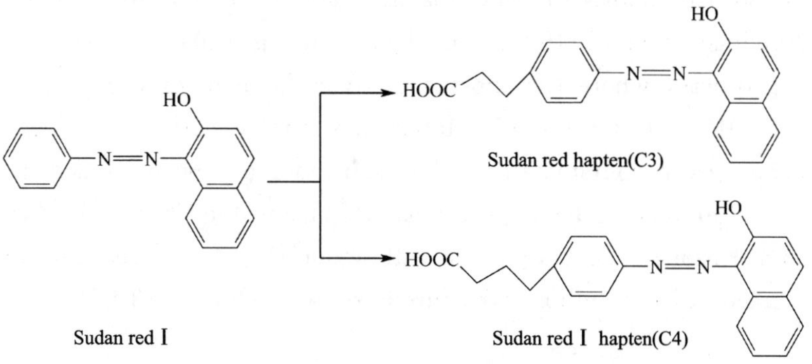

Figure 13-4　Standard curve of Sudan red I for ELISA

Tang[15] reported the reaction of 2-naphthol and diazotized 4-anthranilic acid, which was used for synthesis of derivatives of Sudan red I (CSD I , Figure 13-5). The CSD I was conjugated with BSA for immunization of mice. Then the spleen cells of mouse were fused with SP2/0 cells. Monoclonal antibodies produced by hybridoma were applied in establishment of indirect competitive ELISA. The method sensitivity for Sudan red I , III , and para red was around 0. 1 ng/mL, with IC_{50} values of 1. 03, 1. 13 and 0. 89 ng/mL, respectively.

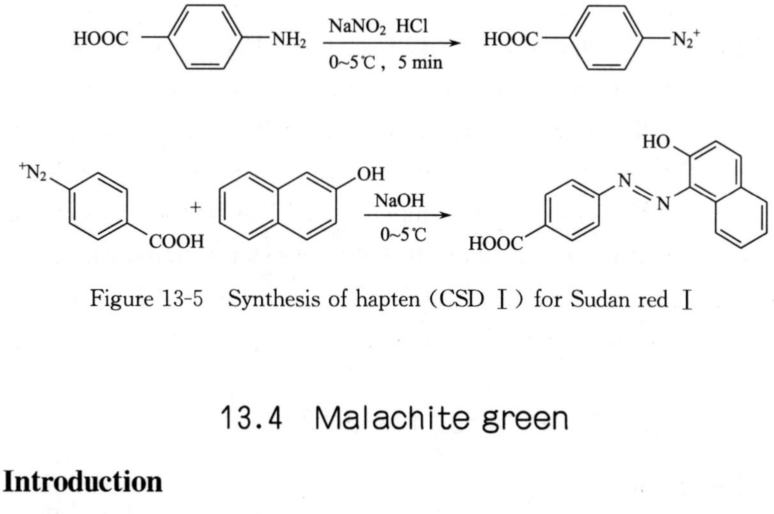

Figure 13-5　Synthesis of hapten (CSD I) for Sudan red I

13.4　Malachite green

13. 4. 1　Introduction

Malachite green is an organic compound that is used as a dyestuff and has emerged as a controversial agent in aquaculture. Malachite green is traditionally used as a dye for materials such as silk, leather, and paper. Although called malachite green, the compound is not related to the mineral malachite-the name just comes from the similarity of color[16]. Malachite green was classified a Class II Health Hazard. Due to its low manufacturing cost, malachite green is still used in certain countries with less restrictive laws for non-aquaculture purposes. In 2005, analysts in Hong Kong found traces of malachite green in eels and fish imported from China and Taiwan. In 2006 the United States Food and Drug Administration (FDA) detected malachite green in seafood imported from China, among others, where the substance is also banned for use in aquaculture. In June 2007, the FDA blocked the importation of several varieties of seafood due to continued malachite green contamination. The substance has been banned in the United States since 1983 in food-related applications. It is banned in China now. When entering in aquatic animal tissue, malachite green will be quickly metabolized into lipophilic leucomalachite green (LMG), whose structure is shown in Figure 13-6.

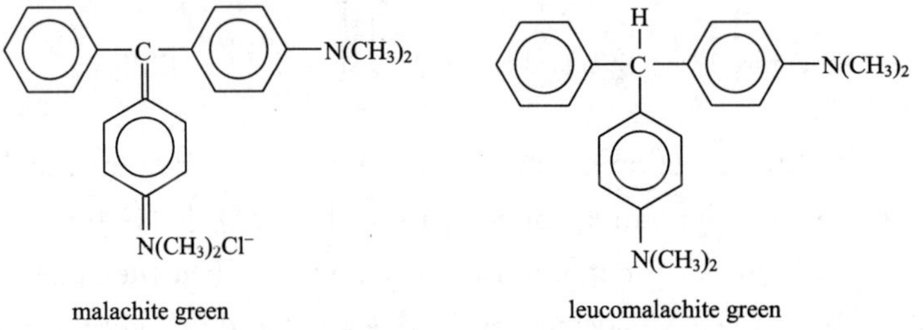

malachite green   leucomalachite green

Figure 13-6   Structural formula of malachite green and leucomalachite green

13.4.2   Immunoassay for MG

Kong[17] used an analogue of MG(para-rosaniline, PA) to couple with bovine serum albumin (BSA) by diazotization reaction. Balb/C mice were immunized with PA-BSA. The positive spleen cells were fused with myeloma cells. Hybridoma cell was screened using two-step method, and finally getting a stable antibody secreting cells for leucomalachite green (LMG)-5E9. An indirect competitive ELISA method was established based on purified ascites. ELISA results showed that no cross-reaction with other analogue of LMG. Indirect competitive ELISA method for detection of LMG kit was initially optimized with limit of quantification of 0.43 ng/mL.

13.5 Melamine

13.5.1   Introduction

Melamine (MEL)is an important class of chemical materials with nitrogen heterocycle, which is also an important post-processing product of urea. It is a white monoclinic crystal, with formula of $C_3N_6H_6$ or $C_3N_3(NH_2)_3$ and molecular weight of 126.12(Figure 13-7). Melamine is a versatile chemical intermediates, and is mainly used as raw materials for the production of melamine-formalde-

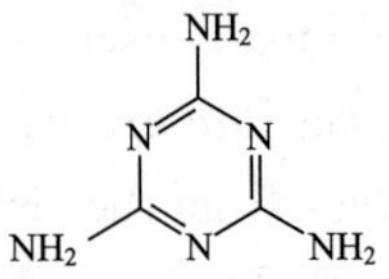

Figure 13-7   Structural formula
of melamine

hyde resin (MF).

In 2004, an outbreak of food adulteration with MEL led to renal failure in dogs and cats in Asia. In the spring of 2007, pet food adulterated with MEL was blamed for the illness or death of thousands of dogs and cats in the USA. In September 2008, the occurrence of kidney stones in thousands of infants across China captured the attention of the world. According to a report from the Ministry of Health (MOH), in China, more than 54,000 infants and young children were hospitalized, and at least six children died in this incident. An investigation revealed that the illnesses resulted directly from consumption of milk, infant formula, or related dairy products adulterated with MEL. So rapid, widely available, and cost-effective methods for detecting MEL in various substances are of paramount importance[18]. Melamine is banned for use in human and animal food in China and critical maximum residual limits (MRL) have been set for milk-based foods (less than 1 mg/kg in infant formula powder and less than 2.5 mg/kg in other foods).

13.5.2　Immunoassay of MEL

Immunoassays and related immunochemical analytical procedures [e.g. ELISA, immunochromatographic assay (ICA), and fluorescence polarization immunoassay (FPIA)] have been widely used to detect various residues in foods and environment. However, ICA and FPIA have not been reported so far for MEL. The use of ELISA for the determination of MEL has been reported in several papers. A comparison was made between three commercial ELISA test kits [i.e. Abraxis Melamine Plate kit (5005B), Abraxis Atrazine ELISA (520005) and EnviroGard Triazine Plate kit (7211000)] for MEL detection and evaluated by Garber[19]. The results showed that only the Abraxis MEL Platekit was specifically manufactured for MEL detection (LOD<20 ng/mL) and did not require extensive sample preparation. Wang and others[20] developed an indirect competitive ELISA based on a polyclonal antibody with an IC_{50} value of 500 ng/mL and applied this method to detect melamine in tissue and body fluid. Furthermore, antibody preparation for the detection of melamine has been reported. Lei and others[21] prepared a polyclonal antibody against melamine and the 50% inhibitory concentration (IC_{50}) value was 70.6 ng/mL. A polyclonal antibody against melamine with an IC_{50} value of 13 ng/mL was developed by Liu and others[22] and showed 59% cross-reactivity with cyromazine. Some reports on the use of a monoclonal antibody against melamine preparations have been reported recently. Yin and others[23] obtained a monoclonal antibody with an IC_{50} value of 22.6 ng/mL and developed an ELISA method for the detection of melamine in foods and milk samples. Zhu and others[24] mentioned a monoclonal antibody to melamine which can selectively recognize melamine with an IC_{50} of 63.7 ng/mL.

13.6　Phthalate(ester of phthalic acid)

13. 6. 1　Introduction

Phthalates (PAEs) or phthalate esters, are esters of phthalic acid and are mainly used as plasticizers (substances added to plastics to increase their flexibility, transparency, durability, and longevity). They are used primarily to soften polyvinyl chloride. Phthalates are being phased out of many products in the United States, Canada, and European Union over health concerns. Phthalates are used in a large variety of products, from enteric coatings of pharmaceutical pills and nutritional supplements to viscosity control agents, gelling agents, film formers, stabilizers, dispersants, lubricants, binders, emulsifying agents, and suspending agents. End-applications include adhesives and glues, electronics, agricultural adjuvants, building materials, personal-care products, medical devices, detergents and surfactants, packaging, children's toys, modelling clay, waxes, paints, printing inks and coatings, pharmaceuticals, food products, and textiles. Phthalates are easily released into the environment because there is no covalent bond between the phthalates and plastics in which they are mixed. As plastics age and break down, the release of phthalates accelerates. People are commonly exposed to phthalates, and most Americans tested by the Centers for Disease Control and Prevention have metabolites of multiple phthalates in their urine. Because phthalate plasticizers are not chemically bound to PVC, they can easily leach and evaporate into food or the atmosphere. Phthalate exposure can be through direct use or by indirect means through leaching and general environmental contamination. Diet is believed to be the main source of DEHP and other phthalates in the general population. Fatty foods such as milk, butter, and meats are a major source[25]. The common PAEs are dimethyl phthalate (DMP), diethyl phthalate (DEP), dibutyl phthalate (DBP), dioctyl phthalate (DOP), dibutyl (2-ethylhexyl) phthalate (DEHP), butyl benzyl phthalate (BBZP), dimethyl isononyl phthalate (DINP), etc.. The most used of PAEs is DEHP, followed by DBP.

13. 6. 2　Immunoassay for PAE

A reliable and sensitive competitive fluorescence immunoassay for the quantitative determination of dibutyl phthalate (DBP) was developed by Zhang[26]. The hapten synthesis was shown in Figure 13-8.

The hapten-protein conjugate was used as an immunogen for polyclonal antibody production. With this polyclonal antibody, a novel fluorescence immunoassay for detection of DBP was described. Under best conditions, DBP can be determined in the concentration range of 0. 1-100 μg/L with a detection limit of 0. 05 μg/L. The cross-reactivities of the anti-DBP antibody to six structurally related phthalate esters were below 9%. Some

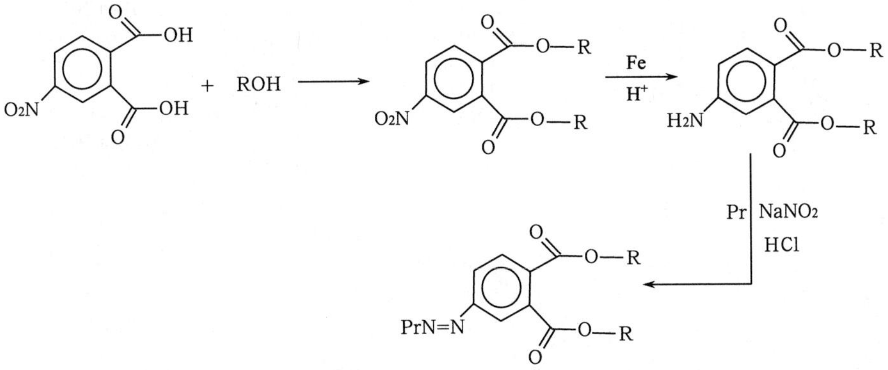

Figure 13-8   Hapten design for DBP

environmental samples were analyzed with satisfactory results. A hapten di-n-octyl 4-aminophthalate (DOAP) was designed by Kuang[27] successfully. It was used to couple with carrier proteins, bovineserum albumin (BSA) and ovalbumin (OVA) by diazotization reaction for immunogen (DOP-BSA) and coating antigen (DOP-OVA), respectively. Rabbits were immunised with DOP-BSA; polyclonal antiserum was raised and determined by competitive indirect enzyme-linked immunosorbent assay (ciELISA). After optimization, a ciELISA was established. The quantitative working range for DOP was 5-75 ng/mL with the detection limit of 1.9 ± 0.1 ng/mL and the IC_{50} of 19.2 ± 1.1 ng/mL. The optimised ELISA had cross-reactivity of 22.6%, 17.6% and 21.2% with di-iso-octyl phthalate (DIOP), di-n-butyl phthalate (DBP) and di-hexyl phthalate (DHP), respectively. The result of the detection of polyvinyl chloride (PVC) samples showed that the immunoassay we developed had high accuracy contrast with high-performance liquid chromatography electrospray-ionisation tandem mass spectrometry analysis and it could be qualified to determine di-n-octyl phthalate (DOP) residue in PVC sample.

Qing[28] prepared complete antigen by coupling dicyclohexyl phthalate (DCHP) and dibutyl phthalate (DBP) with carrier protein (BSA and OVA). The conjugates of DCHP-BSA and DBP-BSA were used to immunize rabbits. The two antibodies were purified from antisera. The anti-DCHP antibody was used to coat enzyme microtiter plates and anti-DBP antibody labelled with fluorescein isothiocyanate was used for detecting antibody for establishment of double-antibody sandwich ELISA. The immunoassay was successfully used for dicyclohexyl phthalate analysis in the tap water and drinking water.

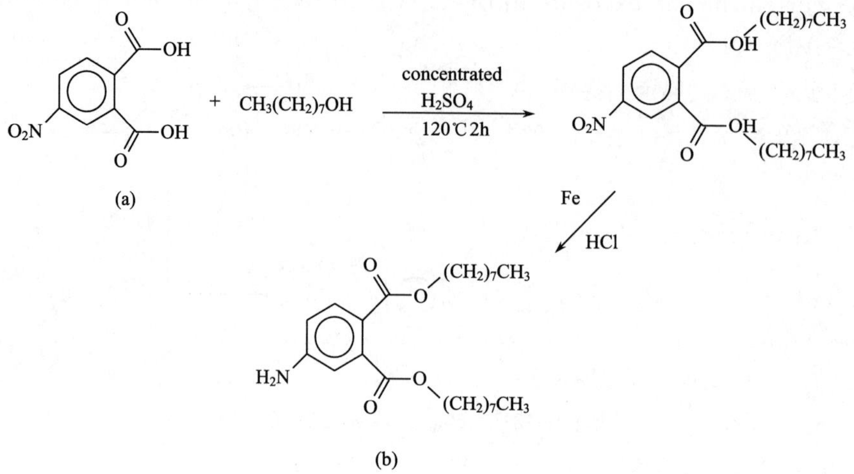

Figure 13-9　Synthetic route for hapten DOAP
(a) di-n-octyl 4-nitrophthalate (DONP); (b)di-n-octyl 4-aminephthalate (DOAP)

13.7　Bisphenol A

13.7.1　Introduction

Bisphenol A, with a scientific name of 2,2-bis (4-hydroxyphenyl) propane, is referred to as bisphenol-propane. Its English name is 2,4-Di (4-hydroxyphenyl) propane or Bisphenol A (BPA), and the molecular formula is $C_{15}H_{16}O_2$. The pure product is white crystal, flammable and with microstrip phenol odour. The molecular weight of BPA is 228.3 and its boiling point is 250-252 ℃ (1.773 kPa). It is soluble in alcohol, acetone, ethyl ether, methylene chloride, benzene and diluted alkali solution, etc., and slightly soluble in carbon tetrachloride, almost insoluble in water. The structure is shown in Figure 13-10.

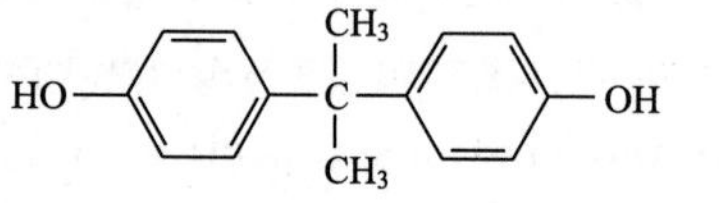

Figure 13-10　Structure of Bisphenol A

Bisphenol A is an endocrine disruptor, which can mimic the body's own hormones and may lead to negative health effects. Early development appears to be the period of greatest sensitivity to its effects and some studies have linked prenatal exposure to later neurological difficulties. Regulatory bodies have determined safety levels for humans, but those safety levels are currently being questioned or are under review as a result of new scientific studies. Concerns about the use of bisphenol A in consumer products have been regularly reported in the news media since 2008, after several governments issued reports questioning its safety, prompting some retailers to remove products containing it from their shelves. In 2009, The Endocrine Society released a statement expressing concern over current human exposure to BPA. A 2010 report from the United States Food and Drug Administration (FDA) raised further concerns regarding exposure of fetuses, infants and young children. In September 2010, Canada became the first country to declare BPA a toxic substance. In the European Union and Canada, BPA use in baby bottles is banned[29].

13. 7. 2   Immunoassay for BPA

Zheng[30] modified Bisphenol A (BPA) by introducing a carboxyl, which was used to couple with bovine serum protein by mixing carbodiimide and N-hydroxysuccinimide method. The artificial antigen was identified by ultraviolet spectroscopy and infrared spectroscopy. The polyclonal antibodies were prepared through immunization of Balb/C mice with titer of 256,000. Indirect competitive ELISA method for the determination of bisphenol A was established and results showed that the linear detection range of BPA was between 10 μg/L and 1×10^6 μg/mL, and the and IC_{50} value was 64. 57 μg/L for BPA with the detection limit of 9. 06 μg/L. The cross-reactions to benzene, phenol hydroxyl, salicylic acid and tert-butyl benzene were tested and the data were less than 0. 01%. The determination of bisphenol A in bottled water, river water and plant water were conducted by this ELISA method.

Huang[31] synthesized two haptens mono-carboxylic methyl ether (MCME) and mono-carboxylic propyl ether (MCPE). The haptens were covalently conjugated to carrier proteins BSA and OVA by the method of active ester or mixed anhydride to prepare artificial antigens MCME-BSA and MCPE-BSA, and coating antigens MCME-OVA and MCPE-OVA. MCME-BSA and MCPE-BSA were used to immunize New Zealand white rabbits respectively for preparing antiserum. The serum titer was determined to be greater than 32 : 1 by double agar diffusion method before collection of whole blood. The anti-serum was purified for establishment of indirect competitive ELISA of BPA. The IC_{50} of this ELISA for BPA detection was 16. 07 ng/mL.

Recently, a rapid, sensitive, specific and simple method based on Fluorescence PolarizationImmunoassay (FPIA) was developed for the quantitative detection of Bisphenol A by Wu[32]. 4, 4-Bis (4-hydroxyphenyl) valeric acid (BHPVA) was selected as the hapten to produce the immunogen. Fluorescein-labelled Bisphenol A derivatives (tracers) with different structures were synthesized and purified by thin layer chromatography. Based on the polyclonal antibody and tracers, an optimized FPIA method was developed with a detection limit (10% inhibition) of 2 μg/L and a linear working range of 20 to 800 μg/L. FPIA was suitable for screening a large number of samples. The recoveries of fortified Tai lake water samples ranged from 91. 85% to 102. 78%, and tap water samples ranged from 90. 36% to 96. 01%. The coefficients of variation were all less than 20%. This FPIA method, which did not require a complicated cleanup process, proved to be very useful for the screening of Bisphenol A in environmental water samples.

13.8   Heavy metal contamination

13.8.1   Introduction

A heavy metal is a member of a loosely-defined subset of elements that exhibit metallic properties. It mainly includes the transition metals, some metalloids, lanthanides, and actinides. Many different definitions have been proposed—some based on density, some on atomic number or atomic weight, and some on chemical properties or toxicity. The term heavy metal has been called a "misinterpretation" in an IUPAC technical report due to the contradictory definitions and its lack of a "coherent scientific basis". There is an alternative term *toxic metal*, for which no consensus of exact definition exists either. Heavy metals occur naturally in the ecosystem with large variations in concentration. In modern times, anthropogenic sources of heavy metals, i. e. pollution, have been introduced to the ecosystem. Waste-derived fuels are especially prone to contain heavy metals, so heavy metals are a concern in consideration of waste as fuel. The main threats to human health from heavy metals are associated with exposure to lead, cadmium, mercury and arsenic. Acute exposure to these heavy metals may give rise to damages of organs. Chronic poisoning is characterized by neurological and psychological symptoms, such as tremor, changes in personality, restlessness, anxiety, sleep disturbance and depression.

13.8.2   Immunoassay

Yuan[33] used isothiocyanobenzyl-EDTA (IEDTA) as chelating agent to bind heavy metal cadmium ions and the resulting products were coupled with carrier protein KLH or BSA. Balb/C mice were immunized with the antigen to obtain antiserum. The antisera was characterized to recognize IEDTA-cadmium well with titer value of $1 : 81,920$.

Yang[34] tried immunoassay development for mercuric ions[Hg(Ⅱ)]. Mercuric ions were coupled to protein carrier (keyhole limpet hemocyanin, KLH) using a bifunctional chelator (1-(4-isothiocyanobenzyl)-ethylenediamine N, N, N', N'-tetraacetic acid, ITCBE)to gain the immuno-conjugate KLH-ITCBE-Hg. BALB/C mice were immunized with this conjugate. Spleen cells of an immunized mouse were fused with myeloma cells, and five hybridoma cell strains (E/H9, E/B2, 1/F11, 1/H7 and B/B11) which can steadily excrete anti-mercury were obtained. The excreted antibodies were both IgM and κ-type, and the titers of ascites of $1 : 6,400$, $1 : 3,200$ and $1 : 6,400$. Antibodies excreted by hybridoma cell B/B11 were selected for ELISA development. The conjugate of

BSA-glutathione-Hg (II) was used as coating antigen at 0. 9μg/mL for microtiter plates. The detection limit was 9. 15$\times10^{-6}$ mmol/L for Hg (II) based on optimized ELISA. Inter and intra coefficients of variation were 1. 2% and 4. 7%, respectively. Recoveries in samples of ultra pure water, tap water, pond water and vegetables were 102. 21%, 96. 92%, 103. 23% and 87. 12% respectively.

Zhu[35] prepared hybridoma cell strains that can steadily excrete monoclonal antibody of high affinity and specificity to lead. Monoclonal antibodies against lead were generated by immunizing BALB/c mice with lead conjugated to keyhole limpet hemocyanin (KLH) via a bifunctional chelator, S-2-(4-aminobenzyl)diethylenetriamine pentaacetic acid (DTPA). Stable hybridoma cell lines were produced by fusion of murine splenocytes and SP2/0 myeloma cells. One of the hybridomas generated from this fusion (4/7) synthesized and secreted an antibody that bound tightly to Pb^{2+}-DTPA complexes but not to metal-free DTPA. The performance for a competitive inhibition enzyme-linked immunosorbent assay (ELISA) incorporating this antibody was assessed for its sensitivity to changes in pH, ionic strength, and blocking reagents. The cross-reactivities in this ELISA were less than 3% for Fe^{3+}, Cd^{2+}, Hg^{2+}, and Cu^{2+} and less than 0. 3% for Cr^{3+}, Mn^{2+}, Mg^{2+}, In^{3+}, Ag^{1+}, Ni^{2+}, Co^{2+}, Zn^{2+}, Ca^{2+}, Cu^{1+}, and Hg^{1+}. The IC_{50} value achieved for lead was 2. 72 $\pm$ 0. 034 μM, showing the detection range of 0. 092? 87. 2 μM and the lowest detection limit of 0. 056 $\pm$ 0. 005 μM. Recoveries from the analyte-fortified tap water and ultrapure water were in the range of 80%-114%. These results indicate that the ELISA could be a convenient analytical tool for monitoring lead residues in drinking water.

References

[1]   Guide for monitoring of air and exhaust gases. Beijing:Chinese Environmental Science Press, 2006, 274

[2]   Jiang X, Qiao PY. Forming process and analysis method of dioxin heilongjiang environmental journal, 2008, 32 (2): 39-41

[3]   Industrial and environmental management association, maruzen Co. , Ltd. Jurassic. Technologies and regulations to prevent-dioxin articles, 2000,(11): 15

[4]   Li X, Wang YP. Molecular structure and properties of dioxin. Journal of Harbin Institute of Technology, 2004, 36 (4): 513-519

[5]   Chen XC, Yu X, Guo JF and others. Toxic effects,health risk and countermeasures of dioxin exposure. Journal of Hebei University of Science and Technology, 2002, 23 (4): 50-53

[6]   Wei N. Hazards of dioxins and control measures. Journal of Health Inspection and Health, 2004, 3 (2): 114-116

[7]   Shigeru K, Hidenobu A, Park JW. Quartz crystal microbalance immunosensor for highly sensitive 2,3,7,8-tetrachlorodibenzo-p-dioxin detection in fly ash from municipal solid wasteincinerators. Analyst, 2005, 130: 1495-1501

[8]   Yukio S and others. Development of a highly sensitive enzyme-linked immunosorbent assay based on polyclonal antibodies for the detection of polychlorinated dibenzo-p-dioxins. Analytical Chemistry 1998, 70: 1092-1099

[9] Zhou B, Qiu YL. Polybrominated diphenyl ethers and their environmental behavior. Environmental Science and Technology, 2008, 31 (5): 57-61

[10] Wei AX, Wan XT, Xu XB. Environmental pollution of polybrominated diphenyl ethers (PBDEs). Chemistry development, 2006, 18 (9): 1227-1233

[11] Weilin LS and others. Development of a magnetic particle immunoassay for polybrominated diphenyl ethers and application to environmental and food matrices. Chemosphere, 2008, 73: S18-S23

[12] Xia HY, Dai ZM. The poison mechanism and calibration method of sudan journal of anqing teachers college. Natural Science Edition, 2005, 11 (4): 53-55

[13] Han D. Development of enzyme immunoassay for Sudan I [Dissertation]. Chengdu: Sichuan University, 2007

[14] Ren LS. Single-chain fusion antibody of Sudan I and lemon yellow and rapid immunoassay method [Dissertation]. Changchun: Jilin University, 2008

[15] Tang Y, Ju CM, Wei ZJ and others. Method of competitive ELISA for Sudan I, III and Para red. Food Science, 2008, 29 (08): 523-525

[16] Li N. Malachite green on the health effects. International Journal of Health Toxicology, 2005, 32 (5): 262-264

[17] Gong PF. Preparation of malachite green monoclonal antibody and initial development of detection kit [Dissertation]. Changsha: Hunan Agricultural University, 2005

[18] Sun FX, Wei Ma, Liguang Xu and others. , Analytical methods and recent developments in the detection of melamine Trends in Analytical Chemistry, 2010, 29(11), 1239-1249

[19] Garber EA. Detection of melamine using commercial enzyme-linked immunosorbent assay technology journal of food protection, 71(3), 2008, Pages 590-594

[20] Wang, ZY, Ma X, Zhang LY and others. Screening and determination of melamine residues in tissue and body fluid samples. Analytica Chimica, Acta, 2010, 662(1): 69-75

[21] Lei, HT, Shen YD, Song LJ and others. Hapten synthesis and antibody production for the development of a melamine immunoassay. Analytica Chimica, Acta, 2010, 665(1): 84-90

[22] Liu, JX, Zhong, YB, Liu J and others. An enzyme linked immunosorbent assay for the determination of cyromazine and melamine residues in animal muscle tissues. Food Control, 2010, 21(11): 1482-1487

[23] Yin WW, Liu, JT, Zhang, TC and others. Preparation of monoclonal antibody for melamine and development of an indirect competitive ELISA for melamine detection in raw milk, milk powder, and animal feeds. Journal of Agricultural and Food Chemistry, 2010, 58(14): 8152-8157

[24] Zhu K, Li J, Wang Z and others. Simultaneous detection of multiple chemical residues in milk using broad-specificity antibodies in a hybrid immunosorbent assay. Biosensors and Bioelectronics, 26(5), 2716-2719

[25] Yao WR. Research survey of phthalates pollutants in food. Food Research and Development, 2004, 25 (6): 21-23

[26] Zhang MC, Wang QE, Zhuang HS. A novel competitive fluorescence immunoassay for the determination of dibutyl phthalate. Anal Bioanal Chem, 2006, 386: 1401-1406

[27] Kuang H, Xu L, Cui G and others. Development of determination of di-n-octyl phthalate (DOP) residue by an indirect enzyme-linked immunosorbent assay. Food and Agricultural Immunology, 2010, 21(3) 265-277

[28] Lang Q. Fluorescence immunoassay method for environmental hormone-dicyclohexyl phthalate and dibutyl phthalate [Dissertation]. Shanghai: Donghua University, 2006

[29] Wang J. Research progress of bisphenol a effect on human body and its mechanism. Phylaxiology Information, 2005, 21 (5): 541-544

[30] Zheng J. Synthesis of bisphenol a artificial antigen and ELISA method [Dissertation]. Guangzhou: Guangdong University of Technology, 2008

[31]　Huang YJ. Chemistry immunoassay of bisphenol a [Dissertation]. Yangzhou: Yangzhou University, 2007

[32]　Wu XL, Wang LB, Ma W and others. A simple, sensitive, rapid and specific detection method for bisphenol a based on fluorescence polarization immunoassay Immunological Investigations, 2012,41(1) 38-50

[33]　Yuan XK. Immunizing antigen preparation and characterization of heavy metal cadmium. Journal of Veterinary Medicine, 2008, 27 (4): 19-21

[34]　Yang FL. Monoclonal antibody preparation and immunoassay methods of heavy metals mercury [Dissertation]. Nanjing: Nanjing Agricultural University, 2007

[35]　Zhu XX. Heavy metal zirconium, lead monoclonal antibodies, single chain antibodies and three-dimensional structure modelling development of antibody variable regions [Dissertation]. Nanjing: Nanjing Agricultural University, 2007

Colored Figures

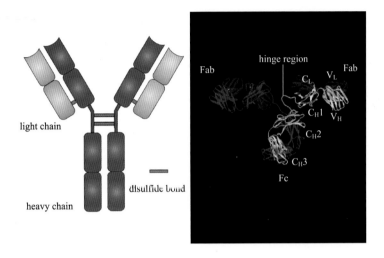

Figure 2-1 Structure model of immunoglobulin (IgG)

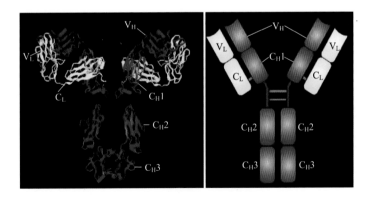

Figure 2-2-A A model of immunoglobulin ribbon

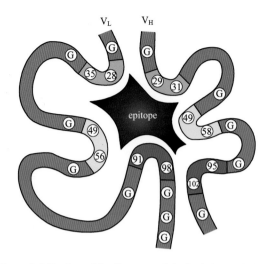

Figure 2-2-B   A model of immunoglobulin hypervariable region

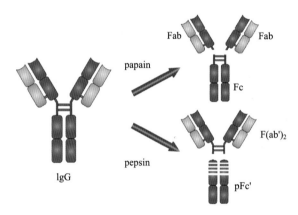

Figure 2-3   Diagram of IgG molecules hydrolysis fragments

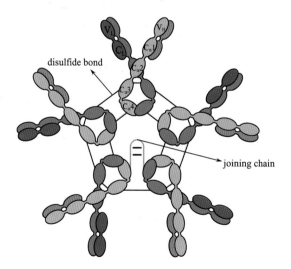

Figure 2-4   IgM pentamer

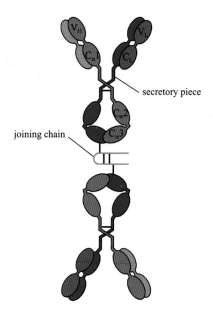

Figure 2-5　Structure of immunoglobulin Ig A dimer

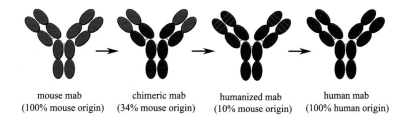

mouse mab
(100% mouse origin)　　chimeric mab
(34% mouse origin)　　humanized mab
(10% mouse origin)　　human mab
(100% human origin)

Figure 4-1　Humanization progress of mouse mab

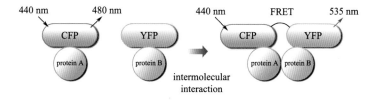

Figure 6-1　Schematic diagram of the FRET principles

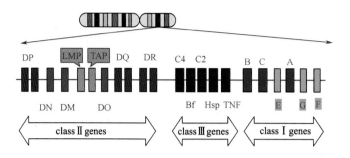

Figure 7-2　Structure of HLA gene

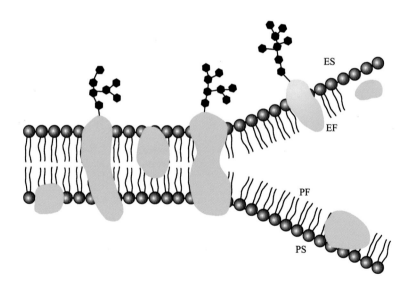

Figure 9-1　Lipid mosaic model and frozen fracture surface of biofilm molecules